Conquest
of the Skies

By Carl Solberg

RIDING HIGH
America in the Cold War

OIL POWER

CONQUEST OF THE SKIES
A History of Commercial Aviation in America

Conquest
of the Skies

A History of Commercial Aviation in America

by Carl Solberg

LITTLE, BROWN AND COMPANY • BOSTON • TORONTO

Solberg, Carl, 1915–
    Conquest of the skies.
    Bibliography: p.
    Includes index.
    1. Aeronautics, Commercial—United States—History.
I. Title.
HE9803.A3S64   387.7′0973   79-15993
ISBN 0-316-80330-8

The author is grateful to Harold Ober Associates, Inc., for per-mission to quote from *By the Seat of My Pants* by Dean Smith.

MV
Published simultaneously in Canada
by Little, Brown & Company (Canada) Limited

PRINTED IN THE UNITED STATES OF AMERICA

Contents

Part I

Prologue

1

To Fly

To fly. The ancient dream has come true in our time. The interdiction that chained us to earth has been lifted. Today we take to the air at will. We have penetrated the abode of angels; we have ascended into heaven.

To fly. Around the globe six hundred million human beings take off in a single year. At 155 major jetports in the United States, airliners soar up thirteen thousand times a day. Americans pay twenty billion dollars a year to ride in them. These silvery argosies cost up to fifty million dollars apiece and consume ten billion gallons of fuel hurtling us through the skies. Every day 657,000 of us go aboard, and fly five hundred million miles — equal to five trips to the sun. In a single year American air passengers fly two hundred billion miles — equal to two voyages to Jupiter and back. Traveling through the air has become an indispensable part of our lives, essential to the conduct of both public and private affairs. We fly — fly! — as an everyday matter of course.

How this came about in just seventy-five years, how the air was won, is a true fairy story, and the subject of this book.

To fly, gravity, primordial, mysterious gravity, has to be overcome. We begin learning the power of gravity the moment we raise ourselves up in the cradle. We persist; we pull ourselves up on our two legs. By dint of such struggling, we join the select company of creatures on earth that go about upright. And how gravity begrudges us this capability! Trudging uphill, falling downstairs, recuperating from our exhaustion every single night, we pay and pay all our lives. Finally, as our strength plays out, gravity wins. At the end of our years we fall as recumbent as we began.

It's enough to make you think.

Think is what Newton did, and in the prime of his two-leggedness he conceived a tremendous idea. He said the force pulling us down is a universal force, governing not only our earthbound steps but all the stars in their courses. He discovered that the attraction exerted between any two bodies is inversely proportional to the square of the distance between them. At a distance of one foot, for example, the attraction between two objects is a hundred times stronger than at ten feet. No wonder we'd rather sit than stand, walk downstairs rather than up. Newton showed how gravity, besides pinning us to terra firma, in fact prevents us from whirling off the surface of the spinning globe. He showed that gravity holds air to earth, enabling us to breathe. And it is this same universal force that draws rain from the clouds, then pulls the water down through streams, lakes, and rivers to the ocean in the never-ending cycle on which all living things depend.

Spurred by these discoveries, others subsequently found out, in their upright cogitations and investigations, about the other energies of the cosmos. We learned to generate them at will — electricity, nuclear power, light, heat, radio waves, X rays. We even figured out how to turn these forces off, increase, decrease, or reverse some of them, with the flick of a switch. But there was no such mastering of gravity. In various ways we could use it — by pulleys, by levers, by hydroelectric dams and dynamos. Yet three hundred years after Newton established that gravity was a force that rules the universe, it remains little understood. By the standards of the other forces, it is puny. Someone has estimated that the gravitational energy of the whole earth amounts to only a millionth of a horsepower. A toy magnet in the hands of a child can be thousands of times stronger. Gravity simply exists, relentless, inexhaustible, the most mysterious force in the universe. We await another Newton to tell us what it is.

Birds have escaped enslavement to gravity. Birds can fly, and the sight of these fellow creatures cavorting in the skies long ago filled our kind with a yearning to do the same — to swing free of the earth's surfaces and escape, if only briefly, the iron dominion of gravity. Daedalus, by legend antiquity's leading technologist, fashioned wings of bird feathers held together by wax. His son Icarus would have escaped imprisonment on Crete with him had he not, as young fliers will, forgotten to fly straight to the next island but climbed instead to stunt so near the sun that the wax melted and gravity pulled him into the sea.

The birds continued to influence flight-fanciers such as Leonardo da Vinci. But the first successful escape from earth was achieved by Frenchmen who exploited a loophole in the law of gravity. In the knowledge that smoke rises, the Montgolfier brothers of Paris encased smoke in a paper ball in 1783, and their friend Pilâtre de Rozier perched in a basket hung beneath the balloon and stayed aloft until the air in the balloon cooled and gravity brought him down again. Much later, when the Germans surrounded Paris in the Franco-Prussian War, balloons carried mail and even passengers out of the beleaguered capital. Léon Gambetta, chief of the French defenders, was one such passenger; he landed safely at Epineuse. Another passenger came down in Norway. That was the trouble; balloons carried you wherever the winds blew. But by that time the gasoline engine had been invented, and the German Count Zeppelin soon found a way to build a cigar-shaped balloon powered by such engines and steered by rudders and other movable controls. As the twentieth century began, lighter-than-air craft offered a promising way to defy gravity and engage in controlled flight.

Of course the invention of a source of power much lighter than the bulky steam engine also excited those who had imagined, designed, and even put together crude machines in the hope of flying like birds. Some went right ahead and simply hitched power to their kites. They came to grief. The German pioneer Otto Lilienthal, insisting that first more must be learned about balance and control aloft, studied the flight of birds. Then he built wings curved like a gull's, attached them to his back, and jumped off hilltops in what we would now call hang-gliding. After making some two thousand flights in which he tried to steer by flinging himself from side to side, he was thrown to the ground when his glider plunged suddenly one day in 1896. His back broken, he died the next day. His last words were, "Sacrifices must be made."

In faraway Ohio two young sons of a United Brethren bishop read the story of Lilienthal's death. Wilbur and Orville Wright, operators of a bike shop in Dayton, were captivated by the ingenuity and daring of the German experimenter. Inspired by his example, they wrote away for his books, studied his data, and like Lilienthal, watched the birds. How effortlessly the wheeling pigeons and soaring buzzards of Dayton kept their balance, compared to the clumsy and inefficient contortions of their hero. By 1899 Wilbur Wright had made a decisive discovery: buzzards "regain their lateral balance . . . by a torsion of the tips of their wings." Thereupon the brothers, aged twenty-nine and twenty-five, made their first important invention — they applied the buzzard's trick to the design of the airplane wing. In the shop one day Wilbur tore off the ends of an empty inner-tube carton so that the box presented in its upper and lower sides the flat wings of a biplane. Then he twisted the opposing corners of the box in opposite directions. That, he told Orville, was how the buzzard contrived to hold steady in gusts aloft; and that, with wires manipulated by an operator lying prone between the wings, was how they could twist the opposing wingtips of a glider to hold it level against the kind of gusts that killed Lilienthal.

This was the idea behind the aileron, and it was absolutely crucial to flying. Years later an Englishman was found to have patented the idea back in 1868. But neither Lilienthal nor the Wrights, nor Professor Langley of the Smithsonian Institution, nor any other pioneer experimenter seems to have known about it.

From that first lancing insight the Wrights went on to the first great invention of the century, the most successful device yet for outwitting gravity. Anyone who thinks these young fellows were just gifted tinkerers is far off the mark. By inspired inventiveness, by pertinacity, and by a matchless combination of theoretical and practical engineering they researched, designed, and built the first airplane — airframe, engine, propellers, and all. They capped this achievement by carrying out the first powered, sustained, and controlled flights. All this they did in their spare time, and at their own expense — in four years and for an outlay of $2,000.

Most of each year the Wrights kept store, working on their latest glider in the back when they could. They thought of gliding as a sport and of their trips to practice on the dunes at Kitty Hawk, overlooking Cape Hatteras, as summer camping expeditions. They flew their 1900 glider as a kite. The following year they climbed aboard and made glides down the dunes of up to 389 feet. These first gliders, with no tail at all, with their elevators out front for fore-and-aft steering, with the operator lying prone amidships and twisting the wingtips by means of a hip cradle, did not look like what we think of as a flying machine. But in their undulating glides down the slopes, the Wrights experienced the classic problems of flight — skids, stalls, groundloops, even spins. The problems, which led them to question Lilienthal's data, drove them to aerodynamic research. Back at Dayton, they rigged up a wind tunnel and tested wing shapes.

The 1902 glider that followed began to look like the real thing. The wings, now spanning thirty-two feet, had less fore-and-aft curvature than Lilienthal's calculations had called for. To counter the yaw that upset them when they twisted the wingtips to bring the machine level, the Wrights had added a fixed rudder behind. When this still failed to give balance on their first tests, they made the rudder fins movable and rigged them so that they moved in conformity with the wingtip controls. After that the machine skimmed the sands like a Frisbee. The brothers made hundreds of perfectly controlled glides; they set a distance record of 622½ feet and stayed airborne as long as twenty-six seconds at a time. Dan Tate, their local helper, said all they needed now was a coat of feathers to make the machine light.

Elated by the success of their third year's gliding, the brothers were now ready to build a powered airplane. Finding no manufacturer ready to supply a light engine, they built one in their own shop, a four-cylinder, 150-pound engine that mustered thirteen horsepower. Understanding already that a propeller blade is also an airfoil, they designed and built props of impressive efficiency. The Wright Flyer that they took to Kitty Hawk in 1903 had the same general design as their third glider — elevator out front; operator, engine, and props between two forty-foot wings; and double rudder now pushed far enough behind to suggest the appearance of a tail.

When assembled at the Wrights' shack on the barren Carolina dunes, the machine rested on skids. In launching the airplane, the brothers planned to set these skids on a yoke on two small wheels that ran along a sixty-foot rail pointed into the wind. One brother would keep the craft tethered while the other, poised prone with his cap reversed, ran up the engine. Then the one whose turn it was to stay on the ground would release the machine and run alongside holding a wingtip until the plane got up speed. In the strong Hatteras winds it would then rise from the yoke and fly. The weight of their contraption was 605 pounds — this compares with the 574 pounds of the modern Piper Cub.

Trouble with cracked propeller shafts kept them from trying their hand at flying the plane till late in the fall. During Wilbur's first go on December 14, 1903, the Flyer stalled out after 105 feet. On the morning of December 17 ice covered the puddles around the Wright shack. A twenty-seven-mile-per-hour wind was blowing. Four men from the nearby Coast Guard station came to act as witnesses, and a local boy named Johnny Ward also showed up. At 10:35 Orville made the first try. He said later:

The machine started very slowly. Wilbur was able to stay with it until it lifted from the track after a 40-foot run.

The course of the flight up and down was exceedingly erratic, partly because of the gusty air, partly because of my inexperience in handling this machine. It would rise suddenly to about 10 feet, and then as suddenly, [when I turned] the rudder, dart for the ground. On one of these darts it hit the ground. I had been in the air 12 seconds, and covered a distance of 120 feet.

The brothers were seen to shake hands. After making repairs,

we went in to warm up. Johnny Ward saw this box full of eggs. He had never seen so many. One of the others said it was the small hen outside. "That chicken lays 8 to 10 eggs a day." He went, looked, returned and said: "It's only a common-looking chicken."

Then Wilbur tried a flight. It too was up and down, again at about ten miles an hour. The third flight was steadier; Orville detected a "sidling to the left," which he controlled, but the flight ended.

At just 12 o'clock Will started on the fourth and last trip. The machine started off with its ups and downs as it had before, but by the time he had gone over three or four hundred feet he had it under much better control, and was traveling on a fairly even course. It was proceeding in this manner till it reached a small hummock out about 800 feet from the starting ways, when it began its pitching again and suddenly darted into the ground. The distance over the ground was 852 feet in 59 seconds.

So, at a faltering ten miles an hour, began the age of flight. But the speed with which the airplane developed after the Wrights showed the world how to build and fly it, was prodigious. Within a year someone took one of their Flyers, put wheels under it, and began flying it right off the grass. In 1909 Louis Blériot crossed the English Channel in a strikingly new type of his own design — a monoplane. The French hung distinct ailerons on the wings and made a central framework called the fuselage, and soon everybody, the Wrights included, put the elevators back in the tail with the rudder.

Its basic shape established, the airplane soared ahead on the crest of the century's technological surge. Two world wars, turning it from the scouting machine the Wrights first proposed into the most mobile and lethal form of artillery ever seen, enormously enlarged the airplane's capabilities. Already, as the first war ended, a British bomber crossed the Atlantic by air. Then, as the second approached, engines, fuels, instruments, and weather aids emerged that enabled commercial as well as military planes to fly unimaginably faster and farther than the Kitty Hawk Flyer. By the time of Orville Wright's death in 1948, airliners crossed the continent in less than ten hours, and only ten years later — still within a single lifespan — jet-powered transports flew routinely across America at speeds fifty times beyond those of the original machine of 1903.

Voyaging in the new element of the air called for individual human qualities of the highest order. One of the primary lessons that the Wrights taught the rest (Wilbur died of typhoid in 1912, and Orville bowed out of flying after that) was the surpassing importance of personal mastery over the machine in flight. Supposing the plane's own power was what mattered, other pioneers had spoken of the operators as "chauffeurs." After watching Wilbur and Orville bank, circle, and cut figures of eight, they realized that to fly meant to *fly*. America's "early birdmen" — Walter Brookins and Arch Hoxsey (trained by the Wrights), Cal Rodgers, Lincoln Beachey — became national heroes recklessly exhibiting the feats of piloting before gaping multitudes. Beachey looped-the-loop and lived, though not for long. Cal Rodgers became the first to fly across the United States — sixty-eight hops in forty-nine days, with fifteen crashes.

No more daring feat could be imagined than taking these flickering moths cross-country. General George Squier said later that in its 1908 tests at Fort Myer, Virginia, when the War Department was considering purchase of the Wrights' invention, Orville Wright was required to fly ten miles around Arlington cemetery and back because "it was the most difficult thing we could think of." In 1910 Glenn Curtiss, the Wrights' great rival, astonished the nation by flying 150 miles from Albany to New York City with only a refueling stop at Poughkeepsie. In a stunning climax, Curtiss guided his tiny *June Bug* out of the Hudson River haze, skimmed boldly around the Statue of Liberty, and landed to delirious cheers at Governors Island.

To fly. Even today only a few fly individually — sky divers, hang gliders, the Californian who in 1977 pedaled a 96-foot wing of cardboard, Mylar, and aluminum tubing into the air off a runway near Bakersfield.

He won the $86,000 Kremer prize administered by the Royal Aeronautical Society for being the first to fly a human-powered airplane around a mile-long, figure-eight course at an altitude of at least ten feet.

To fly. Even today only 783,932 Americans possess licenses and pilot their personal planes through the nation's skies, nowhere near as many as own and drive private automobiles.

To fly. Today as passengers we *all* fly, and it is to the explosive development of mass passenger flying, quite unforeseen by the early airmen, that we must turn. Though piloting as a sport, an experience, a feat of individual achievement will always be an intense and important part of aviation, the rise of scheduled air transport is our theme.

Passenger flying began in 1908 when the French experimenter Henri Farman took Leon Delagrange up from a meadow near Paris. Delagrange in turn took up Mme. Thérèse Peltier as the first woman passenger a few months later. Charles Furnas became the first American passenger to fly when he went up with Orville Wright at Kitty Hawk that summer. Shortly afterward aviation claimed its first victim. As part of the War Department trials at Fort Myer, Orville Wright took Lieutenant Thomas E. Selfridge up for a circuit over the parade ground. A split propeller caused the Wright Flyer to plunge to earth, seriously injuring Wright and killing Selfridge.

The first scheduled airline started up in Florida six years later. By that time Glenn Curtiss had created the flying boat, a seaplane with boatlike hull. Operating entirely off the water, such a craft needed no supporting undercarriage and hence could be built bigger. Thomas Benoist, an auto-parts maker, saw transport possibilities and built one big enough to ferry passengers across the waters of Tampa Bay. With a partner he formed the St. Petersburg–Tampa Air Boat Line, and on January 1, 1914, pilot Tony Jannus carried the world's first scheduled airline passenger, Tampa's Mayor A. C. Pheil, across the bay from St. Petersburg. Twice a day for four months, except for seven and a half days when it was held back by mechanical trouble, the single-engine biplane crossed the bay carrying its capacity load of one passenger. The fare for the twenty-mile trip was five dollars. This earliest of all airlines carried twelve hundred passengers without mishap before it folded at the end of the season.

From these beginnings air transport rose to girdle the earth, bridge the seas, and leap the continents. Abroad, it brought every nation of the world within twenty hours of any U.S. city. At home, it cut the distance between merchants and all markets, between vacationists and all resorts, from weeks and days to a few hours. By 1978 U.S. airlines, scarcely fifty years old, had long since become the dominant means of long-distance transport. For a nation of impatient travelers, air was the only way to go.

Part II

Warm-up (The 1920s)

2

The Post Office Flies the Mail

A *modern* 360-ton Boeing jetliner is made of 4.5 million parts — mostly metal and plastic. The airplane of 1920 was made of wood, cloth, and wire. The wings — it was, of course, a biplane — were fashioned of spruce, covered with coarse cotton cloth, and then coated and recoated with stiffening airplane "dope." The fuselage was also made of wood, mainly plywood, and shaped by a frame of long curved members called longerons, which also held the tail surfaces in place. The longerons were made of long, slender shafts of ash, a wood that would take a lot of bending before splitting.

The flying machine of 1920 was put together in a modest workshop that looked rather like a boat works — indeed, seaplanes and flying boats were among its products. There were a few belt-driven machines on the premises, notably in the machine shop. The tool most often used was a set of pliers. The chief skills needed were carpentry and joining. At the Curtiss plant in Buffalo men in cloth caps and white carpenter's aprons filled with pockets for nails could be seen laying out the wings on long benches.

The main members of the wings were a pair of sizable spruce spars. These had been routed out beforehand, to cut down weight, in the shape of I-beams. They were laid out on the table, one to run laterally near the wing's leading edge, the other to extend parallel to it from the fuselage along the rear, or trailing, edge. At right angles to these the carpenters placed the wing ribs. The ribs were short and light, spaced perhaps twelve inches apart all the way out to the wingtip. Cut to the wing's contour, these strips of spruce were secured in place by small wood braces. More plywood pieces formed the wing's trailing and leading edges. All joints were held together by glue.

Over these featherweight panels, fifteen or so feet long and some five or six feet wide, went the fabric. It was essential that this covering be attached so that there was little chance of its ripping off in flight. It must fit snugly over the contours of the surface — "like a pillow case," it was said — with no dimpling between the ribs. This called for sewing, and was women's work. Sitting or standing on either side of a wing panel propped between them, women in sash-bowed gingham aprons stitched up the cotton wing covering. They used needles ten to sixteen inches long, long enough to pierce the entire wing, since the heavy thread had to be looped entirely around each rib as they sewed.

The air-loading on the fabric might exceed a hundred pounds per square inch, so a smooth and taut surface was much to be desired. On their early, one-hundred-yard hops the Wright brothers had gotten along with wing surfaces of brown paper toughened and tightened by an application of hot fish glue. The French tried silk, but silk leaked air and spoiled the lift. Around 1910 dope arrived. Applied to cotton or to linen, the favorite in Europe, airplane dope was universally used until well after Lindbergh's 1927 flight. It was made of cotton dissolved in either nitric or acetic acid, and its application became if not the most arduous then the most time-consuming part of airplane manufacture. To get a firm, all-weather surface took one coat after another; the first three or four coats were laid on by hand-brushing followed by sandpapering, and only the fifth or sixth, using "pigmented" dope that included varnish and water-proofing material, could be sprayed on.

The airplane of 1920 was very much a creature of the elements. Everything it was made of had to stretch and bend with buffets and strains. This was as true of the metal members as it was of the wood and fabric. There was a newfangled substance around called aluminum but nobody used it yet. William Boeing, in the vanguard of innovation, framed his fuselages with tubes of steel. To fashion a fuselage with longerons and cross-members of metal, his workers set up their forms on sawhorses. Then they cut the tubing to length and machined the ends to fit according to blueprints. There were as yet no master jigs to clamp around the tubing and to hold everything in shape. But once the longerons were in place, welders went to work in pairs, one on each side of the fuselage. As one finished fusing the truss member to the longeron on his side, the other would complete his weld on the other side. So it went, from fore to aft, one weld at a time, until the top links were all welded, and then the process would be repeated for the lower truss. Finally, after the fuselage was assembled, it was laboriously aligned, the plywood interior and canvas covering went on, and the airframe took shape.

After this the men added what they called the accessory parts: the engine mounts up forward, the landing-gear struts underneath, rudders and stabilizers aft, ailerons on the trailing edges of the wings. The propeller, made in the same carpentry shop in those days, was also wood — usually mahogany. The wire-bracing job came last, and called for fine tuning. The control cables, which had to pass over pulleys, were made of cotton cord, some of them with wire cores. But all the bracing in the fuselage and be-

Women install fabric wing panels on a Boeing plane, 1922.

tween the wings was solid steel aircraft wire, up to three-eighths of an inch in thickness. The day had passed when airmen thought a well-made plane had guy wires so numerous that a bird released between the wings could not fly out. But the wires were important to the elasticity of such a contraption, and so were their fittings, the thimbles and ferrules and shackles that fastened their ends, and especially the turnbuckles by which pilot or mechanic could tighten the wires.

Last of all the engine was uncrated and installed. Engines came extra in those days, and indeed right up to the time of the Douglas DC-3 the buyer was always billed separately for a product thought to be quite distinct from the airframe. Either way, it wasn't exactly big business in 1921. Each flimsy little crate took about a month to put together, and cost about $15,-000. A grand total of 302 of them was built in that year, all of wood. Could such be the future Wings of Man? Could commercial flying possibly spring from such tiny roots?

The answer is that it did. The history of commercial aviation in the United States is a true adventure story — but it didn't all come about at once, and it didn't come about in the way you might suppose.

In most places around the world airlines began when enterprisers acquired planes, hired pilots to fly them, and advertised for passengers. That

President Woodrow Wilson awaits the inauguration of the Air Mail Service in Washington's Potomac Park, May 15, 1918. Left to right: Otto Praeger, M. O. Chance, A. S. Burleson, President Wilson.

may seem nowadays the logical, even the only, way to launch air transport. But that is not how scheduled flying got started in the United States. In our country organized air transport grew out of flying the mail.

Of all the arms of government the War and Navy departments first reached out to adopt and exploit the Wright brothers' invention. But the Post Office Department also recognized the flying machine's potential for the peaceful task of delivering the mail. From the moment Earle Ovington, as a stunt at a 1912 Long Island air meet, flew a dozen miles to drop a handful of letters for the Hempstead, New York, postmaster to mail, the Post Office Department began requesting funds to experiment with delivering letters by air. When the Postmaster General finally got his appropriation, six years had passed and World War I was raging. The Army provided the planes and the pilots for the test.

May 15, 1918, was "the first real spring day in this part of the country," reported New York's *Aerial Age*. The Signal Corps and Post Office had made the plans, and everything was supposed to be ready for the launching of the nation's first mail flight — 212 miles between Washington and New York. Military aviation provided as yet only the sketchiest cross-country training, so anything could happen that day — and did.

The flights were organized in relays: the planes — Curtiss biplane trainers with the pilot in the rear cockpit and the mailbags in the front one — were to take off simultaneously from Washington and New York and fly to Bustleton Field, north of Philadelphia, where two other pilots would take over and fly the second leg of each trip. "The much envied lieutenants who shared the adventure" — so *Aerial Age* described them — were Torrey Webb at New York, a recent Columbia graduate from California whose uncle and aunt had come to see him take off; George Boyle at Washington; and Joe Edgerton, the senior pilot, and Paul Culver, who would be waiting at Philadelphia to complete the flights.

At Belmont Park, the New York starting point, the Post Office people rushed the mailbags by railroad car from the city, and in three minutes Webb had them aboard. At that moment dignitaries brought by special train from New York were in front of the hangar listening to a speech by Alan R. Hawley, the amateur balloonist who headed the Aero Club of New York. "There she goes," someone shouted, and the listeners turned to look across the running track. "The mail waits for no man," exclaimed Rear Admiral Bradley C. Fiske, who leaped to his feet and led three cheers as the plane "circled the field, reached the proper elevation — about 400 feet — and streaked off to the southwest." Behind pilot Webb's back an auspicious breeze kept the flags on the Belmont racetrack grandstand fluttering.

In Washington the day was cloudless; a fresh wind blew across the Potomac. To watch the start of the country's first scheduled mail flight President Wilson and congressional leaders broke away from their wartime duties and assembled on a polo field by the river. Postmaster General Burleson made a speech. President Wilson handed the downy-cheeked Lieutenant Boyle a special letter to be auctioned off by the mayor of New York for the Red Cross. There was a delay when the engine of the Jenny, as the Curtiss trainers were called, failed to start. The President was heard saying to Mrs. Wilson, "We are wasting valuable time here." A mechanic discovered the gas tank was empty and filled it. Then at 11:29 Lieutenant Boyle climbed into the cockpit again, shoved his goggles down over his eyes, and took off over the trees.

Alas, as novices will, Lieutenant Boyle lost his way. Instead of heading north he flew south and an hour later brought his Jenny down on a pasture at Waldorf, Maryland, farther away from New York than when he started. Lieutenant Webb had better luck flying the other direction. Following the tracks of the Pennsylvania Railroad almost all the way, he landed safely at Bustleton. Lieutenant Edgerton, picking up the relay, delivered the four sacks of mail from New York to the custody of the Washington postmaster three hours and twenty minutes after it had left New York.

When Lieutenant Culver learned of his comrade's embarrassing mishap in Maryland, he took off for New York with a bundle of Philadelphia mail. From the government field at Mineola, near Belmont Park, two Army scouts flew out to meet his plane. When they were sighted, five more military machines were seen to have joined them and "in V formation like migrating geese swept up behind Culver's plane in a triumphant procession." Somehow most of the guests who had come out from New York had lingered and, "with a large part of the local population, were on hand for the arrival of Lieutenant Culver."

The day's experience, commented *Aerial Age* with a slightly censorious air, showed "how absolutely necessary" cross-country training was and how helpful two engines rather than one would have been in enabling the pilot to "keep the machine in the air until he has found the desired landing place."

The nation's first airmail stamp cost twenty-four cents and was affixed to all 3,000 letters carried; it pictured the Jenny that flew them. By a printer's error 200 of the stamps showed the plane upside down. These became philatelic prizes worth $75,000 apiece years later.

In August the Post Office took over the flying. Soon afterward the war ended. The New York–Washington experiment, although it continued, did not prove much because the planes barely beat the trains, even when they arrived on time. The Post Office Department, recruiting a force of ex-military pilots, laid plans to start a much bigger service — New York to Chicago and San Francisco; and Boston to New York and Florida and on, via Key West and Havana, to South America. That was ambitious, too ambitious. When the Post Office tried to start New York–Chicago service in mid-December, weather forced down every plane and it had to be abandoned.

The official in charge, Otto Praeger, hardly knew a monoplane from a biplane. Pilots said that he would look out the department window, and if he could see the Capitol dome, he expected them to fly that day, whether in Washington or New York. One blustery afternoon in March 1919 pilot John Miller was sent north from Bustleton despite Weather Bureau warnings of forty-eight-mile-an-hour winds. At Trenton he ran into thick snow. Blundering into New York, he just missed hitting the Woolworth tower. Over Long Island his engine cut in and out three times. Unable to see or even control his plane, he finally crash-landed in a field at Great Neck. Unscathed, he handed his mailbags to the local postmaster and, as the official report stated, "returned to Philadelphia by train by order of Mr. Praeger at 5:30 P.M." Making the same trip later that week, pilot E. Hamilton Lee had a harrowing time after running into a wall of fog at Staten Island. As Lee related later: "I flew a few feet above the water following the shoreline as well as I could. And a damn boat was coming up the Narrows right at me. I had to pull up — and there I was, flying in blind fog. I had no instrument except a compass that was spinning with every bump. I tell you" — Lee flew thirty years with the Air Mail Service and United Airlines and was often called the "flyingest airline pilot" —

It was the biggest fight of my life. I flew in that soup for maybe an hour while I tried to shake off the sensation that I was losing level flight when I wasn't. My experience in aerobatics saved me. I had to sit in that soup, letting down slowly, making turns, looking not for sky but for water. When I finally saw the water, I kept enlarging my turns until I saw the shoreline. I saw a field and was measuring it for a landing when I had to pull up over trees.

Back in the soup, I had to do it all over again: careful turns, round and round, until I came out over the water, then wider circles till I saw the ground. This time I went for it. A tree top tore the fabric of my plane — a bough ripped through the fuselage and ripped my pants. I landed — on Staten Island, after all.

I called up Belmont Park. They weren't surprised. They came with a barge from Long Island and found me near shore, and I went back to Belmont Park with them, the plane and the mailbags. They said, "You better not fly tomorrow." But I'd heard too many stories about that. When your nerves are on edge, you may lose it if you don't go right back up. I did, next day, but I shook all the way to Philadelphia.

A couple of mornings later pilot Leon Smith, due to fly to Washington, found fog covering the Belmont racetrack. Smith, known as Bonehead Smith because he had been hit on the head by a propeller and survived, suggested waiting until the weather lifted. Field Manager Harry Powers immediately reported this to Washington. Back came a wire sent by Praeger: "Fly by compass. Visibility not necessary." Smith declined to take off. Powers then called pilot Lee, just back from Philadelphia, to report to the field and take out the mail. When Lee arrived, he supported Smith's opinion. Informed that Lee had refused to fly in zero visibility, Praeger wired: "Discharge both Pilots Lee and Smith."

With that action all six pilots on the run went on strike. For three days not a mailbag moved. Praeger ordered Lee and Charles Anglin to Washington as spokesmen for the pilots. The settlement was a compromise. In future the field manager would make the decision to fly. But if the pilot disagreed, the field manager would take up a plane and make a circuit of the field to show that conditions were flyable; if the manager was not a pilot, he was to sit in the mail bin just behind the engine while the pilot took him up to see how bad things were. The pilots were willing to do this, because the field manager would be laying his life on the line in support of his ruling. There were fewer weather arguments after that, but the rule still prevailed that pilots were paid only when they flew.

By far the largest number of planes left over from the war were De Havilland DH-4s, and the Post Office got a hundred. With these for a starter Praeger was bent on expanding airmail service right across the United States, first between Chicago and Cleveland, then the long jump from Chicago to San Francisco, and last of all between New York and Cleveland, the part where his airmen had come to grief the winter before.

It was before the days of instrument flying, radio beacons, or even weather reports. If a pilot got caught in clouds or fog and lost sight of the horizon, it was not long before he lost control and fell. Ham Lee had barely escaped that fate in the fog off Staten Island. When the clouds were low, you had to fly close to the ground, close enough to see it; the lower the clouds, the lower you flew, dodging steeples and jumping trees and telephone lines. Doing just this, Ham Lee almost rammed into the ship coming out of New York harbor. Elsewhere the fliers followed the railroads from town to town; in bad weather they almost collided with oncoming locomotives.

Service began between New York and Cleveland in the summer of 1920. Along this stretch the pilot could follow rail lines and rivers only until he came to the Allegheny Mountains. Then his route struck across some of the highest and most rugged parts of Pennsylvania — almost exactly, as it happens, along the automobile shortcut known to millions of motorists since 1960 as Interstate Highway 80. The highest mountains were 2,200 feet, but the densely wooded ridges were almost always hidden in clouds. The gaps between them were hard to find, and the twisting valleys through which the pilots felt their way often ended abruptly without exit. This was the Graveyard Run, and right at the center, among the thickly wooded Pennsylvania mountains, was the Air Mail's midway base — Bellefonte.

Pilots had been flying this chancy course daily except Sundays for six months when Dean Smith, six feet four and a war veteran at twenty, was assigned to regular duty flying to Cleveland from the grassy clearing at Bellefonte. He said later:

I took off and headed through the gap in the first ridge. The briefing the chief pilot gave me the night before about the route proved a big help. I made it over Rattlesnake Mountain and followed the river to Clearfield without much trouble. Opaque veils of cloud forced me to twist and dodge, and the squalls grew heavier. By the time I reached the slope leading up to the plateau the clouds were so solid that I had to circle back. I hated to give up. Over Clearfield again the sky looked brighter to the north so I blithely headed that way, ignorant that I was flying over some of the wildest country in Pennsylvania, high and rugged, with few houses and no fields for 50 miles around.

Just then the engine stopped cold. When the timing gear in a Liberty engine fails, one second it is roaring along even and strong and the next there is a tremendous, loud silence. I quickly twisted all the knobs and gadgets in the cockpit but there was no response. To my left was a cuplike basin with a small clearing. One thing I could not know: the clearing was choked with brush and weeds, hiding a three-foot ledge of rock directly in front of my landing spot. The ledge slammed into the undercarriage as I hit. The plane snapped like a popper on the end of a bullwhip. I was catapulted into a long headfirst dive, like a man shot from a cannon. Fortunately I landed in the brush and rolled to a stop in a sitting position. The padded leather ring that rimmed the cockpit hung from my neck like a lei. I was still holding the rubber grip pulled loose from the control stick. My seatbelt lay across my lap. I felt around to determine that I had no broken bones. The wreckage of the plane was piled in a heap, like crumpled paper.

Smith stumbled a mile and a half downhill and located a rural mail carrier who helped retrieve the mailbags. He drove Smith behind his sturdy mare ten miles to Pithole, where Smith caught the train to Cleveland. "Quite a trip, my first flight with the mail," thought Smith. But at Cleveland there was little comment about his accident. Engine failure, forced landings, and crackups were so common on the Air Mail that so long as the pilot was not seriously hurt, no one gave them another thought.

Rube Wagner, who came on as a reserve pilot a little later, learned his first route, the equally hazardous run from Bellefonte to New York, by trailing Smith. Westbound, Wagner trailed Charlie Ames to learn the way. He said later:

It ran from Mineola to the Statue of Liberty, then to the Delaware River as far as to Philadelphia, then west and up the Susquehanna. There was a power line across the river, and you had to be in the middle to go under it. Robby Robinson was killed — decapitated — when he hit those cables the first summer, but we continued to fly that river. Trailing Charlie Ames, who was killed later when he failed to adjust his altimeter after leaving sea level and crashed in fog 100 feet below a crestline near Bellefonte, I got to Ring Mountain. That's a mountain in the Blue Ridge with a saucer in it. Charlie got in this crater, near Sunbury, and flew round and round in it looking for a way out. Finally he found it, got in the clear, and we slipped over and into Bellefonte.

Resolved to show the incoming Republicans what the Air Mail Service could do before they replaced him, Praeger ordered his pilots to fly the mail nonstop across the country. This involved the first scheduled night flying, and in the dead of winter at that. The weather — it was Washington's Birthday, 1921 — was terrible. Four pilots started at dawn, two from New York and two from San Francisco. One of the eastern pilots was soon forced down; the other was grounded by snow at Chicago. A noted veteran, Cap Lewis, flying from San Francisco, crashed to his death in Nevada. By evening the weather worsened and it looked as if the transcontinental night mail was going to be a tragic failure.

But the fourth plane still flew. It was 7:45 P.M. when Frank Yaeger, flying the leg from Cheyenne, taxied up to the Post Office shed at North Platte, Nebraska. Waiting on the line was Jack Knight, handsome, firm-jawed, the very picture in his leather flying suit of what Americans thought pilots ought to be. Knight waited almost three hours at North Platte while mechanics repaired a broken tail skid. Then he was off. At two thousand feet Knight could make out the river through the holes in the clouds, and at several towns along the way people lit bonfires to mark the way. At 1:15 A.M. he was on the ground at Omaha, having a steaming cup of coffee at the all-night stand across from the Post Office shed.

Bill Votaw, the Omaha field manager, informed him that all the other planes were down: the plane that was to have returned east with the Omaha mail was grounded at Chicago. It happened that Jack Knight's father lay ill in Michigan. The pilot accepted the challenge. Having flown the 276-mile leg from North Platte, he agreed to fly the mail 435 miles through the night, over terrain unfamiliar to him, to Chicago.

Snow was falling as Knight took off eastward. All he had to go by was a Rand McNally automobile guide and a Weather Bureau report that he would be bucking a twenty-five-mile-per-hour crosswind. Pointing his machine well to the north to allow for drift, Knight bored eastward until he caught a glimpse of the capital dome at Des Moines. Since the snow was too deep to stop there, he flew on to Iowa City, following the railroad track. There the ground crew, figuring that all flying had ended when the westbound plane out of Chicago was canceled, had turned in. A lone watchman heard his plane and went out with lanterns to help him land. Knight touched down just as his fuel tank ran dry. He got the watchman to gas him up. Then Jack Knight was on his way again, flying over fog until dawn showed him Chicago. He landed at Maywood Field, and with two other pilots completing the remaining relays by daylight, the mail was delivered in New York just thirty-three hours and twenty minutes after leaving San Francisco. Jack Knight was a national hero; Otto Praeger called this first night flight "the most momentous step in the development of civil aviation."

The price of such zealous advance ran high. In 1920 and 1921 nineteen Air Mail pilots lost their lives in eighty-nine crashes. "The forced landing was a way of life," said Dean Smith. "In 1921 we averaged one for every 800 miles — almost one every round trip. We had a fatal crash for every 130,000 miles we flew." At that rate a mail flyer could not expect to live more than four years — indeed, thirty-one of the forty mail pilots first engaged by the Post Office Department lost their lives while flying for the government. Praeger's successor decided too much had been attempted too soon. Congress trimmed the Air Mail's appropriation.

The DH-4s that the Post Office was obliged to use had been known in the war as "flying coffins." Liberty engines, partly because they were flown so much of the time at full power, accounted for a high proportion of the Air Mail's 810 mechanical failures in 1921. Being water-cooled, they leaked. On hot days Ham Lee would spot a threshing rig, land alongside in the stubble, fill up with water from the steam engine's boiler, and take off again. More serious, the gear linking crankshaft with camshaft had a way of shearing. When this happened, the engine stopped and there was a sudden, depressing silence. Compelled on one such occasion to descend abruptly into a barnyard, Dean Smith wired: "Forced landing. Killed cow. Scared me." Later, strong, stub-toothed gears were installed to correct this weakness, but then other things went wrong. Once a connecting rod blew, and Ham Lee's engine caught fire. Said Lee, "I sideslipped the plane and blew the fire out in the air."

When the De Havillands cracked up, they usually burned, which is why they got their grim nickname. The Post Office Department sent sixteen of them to McCook Field to see if the Army could find out what caused the fires. Their water-cooled engines had long exhaust pipes going back along the sides of the plane and fuel tanks in the fuselage between. The Army ran these planes into a stone wall and found that the tanks, breaking up, would spew gasoline onto the hot exhaust manifolds. So they took off these long exhaust pipes and put on short ones, one for each cylinder. "It was my first lesson in human factors," said Jerry Lederer, the pioneer Air Mail engineer who later became a leading safety expert. The Army had tested the short pipes only by day. When the Air Mail flew the planes at night, the flames from the bob-tailed exhaust pipes blinded the pilots. "I had to take them off," said Lederer.

By hard-won experience on the ground and in the air, the Post Office Air Mail Service gradually improved its performance. "Only a tombstone could hold us down," bragged pilot Paul "Dog" Collins. When Collins or anybody else arrived at the end of his 250-mile stint, two planes stood waiting, their engines turning over. Directly after the incoming plane landed, the mail bags were hauled out and flung onto one of the waiting planes; it took off with a delay of less than two minutes. The second plane stood with its engine running until the local manager was satisfied that the first machine was well on its way. If anything showed signs of going wrong within the first few minutes after takeoff, the pilot with the mail turned around immediately and came back, and the mail was transferred to the second machine.

Every pilot was sworn in as a United States Post Office Agent and carried legal power to swear in anybody else with special authority to help him deliver the mail. And he carried a gun. Thus if he had a forced landing in the countryside, he got hold of the first farmhand he came across, swore him in, handed over his gun to him, and set him to guard the machine while the pilot himself made off to the nearest telephone to get help.

Yet because the planes flew only by day, surrendering their mail each evening to be carried forward in the night by train, air service meant little to the public. The Post Office was selecting pouches for the planes at ran-

The Post Office first flew the mail in converted wartime De Havilland DH-4s
with Liberty engines.

Pilot E. Hamilton "Ham" Lee

dom, so if your mail went by air you never knew it. Paul Henderson, the new chief, believed that if the service was to amount to anything, special airmail stamps and rates would have to be reinstituted — and "we will have to fly by night."

This time the service was ready. Following the example of Army experimenters who built a successful eighty-mile lighted airway between Columbus and Dayton, Ohio, the Post Office Department had installed powerful lights at all its landing fields between Chicago and Cheyenne and set up a string of flashing beacons every twenty miles in between. With the aid of these lights the department could schedule flights that started out from New York and San Francisco by day, followed the lighted airway between Chicago and Cheyenne through the night, and then completed their journey the next day. On the first test Ira Biffle at Omaha said, "My God, it's black as the inside of a cow up there." But it wasn't so bad. Rube Wagner said, "You found you could see the horizon same as you could by day, and you learned to fly the same, even with the same landmarks to go by." By the following year the government had lit the entire three-thousand-mile route across mountains, desert, and prairie. The nearly empty mailbags soon filled. Business firms, although still sending duplicates of everything by ground transport, began to send letters by air. Henderson asserted that banks using the Air Mail had cut the time for clearing checks from five to two and a half days. Finally, on July 1, 1925, the Post Office started overnight service between New York and Chicago, and business really began to pick up.

At last the Post Office Air Mail came into its own. When Shirley Short flew the first night mail into Cleveland, a crowd of twenty thousand people gave him a wild ovation and a woman sat right down and wrote a poem, "Hail! Brave Messenger of the Air." All of a sudden, cities like Los Angeles, St. Louis, and Boston decided that they had to have airmail too. The transcontinental service was delivering the mail in thirty hours, and the rate of completions was high in summer if not in winter. The pilots found that thunderstorms, however impressive at night and terrifying from an open cockpit, were not so serious as they seemed. Night flying in the Rockies, where the valleys and passes were big and there was room to turn around if you had to, was not so bad; and if the weather was bad, you didn't fly.

Of course the Air Mail flew when the airlines of Europe, carrying passengers in these same years, took no such chances. The early European airlines knew their passengers must not be scared by being hurtled through bad weather and at night. The British aviation editor C. G. Grey said, "You cannot scare a mailbag." And that was just the difference. Thus it was that the Post Office Air Mail pioneered night flying, its greatest accomplishment.

The achievement was all the greater because it was accomplished by

100 percent contact flying. That is, it relied on the pilot's eyes to follow the lighted beacons that marched single file across the country — and on that extra sense that enabled him to find his way when illumination was not enough. At Chicago's Maywood Field, where fog often overlaid the landing ground with a low, white blanket, ground crews tried putting up captive balloons at the four corners of the field in hopes that the pilots would see them sticking up like bedposts through the fog and so find their haven. They also put out flare pots in the night. Once Dean Smith literally chopped his way down to the ground at Maywood. Discovering that his propeller wash made a gouge in the top of the fog blanket as the plane dipped in, he buzzed diligently back and forth for nearly an hour until he had dug out a ragged trench three hundred feet deep and could just see the end of the runway at the bottom. He landed safely but rolled past the edge of the hole — so dense was the fog he had cut through that he couldn't see to taxi; men had to come out and guide him to the hangar.

Not all the Air Mail pilots were content to fly by the seat of their pants. The most persistent of those who weren't was Wesley L. Smith. Even among the freebooters who flew the airmail, Smith was a character. According to Ham Lee, he was "a warbler, a singer." Dean Smith (no relation) called him "the most stubborn man I ever knew." He was hellbent on flying no matter what. He had more crackups than anybody — fifteen complete washouts of planes. He could be heard muttering, "Nobody will ever fly over me," meaning no other pilot would carry mail through when weather kept him down. One day Dean and Wesley Smith were scheduled to fly first and second sections, respectively, of a mail trip from Cleveland to New York. As Dean Smith told it, "I ran into fog and couldn't get through. I returned and landed. After a bit Wesley Smith came back. First thing he said was, 'Where's Smith?' and shot out his words, 'Gas her up, goddam it.' Then he saw me standing there, and he smiled such a smile. He didn't try to go any further that day."

But Wesley Smith's determination to bull through also made him the Air Mail's most notable pioneer in instrument flying — or blind flying, as they called it then. The first instrument he took aloft, it is said, was a half-empty whiskey bottle, a kind of spirit-level that he placed atop his instrument panel. Then he found out that the Air Mail's instrument mechanic had gotten an old German glassmaker in Chicago to fashion a curved tube with a ball in it, like a carpenter's level. This he fastened on his front panel next to his altimeter — the makings of a turn indicator. He had already been using a clock on the panel to time his turns. But with this latest gadget on his plane it became evident that Wesley Smith could fly when nobody else did. Since pilots were paid only when they flew, this was of considerable interest to Smith's colleagues. Not long after, in Chicago, Shirley Short showed Rube Wagner his new turn indicator, made by

Pilot Wesley Smith

Sperry, and said, "You can fly in clouds with this." Wagner thought about it and answered, "With this, we can think about carrying passengers." By the position of the needle on the turn instrument, which was stabilized by a whirling gyro, a pilot could tell whether his wings were level or tilting. But it would take more than instruments to find one's way cross-country in clouds.

The key to flying cross-country in poor visibility was not what you could see, inside your plane or out, but what you could hear. The Air Mail Service already knew this. Starting early in the 1920s government researchers fixed upon radio as the most promising aid to guide planes through clouds and fog. The idea — it was German — was to use radio not simply for communication but also as a means of navigating through bad weather and good. Radio stations, emitting beams in fixed directions that pilots could follow to and from airports, were established by the Bureau of Standards at the old Air Mail field at College Park, Maryland, in the early 1920s. As early as 1926–27 it was tested in the field. Naturally it was first tried at Bellefonte, and Wesley Smith was right in the middle of it.

The Post Office Air Mail Service launched its frail craft when men knew less about the airways above North America than the Vikings knew of the North Atlantic when they put out in their longboats and sailed west from Iceland. Nobody had attempted to ply these skies before them, and the masses of air into which they flew were vaster, more volatile, and far more subject to perilous change than the waters of the sea. Say it was the valor of ignorance, but someone had to lead the way. Even in the best of weather engines would conk out, and there were times when it seemed that mail planes were dropping out of the skies to alight in some farmer's field almost as commonly as crows. And when the clouds piled up and winds lashed, the mail planes were in hazard as great as any storm-tossed ship on the sea, with a difference — up there it was one man, and one man alone, against the elements.

It was a small band of men indeed — seventy-five in all — who less than sixty years ago blazed the trails for the air commerce of the United States.

Said Dean Smith, who survived to speak a half century later for these unsung pathfinders of the air:

People asked me why I liked being a pilot, when I flew the mail and took such chances of getting killed. I would try to explain, but never could find the words to explain it all. I knew that I could fly and fly well, and this skill set me apart from the run of the mill. I certainly had no wish to get killed, but I was not afraid of it. I would have been frightened if I thought I would get maimed or crippled for life, but there was little chance of that. A mail pilot was usually killed outright. Then, too, sometimes I was called a hero, and I liked that.

One of the most rewarding things about a mail pilot's job was the high pay and the high percentage of leisure time, which made for a merry life, even if indications were that it might be a short one. As a normal thing we worked two or three days a week, five or six hours a day, plus standing reserve perhaps one day a week, which meant only keeping the field advised how they might reach us. I spent my time as unproductively as possible, learning to play golf, chasing girls; investigating dives and joints in the area.

But what I could never tell of was the beauty and exaltation of flying itself. Above in the haze layer with the sun behind you or sinking ahead, alone in an open cockpit, there is nothing and everything to see. The upper surface of the haze stretches on like a vast and endless desert, featureless and flat, and empty to the horizon. It seems your world alone. Threading one's way through the great piles of summer cumulus that hang over the plains, the patches of ground that show far below through the white are for earthbound folk, and the cloud shapes are sculptured just for you. It was so alive and rich a life that any other conceivable choice seemed dull, prosaic and humdrum.

In the workaday world of twentieth-century America, these men were employees, earning pay, doing a job. But what a job! The only remotely comparable employment in American memory might have been galloping over the plains as a Pony Express rider a century before. Sauntering out to their swaybacked crates, wrapped in leather jackets, trailing white silk scarves in the wind, and accepting admiring glances from ordinary, quaking earthlings, these $250-a-month postal employees opened the last of the elements to regular and fruitful traffic.

3

Passenger Lines of the 1920s

As the First World War gave aviation its first great impetus, so also did the war give it a flavor. Say what they might, the proponents of commercial aviation — still called aeronautics, a term that made room for dirigibles and such — had to admit that flying after World War I remained pretty much a stunt. The twenty thousand Americans who learned to fly in the war were thought of by their fellow citizens, and by themselves, as daredevils. When a large proportion of these former war birds bought the government's stock of surplus planes and sought to make a living by flying them, their antics hardly gave aviation a name for regularity and reliability, much less safety. Traipsing footloose from town to town, these itinerant merchants of flight dived, looped, and barrel-rolled to herald their arrival at each town, and when the curious and venturesome flocked to the hayfield where they landed, they took them up in open cockpits at five dollars a whirl. This was called joyriding. The biggest business was conducted during county fairs. Circling over the fairgrounds in an open plane was so much like a ride on the roller coaster below that if by chance the wobbly Jenny happened to straighten out and fly steady for a moment, the pilot would oblige the customer by standing the plane on its wing or even looping it to give him the thrills he expected. If these were passengers, the business of flying them was essentially entertainment. It is estimated that a million Americans first flew in this fashion, and it can be supposed that a very high proportion of them were males of eighteen-to-twenty-five, an age group identified then and now by insurance concerns as the highest-risk segment of the population.

Even when promoters of flying fitted out somewhat larger planes with enclosed cabins and announced scheduled trips between cities like any railroad or steamship company, circus atmosphere was not lacking. On June 2, 1920, Helen McLean climbed into the tiny cabin of a single-engine Curtiss seaplane for Aero Limited's first New York–Boston flight. Flamboyantly publicized as the "first flying saleswoman," she took off from Manhattan's East River with the mission of bringing back more orders for her New York drug firm than the cost of her round-trip ticket. At the price, which was $300, it almost had to be show business. To be sure there was a second passenger, one R. T. Bellchamber of New York, willing to pay this extravagant price to ride with pilot Harry Rogers on what was billed as the first in a thrice-weekly service between New York and Boston.

It is not recorded that Aero Limited, which got its start at scheduled aerial entertainment the summer before by flying passengers between New York and the fleshpots of Atlantic City, ever made another trip to Boston. But this company's operations, particularly in the winter of 1919 when it shifted its seaplanes to Florida and carried some twenty-two hundred vacationers from Miami to Bimini ($80), Nassau ($150), Havana ($100) and back, caught the eye of another marine-minded aerial promoter.

This was a Manhattan Cadillac dealer who had made a small fortune building flying boats for the Navy at a New Jersey plant during the war. Inglis M. Uppercu was a prominent member of the aviation industry, having built airplanes as early as 1915. When cancellation of military orders at the war's end forced him to close down his Aeromarine Plane and Motor Company, he cast about for other ways to put his wartime profits to use in aviation.

Uppercu still had one or two of the twin-engine flying boats he had built for the Navy. Designed for antisubmarine patrol, these boats could fly farther and carry larger loads than most war-surplus planes. In 1919 Uppercu sent one of his F5Ls south to test the possibilities of a route from Key West to Havana. Hugging the shoreline and reading the big white numbers painted on lifeguard stations and lighthouses as he flew ("No. 182 told me I was passing Cape Hatteras, No. 193 meant Cape Fear"), pilot Charles Zimmerman made it safely from New York to Florida. For the Key West–Havana leg he made a strip map, mounted it on rollers in a small box, and unwound it inch by inch as he went, successfully making the eighty-mile crossing.

The first form of organized commercial flying that arose in those days of barnstormers, wing walkers, and assorted aerial vagabonds was the "fixed-base operation." This presupposed that the entrepreneur owned or leased a field and a shed or two, kept a variety of planes there, and contracted to fly people or goods, sometimes quite regularly, to agreed-upon destinations. The term might have described West Indies Airways, which had been carrying fishermen and other pleasure seekers between Key West and Havana, and which Uppercu took over in the summer of 1920.

But Aeromarine West Indies Airways, which Uppercu then formed, was much more than a fixed-base operation. It published timetables and carried passengers. Considering that its first routes linked Miami with the Bahamas and Key West with Havana, these passengers were flying for fun. But on August 30, 1920, Aeromarine won a U.S. Post Office contract to carry mail from Key West to Havana. Four months later the Navy, "to encourage development of commercial air transport," sold Uppercu a dozen of his old flying boats from its war-surplus stores at a bare third of cost. Christened by New Jersey Governor Edwards, the first of these took off for Florida with every one of its eleven wicker seats filled, followed Zimmerman's track southward, and in due course began plying between Florida and Cuba.

The first year was Aeromarine's best. Operating seven flying boats, the line flew 95,000 miles and carried 6,814 passengers. In winter months the airline operated scheduled flying-boat trips not only between Key West and Havana but also between Miami and Palm Beach and Bimini and Nassau. Then in the summer Aeromarine's flying yachts moved north to the Great Lakes and became — in the phrase of President Charles Redden, a former Buick executive — "airborne limousines." Here passengers traveled for business as well as pleasure. Each morning promptly at nine one boat took off from the Detroit waterfront for Cleveland, and another left Cleveland for Detroit. Luxurious houseboats served as passenger terminals; along the way four "signal towers" kept watch and telephoned reports of the flying-boats' progress. Fare was $40 one-way; the distance was 112 miles; flying time was ninety minutes compared with five hours by train. Reservations were necessary, especially on the weekends. "Each seat is numbered," said the *Detroit Free Press*, "and the passenger knows beforehand where he will sit."

In the absence of government regulation of any kind, Aeromarine imposed its own safety standards. Redden told old auto industry associates in Detroit that "building an efficient organization, training pilots and mechanics, and instilling the safety-first idea" turned out to be the firm's real cost determinants, not the fuel bills. Aeromarine's operations boss was Glenn Curtiss's old test pilot; one of its pilots was Eddie Musick, later to be famous for pioneering transocean routes for Pan American Airways. Asked why the line used flying boats, Redden said that people who would not go up in a landplane seemed ready to travel in a plane that was provided with a hull and could therefore always find a safe landing place on water. Also, he said, people who would not dream of traveling at the three-thousand-foot height deemed safe for flying over land appeared quite unworried flying in seaplanes ten or fifteen or even a hundred feet above the water. He might have added that the company was also spared the expense of acquiring and leveling runways for the planes to take off and land.

In its second year Aeromarine expanded. Maintaining a fleet of fifteen flying boats, the line carried 9,107 passengers on some two thousand scheduled flights, all without serious mishap. But its fortunes were adversely affected by an incident that winter.

The flying boat *Miss Miami,* a fixed-base operator's plane hired to carry two vacationing Midwest businessmen, their wives, and another woman from Miami to Bimini, suffered engine failure and came down heavily in the Gulf Stream. While ship after ship passed by unseeing, the five passengers and the pilot bobbed for days atop their waterlogged plane. First the three women slipped off the wing, plunged into the water, and were seen no more. Then the men, parched, weak from hunger, and intermittently delirious under the fierce tropical sun, weakened. First one of the Kansas City executives dropped off into the shark-filled water, soon after, the other. That left but one survivor on the drifting hulk of the *Miss Miami.* On the third day a ship, passing close aboard, saw the outstretched body atop the wing. A lifeboat crew took off pilot Bob Moore. Salt-caked, shriveled, burned as if by fire, he was revived to tell a story that was spread across the nation's front pages and brought outraged editorials on the lack of safeguards for passenger air travel. "There is no constituted authority in commercial aeronautics — the thing is of a wild nature in the eye of the law," wrote Captain Earl Findley, a former officer of the flying corps, in the *New York Times.* He said that in twenty months thirty-six people had died in flying accidents, mostly in stunts, and only ten in "legitimate commercial flying." Not long after, one of Aeromarine's flying boats was forced down by engine trouble twenty miles short of Havana. Two passengers who climbed atop the cabin were swept to their death, and two children went down with the leaking craft.

Continuous performance of commercial flying, day after day on scheduled routes, required planes, pilots, mechanics, managers, airports. But if such a company were to keep going long enough to attract steady patronage, said Captain Findley, one more party must join the venture: "The Banker, with a capital B, must participate — but will he?" Uppercu had found that "you cannot get one nickel for commercial flying." Obviously, the bankers had turned their backs on Aeromarine Airways, which went out of business in 1923. But flying had fans in Wall Street and was not to be denied.

Among all the pioneers of commercial flight, Juan Trippe was Wall Street's own child. Born a commuter's ride away in North Branch, New Jersey, he was the son of William Trippe, senior partner in the Wall Street investment house of White, Weld and Company. The family name traced back to colonial Maryland; an early bearer captained a ship in the war against the Barbary pirates; and four Navy vessels have borne the name more or less continuously since 1812. Trippe seems to have been given his first name after an Aunt Juanita of Spanish descent. This may have been the source of his dark eyes and wide smile.

Trippe's associations with flying went back to the earliest years. His father took him to watch the country's first air race — from the field at Mineola on Long Island, around the Statue of Liberty, and back. With his father he saw Glenn Curtiss triumphantly complete his sensational long-distance flight down the Hudson Valley from Albany in 1910 by alighting on Governor's Island in New York Harbor. As a boy he watched Arch Hoxsey and Lincoln Beachey fly upside down in front of the reviewing stand at Mineola. He met both of the Wright brothers. And when he went to Yale in the war years, he joined the Navy and learned to fly with his classmate John Hambleton in Florida. Returning to Yale for a fourth year, he organized a flying club and took transportation as his chief classroom subject. He also found time to edit the Yale *Graphic;* he hired Sam Pryor, later a Pan Am vice president, as ad salesman and made a profit. There was no question but that he was to go into Wall Street, but when he went to work for the investment house of Lee, Higginson and Company, whatever his father's expectations, his own notion was clear. The work would give him access to cost figures for such transportation industries as railroads, shipping lines, and bus lines, there being no airlines yet. Even in college, Trippe could see far.

Before long Trippe heard from friends that the Navy was disposing of some mint-fresh planes from its war-surplus stores. Calling on some Yale friends to join in and putting up some of the Yale *Graphic* profits for a starter, Trippe bought nine trainers and started New York Airways. Except to fly to Southampton on weekends, Trippe never piloted the planes himself, but he learned a good deal about what it cost to keep planes flying.

As Trippe told it later, some of his best Yale friends came from Pittsburgh, from the very district represented by Congressman Clyde Kelly, chairman of the House Post Office Committee. Trippe and friends, one a railroad president's son, helped persuade Kelly to introduce a bill in 1925 opening the flying of airmail to private contractors. Aviation was of somewhat less importance to Congress in those days than the tariff on pigs' bristles. The bill passed, and soon afterward the Post Office Department advertised for bids to fly mail over eight "feeder" routes, supplementing its own New York–San Francisco coast-to-coast service.

Five contracts were awarded. Air Mail Route No. 1, New York–Boston, was the one on which Trippe and his friends bid. Of the other four, the line connecting Chicago and St. Louis went to Robertson Aircraft Corporation of St. Louis, which proceeded to hire Charles Lindbergh as its chief pilot; the line linking Chicago and Dallas went to National Air Transport, backed by Chicago and Wall Street money; the route from Salt Lake City to Pasco, Washington, went to Varney Air Lines, owned by the former California fixed-base operator Walter T. Varney; and the route connecting Salt Lake City with Los Angeles was won by Western Air Express, an airline formed by the old-time racing driver Harris M. "Pop" Hanshue with backing from some of the richest men in Southern California. These were the true progenitors of the modern airlines —Varney later became part of United Airlines, biggest and most prosperous of modern trunk routes; Robertson was absorbed into American Airlines; and Western Air Express eventually became one of the founding units of TWA. A sixth contract was awarded two weeks later to Henry Ford, the car manufacturer, to carry the mail between Chicago and Detroit and Cleveland and Detroit.

When Air Mail Route No. 1 was offered for bids, Trippe, Hambleton, and various well-heeled Yalies, styling themselves Eastern Air Transport, offered the top price possible. So did a group of Boston bankers who had taken over a Connecticut-based outfit called Bee Line, Incorporated, and reorganized it as Colonial Air Transport. The Post Office looked at these identical bids and said to the Bostonians, "You ought to get together and form one company to fulfill the contract." Trippe and Hambleton met with the men at Colonial Air Transport, and a board was set up with John H. Trumbull, former governor of Connecticut, as chairman. Trumbull was a fine figure of a man and liked the idea of an airline serving the state and doing great things for Connecticut industry, but he knew little about aviation. Airmail service began on June 18, 1926, with Trippe managing it from his New York base.

From the outset Trippe reckoned that Colonial, while collecting a solid $1,000 per trip for carrying the mail, must look for all the passengers it could get. Leonard Raymond, an early passenger, happened into Colonial's tiny Boston office one afternoon, where Trippe, with only a secretary named Peg Lanigan to help him, was busy getting out his first promotional mailing. Raymond lent Trippe fifty dollars to pay the postage. At the New York end Trippe aggressively pushed the passenger part of the business. When Anthony Fokker, the Dutch engineer who designed the best German planes in the war, moved to the United States and opened a factory at Hasbrouck Heights, New Jersey, Trippe was an early and eager visitor to the plant.

Tony Fokker was a kind of Wernher von Braun of World War I. At a time when warplanes were as startlingly new as rockets were in 1942, this young Dutchman fresh from the University of Leyden designed for the German Kaiser the most advanced of all military flying machines. After the war Fokker took his youthful imagination and unrivaled experience to the United States where, much as von Braun did, he became more celebrated for his contributions to design than for any of the machines he built.

Attracted by his ideas for a commercial plane that would provide not only speed but comfort and reliability, ambitious new contract mail carriers like Trippe and Harris Hanshue of Western Air Express journeyed to Hasbrouck Heights, just across the Hudson River from New York. They already knew that only by offering reliable planes could they induce passengers to fly on their airlines. They saw the cantilevered construction of his high-wing monoplane, without struts and without exposed flying wires. They saw the spacious interior of the cabin, with fabric-covered walls and upholstered seats. They listened to Tony Fokker talk about the limitations of wartime engines — the only kind available in the first half-dozen years after the war. To all who said they needed a passenger plane that would not only provide more seats but also attract paying passengers to fill them, Fokker replied that the answer was a trimotor plane. His reasons he stated with mathematical simplicity: "Two engines are more reliable than one, and three more than two by a factor of one-half."

By this time Postmaster General Harry New was throwing contract routes open for bids all over the country. Trippe, having ordered the first three Fokker trimotors for Colonial Air Transport, turned to bid on the best of them all — the Chicago–New York route. But when Trippe sought the approval of his board of directors, the Boston bankers, who wanted no part of such chancy moves interfering with their profits, voted him down. Thereupon Trippe appealed to Chairman Trumbull to call a stockholders' meeting. Though every single shareholder appeared for the session in Hartford, the Bostonians prevailed by a vote of 52 to 48. Trippe and his New Yorkers bethought themselves of a device in Colonial's bylaws by which they could assert control. That night on the way from Hartford they stopped off in Greenwich, held their own rump meeting as a Connecticut corporation in the railway station, elected their own directors, and voted to reinstate the Chicago contract bid. This accomplished, they caught the milk train to New York, where they arrived at dawn.

Then they went on to Washington to inform the Post Office that the bid was reinstated. About ten minutes later the Boston delegation walked into the room, and Second Assistant Postmaster General Harllee Branch had a problem. Though well disposed to Trippe and his friends, Branch said to them, "You'd better see what you can work out." The odds were against the New Yorkers, so Trippe said to Trumbull, "If it comes out against us, we want to buy at cost the three Fokker trimotors we have on order at Hasbrouck Heights." He also asked that if his group lost, the Bostonians buy them out at the level of their original investment. Though Trippe's lawyer, Bob Thach, put up a good fight, those were the best terms he could get — and Trippe was out of Colonial. Juan Trippe came that close to making his aviation career in domestic air transport.

The other airline that plumped for Tony Fokker's trim trimotor in the formative years of passenger flying in the United States was Pop Hanshue's Western Air Express. Western Air had been founded with Los Angeles money, on Los Angeles pride, and on Los Angeles opinion of the city's leading place among the nation's business and cultural centers. Thus the feeder mail contract that Western Air competed for and won was the route connecting Los Angeles with the East via Salt Lake City, not the link with the main line via the city's old rival, San Francisco. That con-

Harris "Pop" Hanshue (in shirt-sleeves) weighs in the mail at Vail Field in Los Angeles in 1926.

tract, which also required flying the mail northward to Portland and Seattle, went to a San Francisco bus line operator named Vern Gorst, whose Pacific Air Transport was badly undercapitalized and had its hands full just delivering the mail.

From the first years after the war, however, there had been more interest in passenger flying in California than in any other part of the country. California had its counterpart of Florida's Aeromarine Airways and the East Coast's flying-boat airlines — Wilmington-Catalina Airways, which began carrying passengers in 1919 between Los Angeles and Catalina Island. A half-dozen small companies sprang into existence to carry passengers between Los Angeles and San Francisco where Pacific Air Transport had only enough planes to carry the mail.

Of all the new airlines, Western Air was the one that from the start made money. The route over the desert was well suited to flying, the Douglas mail plane that Western used performed well, and Hanshue kept his costs down. Mail, of course, was the big secret; an average of five hundred pounds went on every trip. Every single round trip earned a profit of $1,140. Most of the two hundred passengers who flew with the air mail that first year acknowledged that they flew for the thrill; indeed, many of them handed their pilots twenty-dollar to fifty-dollar tips on landing.

Yet Pop Hanshue knew as well as Trippe that an airline could not progress if it remained wholly dependent on mail pay. That was also the view, as it happened, of the Guggenheim Fund for the Advancement of Aeronautics. This foundation had been established from the family's copper fortune at the instigation of Capt. Harry Guggenheim, who had been a naval aviator in World War I. The Guggenheim Fund aided in several aviation gains in the 1920s. It established schools of aeronautical engineering at New York University, Michigan, MIT, and Georgia Tech that trained such leading designers as Kelly Johnson of Lockheed and Wellwood Beall of Boeing. After Lindbergh flew the Atlantic, the fund financed the forty-eight-state tour he made in his *Spirit of St. Louis* to promote the interest in aviation his flight had aroused. It organized and financed the first successful instrument flight. And, in 1928, when airline flying was still almost exclusively devoted to carrying the mail, Harry Guggenheim was casting about for a way to demonstrate that the future of aviation lay in the development of passenger traffic.

This coincidence of views led the fund to advance Western Air Express $180,000 to launch a model airway between Los Angeles and San Francisco. This was where Tony Fokker's trimotors came in. Western Air used the money to buy three of them to fly passengers, and passengers only, between Los Angeles and San Francisco. Since Western Air did not hold the airmail contract between these two cities, it could accept this stipulation. In its first two years of flying between Los Angeles and Salt Lake City, Western Air's planes carried only six hundred passengers; in 1928, when the Fokkers flew on the model airway between Los Angeles and San Francisco, some five thousand passengers traveled by Western Air, without a single fatal accident.

Western Air Express had already assembled the essentials for operating an airline — the planes, the crews, the maintenance forces, the managers. In Los Angeles it operated from its existing base at Glendale, and in San Francisco it used the facilities of Oakland's municipal field. But the federal government, which had taken over the maintenance and development of the flyways used by the airlines under the Air Commerce Act of 1926, had not yet installed lights along the California route, and the Weather Bureau's forecasts left a lot to be desired. The main feature of the model airway experiment turned out to be the weather forecasting, which was organized by the Swedish meteorologist Carl Rossby. Rossby, a genial, wisecracking gnome of a man, had not invented modern air-mass meteorology, but he had studied with the Norwegians, Bjerknes and Sverdrup, who did. Brought over by the Guggenheim Fund, he was responsible for importing the new technology of weather analysis into the United States just when commercial airline operations stood in greatest need of its help. Scoffing at the notion that you could keep track of weather by simply reporting cloud conditions at two or three points along the airway,

Rossby organized a network of observers at thirty-eight points all over California who phoned in reports every ninety minutes of local wind, temperature, cloud cover, and upper-air conditions. At headquarters Rossby's men, tracing their curving isobars across big maps and singling out the high-pressure centers and low fronts, drew up their forecasts. Equally revolutionary, a fancy new gadget called the teletype transmitted the encoded map data to the Western Air operations offices for last-minute pilot briefings.

Harris Hanshue was a onetime six-day bike racer, a stubby man with a brush haircut. He wore gold-rimmed glasses, carried a substantial bay window, and had a whiskey voice that his employees liked to imitate. Behind his back they called him Pop. A short man, he had a tall wife who was garrulous. Hanshue rarely said anything, although when he did it usually packed a punch. And so far as he could help it, he never rode on his airline.

But when Western Air got its Fokkers and flew its experimental service between Los Angeles and San Francisco, Pop Hanshue was somehow persuaded to take a trip. As they were coming in to land at Los Angeles, the copilot went back to see if everything was all right. "Mr. Hanshue, how are you?" he asked. "I can't hear anything," wheezed Hanshue. "Mr. Hanshue, all you have to do is swallow," said the copilot. In a minute everything was all right, but as they taxied up to the ramp Hanshue looked out the window and said, "My God, there's my wife, how do I get back the way I was?"

Herbert Hoover, Jr., joined the airline fresh out of Stanford to help organize air-to-ground radio communications. Hoover hired his college friend Jack Franklin to help him. Once when Hanshue was ensconced in a suite at the Ambassador Hotel in New York, young Hoover was called in for consultations. Always careful to avoid being drawn into situations that could embarrass his father, young Hoover took Franklin along. They were assigned rooms in Hanshue's suite, where there was much going and coming, and a lot of drinking. James Talbott, the Richfield Oil chief on Western's board, was there with others. At two o'clock one morning, Pop came in from a bender and sat down heavily on Hoover's bed. "Herb, you're a good guy — for God's sake don't be like me," Pop wheezed, and lurched out.

Though no fewer than a hundred Fokker trimotors were sold to American airlines, Tony Fokker's airliner was not destined to take the honors as America's passenger plane of the 1920s. That distinction was reserved for a racketing, bucketing old behemoth that came rattling out of the same Detroit shop as the Tin Lizzie of the nation's highways — the Ford trimotor, the Tin Goose. It would be pleasant to relate that this airborne threshing machine that cast such a memorable shadow over the country's flyways was, from its shark-toothed snout to its prairie-plow tailskid, the pure and pristine product of the same native American cut-and-try tinkering genius that gave the world Eli Whitney's cotton gin, Sam Morse's telegraph, and Henry Ford's Model T. On the contrary, almost everything allegedly unique about it, including its ugliness, seems to have been borrowed from abroad.

The plane got its Ford name because a marvelously engaging talker with a shock of hair and a Chaplinesque wisp of mustache had induced Henry's son Edsel and various other motor magnates to put up money for a radically new kind of flying machine — one made almost entirely of metal. The men in Detroit listened entranced as William E. Stout, former editor of the boys' page of the *St. Paul Dispatch,* wartime plane designer at Packard Motor Company, and all-round idea man, proposed kicking the lumber kiln and dry-goods technology of the Wright brothers right out of the airplane business. Instead, Stout said he would construct a plane almost entirely of a new metal called duralumin — almost as light as aluminum and more than twice as strong, as strong as cold-rolled steel but weighing only a third as much. .

This was the kind of notion that appealed to Henry Ford, who had achieved his great breakthrough to the low-price Model T in part by fabricating engine parts out of a then little-used metal called magnesium. In the flurry of decisions, nobody mentioned that the corrugated metal of Stout's design was just like that on German Junkers planes, or that the exceptionally thick wings also bore close resemblance to Junkers models. When Stout's plane flew, powered by a single Liberty engine, Ford not only bought Stout's plant and made Stout manager but also launched Ford Air Transport, using Stout's all-metal planes to tote spare parts on regular daily schedules between Detroit and Ford assembly plants in Cleveland and Chicago.

When the Post Office invited bids from private operators, Ford bagged the contracts to fly between these two cities; since his planes and crews had already been put through their paces flying cargo, Ford Air Transport was the first, on February 15, 1926, to carry domestic contract mail. But Ford, like Trippe and others, wanted a plane that could carry passengers as well as mail. Stout therefore went back to his drawing board. Following Fokker, he too would produce a trimotor with a commodious cabin — but it would be all metal.

The sequel was much like the story of what happened when the men of Detroit set out in 1917 to design their Liberty engine for World War I. At a very early stage Bill Stout, like the Detroiters before him, seems to have come a cropper. The three-engine aircraft known as the 3 A-T was evidently such an awkward and impractical design that Henry Ford decided to take Bill Stout off the job. Not only that, he saw that a timely plant fire wiped out the offending airplane and all the blueprints and papers con-

nected with it. The first three-engine aircraft produced on the Ford premises could therefore be declared never to have existed.

The next version, the 4 A-T, otherwise known as the Ford trimotor, became famed in song and story as the Tin Goose. The machine became
America's idea of the airliner. However, Bill Stout, who proposed and
thought to build it, had very little to do with the successful version. Once
Stout had been sent to the showers, Ford turned the ball over to William
Mayo, his chief engineer, and a team of designers that included James
McDonnell, later boss of McDonnell-Douglas. Under imperious orders to
turn Stout's monstrosity into a winner, the harried automen took help
wherever they could find it. One day Admiral Byrd dropped in at Dearborn with his new Fokker F-7.

The story is that the Admiral was given a warm welcome, and his
Fokker was hangared for the night. All that night the Ford team buzzed in
and out of the high-wing Fokker monoplane, bending copper into tem-
plates as they measured every foot of its length, breadth, and height. Only
a few people were told of the all-night measuring party, and certainly not
Henry Ford. But those who heard were not surprised to discover a remarkable similarity between the metal airfoil of the Ford 4 A-T and the
airfoil of the plane that spent that night at Dearborn.

The Ford 4 A-T was built like a brick wall. Its very appearance — the
absence of bracing wires along its cantilevered wing, the suggestion of
great strength in the thickness of the airfoil, the sense of reassurance in its
three motors and all-metal construction — had a big psychological impact on passengers. Even before many of them had sampled its spartan
comforts and found out about its performance, the Tin Goose was the
plane for them. Timing, as well as the Ford name, had a lot to do with its
quick acceptance: the Ford trimotor came along in 1927 just when the
Lindbergh flight stirred the first real public interest in flying. Its speed
was something like 100 miles an hour. It could fly 250 miles at a clip. It
could, and often did, land and take off in a pasture. It could carry twelve
passengers in a cabin high enough for a medium-sized man to walk down
the aisle without stooping. There were two pilots, and starting in 1929 the
Ford trimotor carried a third crew member, a flight attendant called a
steward.

The Tin Goose had one family resemblance to the Tin Lizzie: it had to
be hand-cranked. Novelist Ernie Gann, who copiloted the critter for
American Airlines in the 1930s, has provided a nostalgic description of
this foible of the winged flivver:

The three engines on Fords were activated by inertial starters. On a signal, a
mechanic or the copilot inserted a crank in each engine separately and commenced his labors. He wound slowly at first and then faster as his efforts spun a
heavy flywheel, which was geared to the crankshaft. Once the flywheel was

whirling at what sounded like full speed, the cranker pulled a cable that engaged a spring-loaded clutch, thereby transmitting the energy of the spinning wheel to turn the engine. The squealing sound it made was almost identical to the shriek emitted by a pig when kicked in anger. If the oil was not too cold, the propeller would turn at least three or four revolutions. Usually, this was enough to start the engine, but if the pilot was not alert and missed the moment of truth with mixture lever and throttle, the cranker was obliged to start his labors all over again. Winding the crank was very hard work, and at times when the engines proved unaccountably balky, there were impolite exchanges of opinion between cockpit and ramp.

Planes, people, capital, airports, flyways — even after they had assembled all these rudiments of scheduled flying, the pioneering airlines had to do what the railroads and steamship lines never did: they had to overcome mankind's inexperience with voyaging in an entirely new element. The worst of it was getting used to the movements, heat, and moisture in the air aloft. Even when the atmospheric disturbances were not severe enough to cause storms, they laid down blankets of cloud and fog that blocked the earth from view.

From the earliest days airmen had relied on two instruments aloft that had been originally developed to cope with similar problems in surface travel. These were the compass and the barometer. For use in airplanes the barometer was adapted and linked to a gauge whose needle registered the drop in atmospheric pressure in terms of linear elevation as the plane flew higher. This was the altimeter, first developed by the Europeans for use in World War I. The magnetic compass was of limited value in an airplane. It functioned well only in straight and level flight. In disturbed air, or if the pilot made a turn, it began to spin, and kept on spinning. It did not steady down for some time, and the trouble was that it became erratic just when a pilot needed it — in turbulence or in a turn.

By the time the Post Office airmail service flew the nation's airways there were two more standard instruments on the panel in front of the pilot's seat. One was a clock that some pilots used to time their turns. The other was the airspeed indicator, a dial registering in miles per hour the wind pressure in a tube located on the outside of the plane. There were also gauges that showed fuel and oil pressure and engine revolutions per minute, important information about the state of the plane's power plant but of course no help in navigating.

By the late 1920s the instruments available for navigating through a disturbed atmosphere were really only three — airspeed indicator, altimeter, and a little one called the turn-and-bank indicator, which was first produced by Sperry after 1922. This was a gyro instrument. It had a needle that pointed up; if the pilot flew the plane straight, the needle stayed upright. Below the needle was a curved tube containing a bubble, and as long as the plane flew properly, whether in a turn or flying straight and level, the bubble remained in the middle. If the airplane began to slip (to drop one wing lower than the other) or to slide one way or the other, the bubble moved off center. This told the pilot the plane was not flying true to its direction.

Pilots used these three instruments to maintain airspeed and altitude, keeping the bubble in the center of the turn-and-bank indicator. Pioneering pilots in the Air Mail Service like Wesley Smith would practice flying for long spells solely by these three instruments without ever looking outside their cockpits at the horizon. Others, like Walter Addems, who was flying for National Air Transport in 1926, made a mask that kept them from looking outside while concentrating on their instruments.

By the end of the 1920s instrument makers like Sperry, Kollsman, and Bendix had designed and developed more and better instruments, and better systems for operating them. Both the Ford and Fokker trimotors had quite a complete array of good basic instruments. The Ford had an airspeed indicator, an altimeter, a turn indicator, and another pressure instrument called the rate-of-climb indicator. It also had a new and beautiful instrument called the artificial horizon.

On the face of this instrument was a form resembling the wing of an airplane. This wing would reflect the lateral position of the plane as the pilot flew it. If he flew level, the little wing in the instrument stayed level. It not only told him whether he was keeping the plane's nose level with the horizon, more important, if his plane went into a bank or a turn, the figure on the face of the instrument would indicate that bank or turn. By referring to the ball in the turn-and-bank indicator, the pilot could see whether his plane was slipping, sliding, or flying true.

That was the state of the art of instrumentation by the year 1928, a quarter of a century after the Wrights first ventured into the new element. These instruments were still not fully exploited because the pilot lacked the technology to fly when clouds or darkness prevented him from keeping the horizon in sight. But they helped at certain moments when the pilot lost his bearings — at dusk, for instance, when he lost his natural horizon, when all light blended together in the dimness and he could not tell whether he was flying right side up or not. Airplanes were not ready, navigating systems were not ready, communications were not ready, and pilots were not ready to do the kind of flying that these instruments made it possible for them to do.

At this point the airlines were just beginning to get the hang of radio communication. Of course radio's great potential for air communications had been seen from the first. The military tried dot-and-dash communications in some planes in the First World War; it also began experimenting with radio telephones in those years. European airlines used air-to-ground radio on their short hops after the war. On the first Atlantic crossing in 1919 Alcock and Brown carried a three-hundred-pound set, but it conked out. Because the sets were so bulky the Post Office never tried radio on its De Havillands, but after private contractors began flying the mail, the government offered to pay them a premium on their deliveries if they used two-way radio in their planes.

Both Boeing Air Transport and Western Air Express were very interested. As United people tell it, one of their planes went down in a Wyoming storm one night in the 1920s and Bill Boeing was pacing the floor in Seattle.

"Everybody knew about the storm except the pilot," exclaimed Boeing. "If only we had some way to talk with our pilots when they are in the air."

Boeing's brother-in-law Thorp Hiscock, an amateur tinkerer and sometime pilot who had a ranch in nearby Yakima, happened to be present. "What about radio?" he asked.

"The companies say it can't be done," said Boeing.

"It can too — I'll do it," said Hiscock.

After furious effort Hiscock and his fellow handyman Bill Lawrenz put together a crude set. The static aloft was ear-splitting. Western Air Express had its Edison in young Herbert Hoover, Jr., who also engaged in early air-to-ground radio experimentation with his Stanford fellow student Jack Franklin.

These were talented and persevering young men — Hiscock went on to invent the deicing gear that helped keep airliners flying in the 1930s, Hoover to become Eisenhower's Undersecretary of State, and Jack Franklin to be TWA's chief engineer — but the determining factor in their success probably came from elsewhere. Through the National City Bank, the group that controlled United put pressure on Bell Telephone Company to invest $2 million to refine Hiscock's equipment.

Bell Labs took a hand, bought a Ford trimotor, and put some of its crack researchers on the job. They made fundamental studies of the frequencies used — and incidentally made one important mistake. They decided on a single frequency for the planes to use day and night. It is now well known that waves in the three-to-ten-megacycle range exhibit the "skip effect." By day they are reliable but at night they skip, often right over the intended receiver. United and Western Air began using the equipment the telephone company developed; they soon learned that a single frequency would not do the job. United worked out major modifications to permit use of two frequencies — higher by day, lower by night. These modifications were turned over to Western Electric, Bell's manufacturing arm, for fabrication; in a very short time airline pilots were talking to ground stations, and ground stations were relaying storm warnings and other vital information to pilots in the air. It was a great coup for the airlines, especially United. But it wouldn't have been possible if the mighty telephone company hadn't been drawn in. Said Jack Franklin later, "Here was Western Electric, used to filling orders for fifty thousand telephones at a crack, getting orders from Western Air for six microphones and one radio transmitter. The Bell people lost their shirts."

4

A Reliable Engine

D*ayton*, Ohio, was not only the hometown of the Wright brothers, it was also one of the most prolific breeding grounds of twentieth-century adventure capitalism. The whole Miami River valley of which Dayton was the center was dotted with mill towns where strapping youths off farms and stolid immigrants from Bavaria and Bohemia hammered iron and turned lathes. A factory boss said later, "Everything in this part of Ohio centered on machinery." Dayton's biggest business was the National Cash Register Company. "The Cash," as locals called it, was the dominant cash register company of the world. By some accounts its hard-driving founder and president, John Patterson, invented modern high pressure salesmanship to make it so. Enterprisers of pith and moment learned in the Cash school. One of its vice presidents, Thomas J. Watson, went on to start an even bigger business machine company — IBM. Others later rose high in the automobile industry. And a family that shared richly in the Cash flow was the Rentschlers, who made castings for the machines in nearby Hamilton.

Old Adam Rentschler, an immigrant from Württemberg, was mustached, stolid, swarthy, and strong as an ox, and he ruled his hundred German-speaking moulders and puddlers with an iron hand. A hard man on the foundry floor, he was also a stern taskmaster to his sons. As fast as they returned from Princeton, he put them to work learning the business. Big and strong as their father, the sons tossed pigs, ground castings, mixed the furnace charges, and cursed and sweated with the factory hands the whole twelve-hour day. Machinery was bred in their bones. Robert lost his life in a forge accident, but Gordon, the next eldest, showed an early flair for front office management and was soon starting new departments and pushing sales so adroitly that the forceful Rentschler style began to be renamed "dictatorial diplomacy." Fred, two years younger, liked to play poker, drink beer, and drive a car at breakneck speed. When the old man would say, "Only two things are worth having, money in the bank and pig iron in the plant," young Fred would dispute him. Yet Fred, more than the others, showed his father's sure hand with machines.

Of all the big men at the Cash, none was closer to the Rentschlers than Vice President Edward A. Deeds. A farm boy from Granville, Ohio, who attended Denison and Cornell universities and became the Cash's chief engineer, Deeds thought it would be a great idea to electrify the cash register. An inventive young fellow hired off the Ohio State campus, Charles Kettering, knew how to do it. The principle of Kettering's invention — that a little motor could carry a huge load if the burden were only temporary — seemed such a good one that Deeds thought it should be applied to the ignition system of automobiles. In 1908 the two formed the Dayton Engineering Laboratories Company (Delco) and produced the self-starter, the invention (first used in the 1912 Cadillac) that domesticated the automobile. Their voracious fancy for machinery that could make money led them to make friends with the Wright brothers, devise portable Delco light plants for farms, and form ever closer ties with the automobile industry. Deeds went on to help design the first eight-cylinder car (the 1914 Cadillac) and Kettering to become General Motors' greatest gadgeteer, but before that the two took a brief flyer with the Rentschlers in making a car in Ohio. Young Fred, the rebellious racing buff, got a start in engines working on the first few four- and six-cylinder Republics.

In 1913 the Miami River rose in flood and inundated homes and factories — including the Rentschlers', where, proving the old man's adage, pig iron was about all that survived. This happened just when another flood, a wave of antitrust feeling, swept over the land. The Supreme Court the year before had struck down Rockefeller's Standard Oil monopoly, and shortly thereafter, Patterson, Deeds, Watson, and other top Cash men were tried, convicted, and sentenced to prison for conspiracy to maintain a cash-register monopoly. They had just appealed their conviction when Dayton suffered its disaster. Seizing the occasion to show their citizenly merit, Patterson took charge of Dayton flood relief and Deeds, rallying valley enterprisers, organized a total rebuilding program. The Miami Conservancy District, the power and flood-control scheme Deeds led, was a considerable effort. Its chief engineer, Arthur Morgan, subsequently became first head of TVA. Another close Deeds associate, Harold E. Talbott, built the dams, and young Gordon Rentschler, volunteering to sell the bonds, performed his mission to Wall Street so efficiently that Charles Mitchell, president of National City Bank, brought him into New York as his number-two man and eventual successor.

After the court of appeals in Cleveland overturned the Cash men's convictions on a narrow technical issue, Deeds did not rejoin the company but busied himself with his automobile and other interests. The First World War broke out, and at once the Wrights' invention became a weapon of war. Seeing an opportunity to make flying machines for war, Deeds, Kettering, and Talbott thereupon organized the Dayton-Wright Company, with Orville Wright as vice president and consultant; when the United States was drawn into war, Deeds went to Washington. The Aircraft Production Board, of which Howard Coffin of the Hudson Motor Car Company was chairman and Deeds a dollar-a-year member, was given $640 million and drew up a plan to deliver 50,000 planes to France in a year's time. Then Deeds, appointed a colonel and put in charge of all aircraft procurement, decided that America should concentrate on a single engine. Locking two automobile engineers in a Washington hotel room, he obtained the design in five days.

Their engine had automobile stamped all over it — eight cylinders, water-cooled, with Delco battery ignition instead of magnetos — and was dropped. Redesigned, it was made into a twelve-cylinder engine like the British Rolls. The revamped engine passed the test. This was the Liberty engine that became America's number-one contribution to aviation in World War I. By Armistice Day, 20,478 Libertys had been manufactured by Packard and others, and perhaps half had been shipped to Europe for installation in planes that saw front-line service. But most of the $640 million was spent, and nothing like 50,000 American-made planes reached the American Expeditionary Force. In early 1918 a shocked President Wilson was told that profiteering was going on while not a single plane was getting to the front. In fact, only 248 American-made warplanes ever flew at the front. Charles Evans Hughes, asked by the President to investigate, unearthed evidence that Deeds had been passing inside contract information to his friends in Dayton-Wright and recommended that Deeds be court-martialed. Although Secretary of War Baker appointed a review board that exonerated him, Deeds left Washington under a cloud.

Because the surplus Libertys and the De Havilland DH-4 observation planes finally produced by the Dayton-Wright Company became the standard equipment with which the U.S. Post Office Department pioneered airmail service, Deeds had a shaping influence on the rise of scheduled flying in America. But Deeds left still another legacy from his high times in Washington, and this was to prove even more important in the emergence of commercial air transport.

Not quite all the engines manufactured in World War I were Libertys. As a hedge, the businessmen who organized wartime production also backed manufacture of the best French engine, the 150-horsepower Hispano-Suiza, in the United States. And when young Fred Rentschler sought a wartime commission in the U.S. Air Service, Deeds helped the motor-loving son of his old Ohio partner to find a place as inspector at the Wright-Martin Engine Company's plant making Hispano-Suiza engines under license in New Jersey. That, said Fred Rentschler long afterward, was when he fell in love with airplane engines. It was an affair that changed U.S. aviation history.

After the war the automen, licking their wounds, went back to what they did best — mass-producing vehicles that never left the ground. But Fred Rentschler told his brother, "I'm staying in aviation, come hell or high water." Gordon's advice was that the family business in Hamilton could not sustain an aviation venture. Fred Rentschler therefore joined what was left of the Hispano-Suiza business after the main backers pulled out, and found a new plant for it at Paterson, New Jersey. The company, renamed the Wright Aeronautical Corporation, proceeded to turn out Hispano-Suiza engines in drastically reduced numbers for the Army and Navy.

Who could have foreseen that at this absolutely lowest point in aircraft manufacture, when the war was over and the captains of industry had hastened away, a movement was starting that would lead to the next big advance in aviation technology in America? Looking back, it now seems perfectly plain that it takes government to determine the course of technology development, and so it was with the rise of the air-cooled engine.

Most of the major engines of World War I were water-cooled, like the Hispano-Suiza and Liberty, and in 1920 most airmen thought the future lay in such engines and in flivver planes. But the military thought otherwise. By withholding contracts for Rentschler's Hispano-Suiza, the Navy forced his firm to buy out a small Manhattan company that made the country's only air-cooled engine but lacked the means to develop it. Rentschler and his able lieutenants, rising to the challenge, eliminated the production bugs and perfected the engine, the 200-horsepower Wright Whirlwind. Ordered by both Army and Navy, this 500-pound engine powered the famous long-distance flights of the mid-1920s, including Lindbergh's.

Now sure where the future lay, Rentschler urged the Wright board to plow money into research on building more powerful air-cooled engines. Overruled, he resigned in the fall of 1924 and, joined by key aides, set out to outdo Wright in building a much bigger air-cooled engine. He wanted to run his own show.

At the Bureau of Aeronautics in Washington, Admiral Moffett declined to offer any contract in advance. But he waxed enthusiastic at the prospect of a second company competing in air-cooled engine design. Rentschler went away with the definite impression that for the first fleet aircraft carriers then in the planning stage, planes with much more powerful engines would be required. This was reinforced by what an old friend, Chance Vought, the brilliant plane designer, confided. If Rentschler could provide a 400-horsepower engine weighing no more than 650 pounds, Vought said, he could build a new carrier plane that could win them both contracts.

Rentschler had already made sure that his ablest associates at Wright — including top engineer George Mead and designer Andy Wilgoos — would join him. But first he had to raise money and find a plant. Very likely investment bankers would have backed him, but to invite them in would put Rentschler right back in the position from which he had just resigned. He therefore went to consult his brother Gordon, now executive vice president at Wall Street's biggest commercial bank. "Lying on Gordon's desk," as Fred later recounted, was the report of the Niles-Bement-Pond Company, a machine-building combine that Gordon had just helped reorganize. Gordon drew his brother's attention to the fact that the combine included the Pratt & Whitney Tool Company of Hartford and that this old-line firm still had a sizable war-earned cash surplus.

The combine's new chairman was none other than Colonel Deeds, and its president was another old neighbor who had managed the combine's Niles works next door to the Rentschler foundry back in Hamilton. Fred talked to these old friends and then worked out a deal with the people in Hartford: Pratt & Whitney Tool would stake him to $250,000 and factory space to get started, and a further $1 million to set up production. The company took the preferred stock and some of the common, but Rentschler treated it as a business loan and subsequently paid it off. His group, which included his brother and Deeds's son Charles, kept almost all the nominally valued common stock. Colonel Deeds became Pratt & Whitney Aircraft's first chairman.

Even before the idle Hartford plant was occupied August 1, 1925, Mead, Wilgoos, and a few other key people quit Wright and started designing the new engine. The one thing they had to go on was that the Navy wanted an air-cooled radial engine of 350 horsepower, and if possible, 400 horsepower. Exercising his brand of "dictatorial diplomacy," Fred Rentschler ordered them to shoot straight for the upper target — 400 horsepower. For the rest, he said, "You can design anything so long as it's economical." So Mead and his men laid on all sorts of weight-saving and power-boosting tricks: a forged crankcase instead of a casting, a split crankshaft instead of a solid one, cylinders ringed with many more cooling fins than had ever been tried before.

In late summer tools were set up and fast as the blueprints flowed out of the designers' cubbyhole, thirty men cut, trimmed, and bolted. Barely six months after the engine had been conceived, the job of assembling began. From Cuba Colonel Deeds cabled that if the engine were completed by Christmas Eve he would personally give each man a turkey. The last nut was screwed tight; the birds were borne home to Christmas tables. Mead said, "We still have to see if it will start." On December 29, 1925, the engine was started and made a gratifying din. Mrs. Rentschler came, listened, and gave it its name: Wasp.

Frederick Rentschler in 1932

On January 15, 1926, the thirty men disassembled their engine and proceeded to rebuild it, drastically. This was the way engines evolved. By March the Wasp was ready again, this time for the crucial test-stand run. It swept through the fifty-hour test, chalking up 415 to 420 horsepower at full throttle.

Having given Rentschler a contract for six engines to be built for $90,-000 development money in 1926, the Bureau of Aeronautics now ordered a half-dozen more. Then the Wasp was put in the air, in planes never meant to take its bulk and power — and the Navy really liked what it saw. From Admiral Moffett came an order for 200.

At this point Rentschler told Pratt & Whitney Tool he needed his $1 million and took over not just the top floor but the whole building for quantity production. The record shows that in 1926 Pratt & Whitney Aircraft delivered 19 Wasps, 12 of them experimental; in 1927, 260; in 1928, 665; in 1929, 1,666. They went on making Wasps until 1960 — by which time a grand total of 34,966 had been launched on the world's wings.

Even before they got the Wasp on the test stand, Mead and Wilgoos began designing a bigger air-cooled engine. This was the Hornet, which produced 525 horsepower and was only slightly less successful than the path-breaking Wasp. To Admiral Moffett's satisfaction, Wright Aeronautical recovered from the loss of its top staff and, after merging with Curtiss, produced the 500-horsepower Cyclone, an engine that gave Pratt's new family of radials stiff competition.

Both the Wasp and the Hornet, it is important to note, were built as military engines. So was every successful large commercial engine ever designed in this country. "The engine is the key to air supremacy," was one of Fred Rentschler's favorite sayings. When the first big flattops, the *Lexington* and the *Saratoga*, joined the fleet in 1927, all 160 planes aboard mounted Wasp or Hornet engines. Some, as Chance Vought had predicted, were his stubby Corsair fighters. But many were Boeing F2Bs, and this marriage of Rentschler engines and Boeing planes proved significant not only for military but for commercial aviation history.

5

A Winning Combination

Rentschler, Rickenbacker, Kindelberger, Loening — the list of German names in American aviation is a long and notable one. William E. Boeing of Seattle, the second big figure to emerge in U.S. commercial aviation in the 1920s, was yet another who grew up with the sound of German accents. His Austrian-born father, Wilhelm Boeing, emigrated to Michigan, opened an office on Detroit's Woodward Avenue, and by 1880 grew rich buying and selling North Woods pinelands and iron-ore properties. When the old man died, his wife remarried. Young Bill was left as their only son. The minute he became twenty-one he ended his engineering studies at Yale and took off. Bidding goodbye to Detroit in much the same way that Howard Hughes later abandoned Houston, this heir to a fortune went West to establish his independence by investing in Washington spruce as profitably as his father had in Michigan pine.

Tall, mustached, with thin-rimmed glasses and a markedly professorial air, Boeing was anything but a playboy. He was one of those capitalists the popular press of the first years of this century called, for want of a better word, a "sportsman." Such men were rich nobs who did not have to keep regular work hours like ordinary money-grubbers, and yet expended their days so energetically that their doings often came to public attention. Boeing seems to have been a rather aloof, stiff, and cold fellow. If he resembled the young Hughes in any other way, it was in his urge not only to master the pleasure vehicles he acquired but also to improve them by tinkering. In his formative years Boeing, like other early aviators, appeared to be trying to prove something, and making money from his self-ordained initiation rite was part of the show.

Boeing first became fascinated with flying when he saw Louis Paulhan defy the laws of gravity by whirling his Farman biplane around a 1.7-mile course forty-seven times at the 1910 San Diego International Air Meet. He stalked through the crowd to ask a ride, then waited vainly through four days for the French champion to keep his promise. But in Seattle he met one of the most compelling figures in early American aviation. This was Conrad Westervelt, newly graduated from Annapolis and posted to the Puget Sound Navy Yard. Westervelt, a first class engineer who later codesigned the first plane to fly the Atlantic and organized the first commercial airline in China, was also a conversational volcano, the author of novels and a Broadway play. Although Westervelt was already one of the Navy's most outspoken proponents of aviation, boats first brought them together — Boeing not only owned one but also had bought a boatyard with the idea of building a better one himself. Bachelors both, they cruised the Sound, and Westervelt fairly erupted with red-hot talk of hulls, power plants — and wings. Boeing said less. But one day the two of them went for a ride in one of Seattle's first seaplanes, and afterward Boeing said, "I think we can build a better one." So they went to talk with Herb Munter, an oil-spattered birdman who had built three planes and was hammering together a fourth in a leaky lakeside shed. It was 1915, and Westervelt had been hurling fire-and-brimstone about airplanes as weapons and about a war that was drawing closer and closer to America. With Munter's help they started to build two floatplanes to be called B & W, for Boeing and Westervelt, at Boeing's tiny shipyard. Boeing traveled to Los Angeles to buy a seaplane from Glenn Martin and learn to fly it.

Before their first B & W was finished, Westervelt was ordered East. It was Boeing himself who flew its test flight in June 1916. A month later, putting up $100,000 capital, he organized a manufacturing company. The second plane he built, with certain improvements, was the Model C, which made its first flight just before America entered the war. Now at the center of military procurement in Washington, Westervelt urged Boeing to apply for a contract. The Model C passed the Navy tests, and the "sportsman" in Seattle looked around for help. The University of Washington engineering department sent over a couple of fresh graduates, Claire Egtvedt and Phil Johnson, and Boeing's company turned out fifty trainers for the Navy before the war ended.

After the war Egtvedt and Johnson scrounged small orders from the Army and Navy. Boeing paid the bills and held the company together. He even built his first commercial plane. Eddie Hubbard, an early test pilot, convinced the U.S. Post Office that there was need of an airmail service between Seattle and Victoria, British Columbia. Hubbard's idea was to pick up mail that had missed outbound steamers at Seattle and put it aboard the ships when they called at Victoria before setting out across the Pacific. The flying boat would then scoot home with mail from any incoming ship that had paused at Victoria. Had the Post Office tried to extend its own routes to provide such services it would have gotten into border complications. But there was nothing to stop the U.S. and Canadian governments from paying a private operator for the work.

Hubbard and Boeing made the first trip March 3, 1919, in a Model C.

But the new B-1, an open-cockpit flying boat designed by Egtvedt, was soon ready and Hubbard, occasionally spelled by Boeing, flew back and forth eighty-four miles on the country's first international contract airmail service. This odd little exercise (there was another airline that briefly performed the same job between New Orleans and steamers calling at Pilot Town at the mouth of the Mississippi River) had a significance quite out of proportion to its size and remoteness. First, it gave Boeing a chance to show, by the B-1's steady performance, how rugged and dependable the planes he could build were. Second, it gave Hubbard and Boeing a small but unique experience with scheduled flying, maintaining flights week after week and month after month. And it led on to a far bigger venture. It is true that after several years Boeing disposed of his share and Hubbard ceased to perform the trips. Therefore, since the antiquity of commercial air transport is customarily measured by the time of continuous operation, Hubbard Air Transport — or the Seattle-Victoria Air Mail Line, as it was also called — counts officially as a mere precursor of the airlines of today. Yet the true origins of the biggest airline of them all, United, trace straight back to this solitary B-1 putt-putting eighty-four miles up and back between the fir-lined shores of Puget Sound.

One day in the fall of 1926, after the Post Office Department had begun to let private contractors fly the mail on various feeder routes in east and west, Eddie Hubbard burst into Boeing Airplane Company offices in great excitement. The word was out, he told Claire Egtvedt, that the Post Office was going to put up its transcontinental airmail route for private bidders. The Chicago–San Francisco segment would be offered first, in November. "This is the opportunity of the century, Claire," he said. "I've got all the figures on mileage and pounds of mail carried. If you can produce some mail planes, I know we can operate them successfully."

Taken aback, Egtvedt asked questions — about the tremendous distances, about the terrible winter weather, about flying at night. Although the company had been turning out military planes almost exclusively, Boeing had just built the experimental Model 40 mail plane for the Post Office Department. The trouble with the 40, Hubbard said, was that it carried the same old World War I Liberty engine. To this Egtvedt, who had been designing new Navy fighters to carry the latest engine, had an answer. "How would you like an air-cooled engine?" he asked — and his mind really began to turn. The Wasp was 200 pounds lighter than the Liberty. With a Wasp for a power plant, the Model 40 could carry 200 more pounds of mail than Hubbard had figured.

Together the men worked out the costs anew — how many people would be needed, what facilities would be required, how much it would cost to operate the plane. They walked in on Bill Boeing. Boeing, said Bill Allen, his lawyer and later his successor, "was an austere man, basically shy. He did his own thinking. He was able to pick people and build an or-

William E. Boeing (right) and Edward Hubbard flew the first international mail route, from Vancouver to Seattle, on March 3, 1919, in a Boeing Model C.

ganization around them. But he was difficult to approach. And he was never a man of words." Listening, he sat silent a long time. Then he said, "This is something foreign' to our experience." When Hubbard pointed out that he had flown the Victoria route and made money, Boeing raised a hand. "This is different," he said, "mighty large, mighty risky."

But Boeing had been watching the first feeder-line operations. He knew the risks but he saw that the right kind of operation could be profitable. With good management, for instance, Pop Hanshue's Western Air Express was making a satisfactory profit. Furthermore, in Boeing's mind an airline might be an outlet for the planes he built. Next morning when Egtvedt came to work at 8:00, he learned that Boeing had been trying to reach him for an hour. By the time Egtvedt and Hubbard joined him, Boeing was pacing the floor. Their big chance, as he said afterward, was the competitive edge that would be gained by putting the Wasp engine in a commercial plane. "It kept me awake all night," he said.

The Boeing B-1E was used on the Vancouver–Seattle route.

Boeing wanted to go over the figures again. He called his production chief, Phil Johnson, in Washington, and Johnson, then discussing deliveries of Wasp-powered Boeing fighters to the Navy, said go ahead. Boeing turned to the others and said, "Those figures look all right to me. Let's send them in." It was one of the pivotal decisions in the history of air transportation.

When the bids were opened in Washington on January 15, 1927, everybody in aviation was shocked. Boeing and his old partner, Hubbard, had offered to carry the mail halfway across the United States for no more than the Post Office was paying others to tote the mailbags between New York and Boston. That was how much Boeing's bid of $1.50 a pound for the first 1,000 miles and fifteen cents a pound for each additional 100 miles added up to. Harris Hanshue's Western Air Express, which had been flying the Salt Lake City–Los Angeles feeder line for $3.00 a pound, had bid $2.24 and twenty-four cents respectively, and fully expected to win the job. But Hanshue was bidding for profits, not for a manufacturing outlet. He and other rivals protested bitterly that Boeing would ruin the whole private-contract system, could not operate safely on such an income, and was letting his factory bear the brunt of the cost.

To fulfill the contract, Bill Allen drew up papers for a new Boeing Air

Transport Company. Equity consisted of common stock of nominal value plus $750,000 in preferred, all taken up by Bill Boeing himself. The new entity contracted with the Boeing Airplane Company to build twenty-five planes, and with Phil Johnson as president of one outfit and Claire Egtvedt chief of the other there was, of course, no haggling. But the crucial part of the deal had been accomplished in Washington where Boeing already maintained a five-man lobby with an $85,000-a-year payroll. The Navy agreed to drop back twenty-five places in Pratt & Whitney's Wasp delivery schedule. Beginning in February 1927, Pratt & Whitney shipped five Wasps a month to Boeing. As a result, with some timely redesigning by Claire Egtvedt (at Boeing's suggestion, he added two seats in a tiny enclosed cabin between the wings, "for a mechanic and a returning pilot"), twenty-five Boeing 40As were on station along the route by the July 1 starting date. To keep these planes flying, Eddie Hubbard had simply signed on the Post Office Department's veteran pilots and grease monkeys.

The new airline made money from the start; within six months its debts were all paid off, and it began to expand. One of the best feeder lines in the West was the route linking San Diego and Seattle. This coveted franchise had been won by a former bus driver named Vern Gorst, whose pioneering efforts to get a service going and to maintain schedules through Pacific Coast fogs and storms are a legend. In the absence of any publicly maintained route markers Gorst bought old ship searchlights and installed them on hills north of San Francisco. He often had to pay his pilots in stock. In late 1928 Gorst went to the Boeing plant to see about buying some new planes for his line, Pacific Air Transport. The 40As cost $25,-000, an unheard of price at the time. Gorst, still struggling to finance his line, was unable to raise the money for the planes. Moreover, competition had reared its head on the California sector of the line; Maddux Airlines had begun flying Ford trimotor planes and another upstart, West Coast Air Transport, owned by Western Air Express, had introduced three-engine Fokkers. They carried only passengers because Pacific Air Transport had the mail contract.

To get the airmail contract, Harris Hanshue of Western Air offered to buy Gorst's controlling stock for $250 a share. A young loan officer at the Wells Fargo Bank in San Francisco, W. A. Patterson, had extended Gorst his first credit and had since become the line's voluntary financial adviser. Patterson pointed out that Hanshue's scheme made no provisions for other stockholders and left the line's employees out on the street. In the end Patterson suggested to Boeing that his airline buy all the Pacific Air Transport stock, voting and nonvoting alike, and merge the two airlines. Bill Boeing offered $200 a share for the stock, and Gorst accepted, mainly because Boeing agreed to keep all the employees on the payroll and to buy out all the other shareholders on the same terms or better.

Shortly after concluding this deal, Boeing brought Patterson to Seattle as assistant to President Phil Johnson, and thus the man who was to guide United Airlines through its greatest years entered the air transport business. One of young Patterson's first jobs was to round up all of the outstanding Pacific Air Transport stock that Vern Gorst had peddled up and down the Pacific coast. Pat, as he was already known, began buying it at $200 a share. Pilots who thought the stock was no good began shaking certificates out of old flight jackets and fishing them out of their lockers. The price began to climb when word spread that Bill Boeing wanted to buy in every share. The price skyrocketed to $666 a share for the once worthless certificates. At this price the syndicate sold out. Patterson picked up the last two outstanding shares from a Seattle prostitute, who showed sound business judgment by holding out for $666 a share.

The second feeder line acquired was Varney Air Lines. Walter T. Varney was a World War I pilot who ran a flying school in the San Francisco Bay area and built up scheduled flying experience operating an air express service between San Francisco and two or three inland towns. When the Post Office called for bids to carry mail on feeder lines in 1925, Varney shunned the more attractive routes and applied for one linking Pasco, Washington, with Elko, Nevada, because he figured no one else would want it. Sure enough, his bid of $1.28 a pound was the only one entered. On April 6, 1926, some nine days before Western Air Express began flying between Salt Lake City and Los Angeles, Varney took to the air — first of the private contractors to do so. But the six tiny, ninety-horsepower Laird Swallows that Varney had bought were lamentably underpowered for the task, and Varney had to ask the Postmaster General for sixty days' grace while he sought bigger engines for his planes. Then Varney got three Ryans, from the same plant that built the *Spirit of St. Louis,* and things began to look up. The Post Office boosted his mail pay to $3.00 a pound, and he was authorized to shift his southern terminus from Elko to Salt Lake City. Then the Post Office decided it was all right to fly over the Cascades into Portland and Seattle, and Varney landed that contract. This put him into some sizable cities and gave him the short route from the East to the Pacific Northwest. With mail revenues of almost a million a year, Varney ordered nine 40B-4s, a version of the Boeing plane that had seats for four passengers. Something of a high flyer, Varney was by now bidding on routes all over the place, even in Mexico, and giving less than total attention to operating details. His passenger setup consisted of a coffee percolator at Pasco, another at Boise. Tickets were mimeographed. He kept running out of money and in 1930 sold out to Boeing.

Making every allowance for the first-class men he gathered around him, the secret of Boeing's airline success, as he himself said, was that his planes were carrying mail instead of radiators and water over the mountains. For a year after Boeing started operations, only he had this advan-

The Boeing 40B-4 carried a Pratt & Whitney Hornet engine and had seats
for four passengers.

tage. Nobody else had air-cooled engines, nobody else had Wasps. The
self-evident need of Vern Gorst's Pacific Air Transport for planes with radial engines that permitted an increase of payload from one thousand to
two thousand pounds and added a potential four hundred dollars of income to every trip played its part in Boeing's takeover of his airline.

Wasp engines in Boeing planes were a money-making combination.
Curtiss Aeroplane and Motor Company, the longest-established aviation
manufacturer, was another such combination. Curtiss built both the airframe and the engine (liquid-cooled) for the military planes that won
Schneider Trophies for the United States in 1925 and 1926. As for Boeing
and Rentschler, their ties went back to the days when Boeing's warplanes
were powered by engines produced by Rentschler and Co. at Wright
Martin and Wright Aeronautical. Whenever Boeing traveled east, he and
Rentschler got together to swap ideas. When Rentschler moved to Hartford, it was natural for Boeing contracts to go with him. In late 1928,
when Boeing and Pratt & Whitney were collaborating profitably on both
military and commercial planes, the idea bubbled up: why not pool the
resources of the Boeing airline and factory with the Pratt & Whitney en-

gine works to form the nucleus of a well-rounded aviation holding company?

Mergers were in the air then. Interest in flying was aroused by the Lindbergh flight, the public had developed a sudden and voracious appetite for aviation stocks. Late in 1928 Boeing Aircraft and Transport Company, a holding company, took over the stock of Boeing's factory and airlines. Then, early in 1929, the two enterprisers set up a new holding company, known as United Aircraft and Transport Company. Rentschler swapped ownership of Pratt & Whitney Aircraft for 800,000 shares in the new concern. Boeing then turned over his shares in his companies for the new United stock. So did Chance Vought, who had been present during the first Boeing-Rentschler merger discussions and at once declared that he wanted to be included. And so did the owners of Hamilton Aero Manufacturing Company, which Rentschler had identified as important for such a combine because it made the propellers for the planes.

Although Boeing became chairman, Rentschler, as president, took the more active part. Having brought Hamilton into the original merger, he now rounded up Sikorsky, Northrop, Stearman (all plane makers), Standard Steel Propeller, and Stout Airlines, an air transport service that lacked airmail contracts but had five years' experience in carrying passengers.

The style of corporate combines set years before by J. P. Morgan, who formed the giant U.S. Steel Corporation out of many lesser firms by an exchange of stock in which all profited, was followed in Rentschler's maneuvers, and his timing was deft. It was the crest of the Wall Street stock boom of the 1920s. Just before they formed the trust in November 1928, Rentschler and his fellow Pratt & Whitney directors, who had acquired their common stock holdings three years before at a nominal twenty cents a share, voted a 78-to-1 stock split. Then, when the combine came into being, they exchanged these shares for 2.2 shares in the new company.

An individual who had paid $18 for his original 90 shares of Pratt & Whitney Aircraft, and in 1926 paid another $22 for 110 shares more, would have found these shares increased by the board's 1928 action to 16,000 shares. And after the exchange of these 16,000 shares for the 34,-720 shares of United Aircraft, his $40 total investment would have become worth $3,367,840. That was precisely the experience of Colonel Deeds's son Charles, who had been installed as treasurer when Pratt & Whitney was formed. Founder Fred Rentschler's original cash investment of $253, for which he received 1,265 shares, blossomed in the same way into $35,575,848 by May 1929. Gordon Rentschler, on the secret "preferred" list to whom stock was offered at a discount, merely doubled his investment.

Bill Boeing also gained by the consolidation. He had already put some $480,000 into Boeing Aircraft and Transport, but when the merger was in the works he picked up a further 4,319 shares of Boeing Aircraft and Transport at six cents a share, and then exchanged these for United Aircraft stock at the rate of 12.73 for 1. So high did the new holding company's stock sell on Wall Street that between March 1927 and September 1928 Bill Boeing made $5,332,284 from $259.14.

6

Lindbergh

Two men ushered America into the Air Age. One was Charles A. Lindbergh, who awakened the country to flying by piloting the *Spirit of St. Louis* alone nonstop from New York to Paris. The other was his father-in-law, Dwight Morrow, who two years before had set the broad pattern and laid down the rules by which America thereafter took to the air.

Up to 1927, although hundreds of thousands of Europeans had done so, practically nobody in the United States had traveled by air. In all the country there were but thirty airliners, all together holding seats for no more than two hundred passengers. Despite expectations, there had been no aviation boom in the land of the Wright brothers after World War I. The automobile men who had so rudely taken over plane production from the Wrights and the Curtisses and the Glenn Martins in the 1917 emergency had as abruptly returned to the more profitable business of making motorcars, leaving thousands of surplus warplanes to glut the small aviation market.

By 1925 there was a new stir in the air. Most of the war-surplus planes had crashed or otherwise had their day; the Liberty engines that powered them were mostly used up. Early in the year Congress passed the Kelly Act, which turned flying of the Post Office's airmail over to private contractors. This brought into being the very first commercial airlines, all in the market for planes. Aviation in America was ready for a new start.

But first, if the industry were to grow, the nation had to establish an aviation policy. So in the latter part of 1925 President Coolidge appointed a board to provide one. The chairman he picked, his Amherst classmate and friend Dwight Morrow, seemed just the man for the job. In an era in which big business was calling the shots, Morrow was the biggest of big businessmen. Able and politically ambitious, he was already senior partner of the Wall Street banking house of J. P. Morgan, the most venerated institution on Wall Street. In a decade when the President himself could say, "The business of America is business," the Morgan Bank stood at the pinnacle of the whole American establishment.

Easy and assured, the practiced master of corporate destinies, Morrow heard testimony on all aspects of the business. General Billy Mitchell had raised a storm with his demands for an independent air force, so Morrow paid particular heed to the general's charge that the Army and Navy were stifling aviation. He took sage note of the observation of a leading designer, Professor Jerome Hunsaker of MIT, that the flying machine had evolved for twenty-five years without any startling breakaways from the basic creation of the Wrights. He allowed the pleaders and promoters of the infant industry their clamant hours on the witness stand.

At the end Morrow drew up a set of recommendations. It was the voice of Morgan speaking, practically ex cathedra, yet consensual, in the way that the director of many large corporations may be expected to speak for his fellows. The signals thus sent down were heard with instant comprehension in both Wall Street and Washington. The rules had been laid out and would be quickly enacted into law.

First, said the Morrow report, aviation was vital to the national defense. Its instrument, the flying machine invented by the Wrights, demanded continuous improvement. To that end, the country must invigorate its aviation manufacturing industry. At the board's urging, Congress voted a five-year Army and Navy program of aircraft procurement.

Second, said Morrow, private flying serves national purposes. To help it grow, the government should take a hand, but not a heavy hand. Out-and-out subsidies of the kind prevailing in Europe would be un-American. But without a certain stimulation from the federal treasury, scheduled flying would languish. To regularize the business and give the public some confidence in flying, the government should set up federal licensing standards for planes and pilots. To facilitate organized flying around the country, the government should establish and maintain flyways — the same as it maintained waterways and highways. Finally, the government should enlarge and expand its aid to contract airmail carriers flying these routes.

At the board's prompting, Congress enacted the Air Commerce Act of 1926, which embodied the first two of these recommendations. By amendments in 1926 and 1927 to the Kelly Act, which nearly doubled the indirect subsidy to airmail carriers by changing the method of compensating them, Congress adopted Morrow's third recommendation as well.

The stage was now set for the Wall Street aviation boom of the late 1920s. It might never have taken place, however, without Lindbergh's 1927 flight. It was the first of two great milestones in the history of commercial air transportation — the introduction of jet airliners a quarter of a century later was the other. Of course Lindbergh's little plane was anything but an airliner, and it carried neither passengers nor mail. Nor is the importance of Lindbergh's flight due only to its place in the history of aircraft development, although no doubt it was a needed proof of the range of the machines and the reliability of their engines. The importance of Lindbergh's solo conquest of the Atlantic was, quite simply, that it awakened America at last to flying. Lindbergh — with his father-in-law — launched the Air Age.

Charles Lindbergh was the greatest American hero of the century. And in looking back long after, it should not be surprising that he was the product of a broken home. His father was a Swedish immigrant's son who farmed, practiced law, and served two terms in Congress as a radically anti–Wall Street Republican congressman. Lindbergh lost his seat rather than give up his opposition to United States entry into World War I. In his love for the outdoors, young Lindbergh took after his father. His flair for the mechanical can probably be traced to his mother, a high school chemistry teacher and cousin of Edwin H. Land, who invented the Polaroid camera.

Evangeline Land Lindbergh thought her only son should study at the University of Wisconsin, but he was more interested in tearing around Madison on his motorcycle. He dreamed constantly of flying, so intensely that he abandoned his engineering classes to do it. Skimping, saving, working at whatever job offered him a chance to be near airplanes, Lindbergh finally soloed. His dreams were often nightmares; until, in June 1922, he made a parachute jump. That was catharsis. "I'd stepped to the highest level of daring, a level above even that which airline pilots could attain," he wrote later. Never again did he dream he was falling through space, sick with terror. Actual experience had cured his unconscious dread.

There followed a period of barnstorming through the country for a living. His fearlessness became legend. He developed an act in which he stood on the top wing of a looping plane, to the delight of the crowds and the horror of the pilots. At Kelly Field he won the silver wings of the Army Air Service, parachuting when his plane, in mock combat, tangled wings with another. From there he joined the Robertson Aircraft Corporation as pilot, flying the mail between St. Louis and Chicago. At twenty-five he was chief pilot.

Flying high over the prairie at night, he conceived an idea: "I could fly nonstop between New York and Paris." Raymond Orteig, a Frenchman who owned the Brevoort and Lafayette hotels in New York, had already offered $25,000 for the first person to do so. Lindbergh had saved money. He had never spent a dime on liquor or women. He had become one of the best pilots in the country. He had logged two thousand hours under every kind of flying condition. "Slim," as he was called, went to a few wealthy flying enthusiasts in St. Louis who agreed to back him. The Ryan Airplane Company in San Diego agreed to build a plane that would carry him — alone — more than three thousand miles through the air.

Knowing that others were about to try the same stunt, they built the plane in sixty days. Lindbergh, who had quit his job, worked at the plant daily alongside Donald Hall, the designer. No happy-go-lucky flyboy, he had analyzed the task. He knew the Wright Whirlwind engine had the endurance — that was proved when Clarence Chamberlin and Bert Acosta kept a Whirlwind-powered Bellanca aloft for fifty-seven hours in April. Lindbergh had calculated that with extra tanks in the wings and fuselage, he could fly the distance with some fuel to spare.

He took his biggest chance in deciding to fly alone. A half-dozen rivals readying for takeoff on the East Coast and in France assumed they had to take somebody along to figure out the navigation. Lindbergh chose to take more fuel instead and do his own navigating. Since there was so little room in the cramped cabin to take star sights, he planned to fly the whole way by dead reckoning. He worked it all out in San Diego with maps from the local library. He laid out the great-circle route from New York to Newfoundland to the mouth of the River Seine in thirty-three 100-mile segments and marked alongside each segment the heading to be followed as he flew it. If he could only keep to these indicated headings, making accurate allowance along the way for the drift of the plane caused by wind, he would strike the Irish coast at a place called Dingle Bay.

In San Diego he waited impatiently for the machine to be completed. News bulletins kept arriving of the preparations of his competitors. Furthest along were the Frenchmen Nungesser and Coli. But Chamberlin and Bertaud in the Bellanca were nearly ready. Admiral Byrd, with his Fokker trimotor, was best financed. Two Navy contenders, Noel Davis and Stanton Wooster, were taking practice flights. But the Navy fliers crashed and were killed. Charles Levine, the Jersey scrap dealer who had bought the Bellanca front-runner, kept changing crews and became snagged in a lawsuit brought by one of them.

Lindbergh's plane, named the *Spirit of St. Louis,* was wheeled out May 11 for its first flight tests. It was a good plane, a high-wing monoplane just under twenty-eight feet long with a wingspan of forty-six feet. It was built of steel tubing, wood, piano wire, and cotton fabric. In the air Lindbergh made careful notes of every detail, and Hall made certain changes. With each flight they loaded on more and more fuel. Finally Lindbergh took off with such a big load that Hall told him he saw smoke coming from the bearings on the landing wheels. They decided there was no sense in trying to carry the full 450 gallons of fuel until the actual flight.

From San Diego to New York is almost as far as from New York to Paris, and if Lindbergh had been as cautious as Admiral Byrd he might have had the *Spirit of St. Louis* shipped across the country. It never occurred to Lindbergh to do so. He was in a hurry lest others get ahead, and he made the long cross-country trip part of his test program. Never had he flown so far in the night. Yet, deliberately avoiding looking for landmarks along the way, he found St. Louis without veering as much as fifty miles off line. There he heard that Nungesser and Coli had taken off across the Atlantic, were overdue, and presumed lost. Without even pausing for lunch with his backers, he pushed on to New York in a second nonstop flight, breaking the transcontinental record by five and a half hours.

There, waiting on the Long Island airfield, was the Bellanca, its takeoff still delayed because of the court squabble between Levine and his airmen. Waiting also was Admiral Byrd's big trimotor, *America,* delayed by damage on a test flight a few days before and by the admiral's fastidious determination not to take a chance until everything was in order.

Meanwhile, the barrels of special gasoline that Lindbergh had chosen to give him the longest possible range arrived from California. The Pioneer Instrument Company, which had provided the plane's basic instruments, placed a newfangled earth-inductor compass aboard, with a needle on the front panel. A new model of the magnetic compass, on which Lindbergh tended to rely more, was also installed. Because of all the iron in the gas tank just forward of the cockpit, the compass had to be installed behind the pilot's head. Somehow the Pioneer people forgot to put up the mirror that Lindbergh would need to read the instrument behind his head. A young woman in the curious crowd handed Lindbergh a little mirror from her purse. This was mounted on the panel so he could see his compass.

For four excruciating days and nights Lindbergh had to wait for the weather to clear. The Wright Aeronautical people, helping get his engine in top shape, undertook to look after Lindbergh, who was staying at a hotel in Forest Hills near the airfield. On the fourth night they said, "Why not see a show for a change?" It was still raining after dinner, but Lindbergh thought he had better check with the Weather Bureau. One of the Wright men jumped out along Forty-second Street and went to a phone. He came back to report: "Weather over the ocean is clearing — it's a sudden change. . . ."

Lindbergh called a halt to the theater party. "We're going," he said. But he needed to get some sleep. He went back to bed at his hotel with orders to be roused at 2:00 A.M. He had just dozed off when his friend George Stumpf, who had come from St. Louis to share his room and stand guard outside to keep others from disturbing him, shook him awake to ask: "What's to become of me now?" Lindbergh never got back to sleep. At 2:00 he started for the field.

It had stopped raining, but the ground was soft and muddy. Lindbergh had much to do. First the plane had to be moved from the hangar at Curtiss Field to Roosevelt Field a mile away. He had thought he might have to fly it, but a way was found to drag it, tail first, behind a truck. Lindbergh followed in a car. At Roosevelt Field Lindbergh found, to his relief, that his rivals had made no plans to start that day. A crowd was there.

Chamberlin, Acosta, Byrd, and Juan Trippe wished him well. Byrd asked if he might fly off his *America* for a test while Lindbergh was making his preparations. "Of course," said Lindbergh. The photographers asked for a picture. "One more," said Lindbergh, "I've got things to do." The laborious task of pouring the special California gasoline from cans continued. Lindbergh was helped into his flying suit, which went on over his suit, white shirt, and tie. Carl Schory of the National Aeronautics Association fastened the barograph, which marked time and altitude on a slowly revolving cylinder, to the fuselage — without it the record of the flight would not be officially accepted. Lindbergh took aboard a canteen of water and a brown paper sack of five sandwiches and strapped himself in his wicker seat. Finally he was ready. It was 7:55 A.M., May 21, 1927.

The plane was laden with 451 gallons of gasoline — half its total weight. Because of the big tanks built into the fuselage in front of his seat, Lindbergh had no forward visibility. Only by looking through two side windows could he see out at all, but that was a disadvantage he had deliberately accepted so as to carry as much fuel as possible. He carried a simple hand periscope that he could stick out the side window to track his course. There is a favorite story told in Lindbergh's family about this. When his daughter Reeve was taken to the Smithsonian Institution to view her father's airplane, she asked, astonished, "Did he really fly it without being able to see ahead?" "Sure," said her brother Jon. After a long pause, she said, "The damn fool!"

Now came the first big test of the whole enterprise. All the contestants who had failed up to this time had failed because their planes were overloaded. Lindbergh and Hall had calculated that the *Spirit* could get off with the load it was carrying, but they had never actually tried it before. Adding to the margin of uncertainty was a deliberate decision Lindbergh had made before takeoff. The plane's propeller, adjusted before takeoff by Standard Steel Propeller, its maker, might have been pitched at an angle to ensure a good grip on the air during the takeoff run. But, as Lindbergh wished, it was adjusted to an angle that would serve most efficiently during the long hours of cruising — after the takeoff.

The load was huge; the wheels sank into the soggy dirt. Other heavily laden planes attempting takeoff from this same runway had failed to reach sufficient height after long runs, and crashed.

Here too Lindbergh had worked out his plan in advance. He had marked the point where he would cut the throttle if he judged the plane would not get off. But this was not necessary. The tiny plane gathered speed slowly, but the tail went up as Lindbergh and Hall knew it would. Perhaps the plane stayed on the ground for a thousand yards that gray morning. It rose once and touched wheels to the ground again. Then it lifted off the ground and pulled slowly into the air as it reached the fence.

It cleared the telephone wires at the end of the field by twenty feet and passed close over the heads of a foursome on the Hillside golf course.

Never in his life had Lindbergh flown over any body of water as broad as Long Island Sound. Throttling his engine back to the three-fourths mark and keeping it there, he crossed it and set his course by his compass card. It was a long way at 100 miles an hour to Newfoundland. He made his landfall at dusk in good weather and turned aside briefly to fly over the capital city of St. John's to make sure he was right. Then he pointed his plane on a course of 065 and flew out over the Atlantic. It was 7:52 P.M.

Throughout his life, Lindbergh never ceased to think about the next hours. He kept notes for years and finally published an inimitable account of them in 1953. The only instruments for guiding him over the ocean were a compass and a sextant. All he had was the line on his chart and the marks he had inked in beside it to show him the way. He had been flying for eleven hours but it was thirty-six hours since he had had any real sleep.

The night was dark. He wrote later:

Is aviation too arrogant? I don't know. Sometimes flying feels too godlike to be attained by man. Sometimes the world from above seems too beautiful, too wonderful, too distant for human eyes to see — like a vision at the end of life forming a bridge to death. Can this be why so many pilots lose their lives? Is man encroaching upon a forbidden realm? Is aviation dangerous because the sky was never meant for him?

It was not long before he ran into the North Atlantic's feared and famous weather. First it was fog, and he climbed to keep his eye on the constellations, but the stars kept blinking out. In the fourteenth hour he leveled off at 10,500 feet and hit sleet, then snow. Thoughts of turning back crossed his mind, but Lindbergh kept going. Sleep was his worst enemy, and Lindbergh, who had thought of just about everything, had asked the engineers to build a slight instability into the aircraft. It was his way of building in a compelling reason to stay awake.

In the seventeenth hour he passed the point of no return. From that moment it would be shorter to fly to Ireland than back to the American shore. He took a drink of water. It was so relaxing that his eyes closed. He shook his head violently and pulled his lids open with his fingers. He leaned out, gulped in cold air, and concentrated on the thought that if he did not stay awake he would die.

Dawn came in the nineteenth hour of his flight, but at first it did not cheer him. "For the first time in my life I doubt my ability to endure," he said. He flew in fog, and when it broke up he had to blink away notions of seeing things — "shorelines, trees perfectly outlined against the horizon. . . . I'm so far separated from the earthly life I know that I accept whatever circumstances may come."

In the twenty-seventh hour he snapped out of it and began to see things for real — gulls, some driftwood, then two fishing boats. Circling one, he saw a man, cut his engine, and shouted, "Which way is Ireland?" The fisherman made no reply. Then, not much later: "Land! Below me a great tapering bay — a long bouldered island, a village. Yes, there's a place on the chart where it all fits — line of ink on line of shore — Valentia and Dingle Bay on the southwestern coast of Ireland. True! I'm almost exactly on my route. And over two hours ahead of schedule!

"The wish to sleep has left."

The word had already gone out from Ireland that Charles Lindbergh had passed over. In the thirty-first hour of his flight he was sighted over Cornwall. Just as the sun set, the *Spirit of St. Louis* passed over the French coast. The weather was clearing as he crossed the English Channel to Cherbourg. He had first thought, airmail pilot that he was, to follow the Seine to Paris. But Charles Colvin, the instrument maker, had told him in New York the river was too winding and had given him a map of the direct route. The Parisians rushed out expectantly to Le Bourget to see the young American and his little plane for themselves.

At 10:24 P.M. Paris time, thirty-three hours and thirty minutes after takeoff, the *Spirit of St. Louis* landed, and crowds of Frenchmen broke through lines of soldiers and police to surround the plane and the man.

"Well, I made it," said Lindbergh — and with eighty gallons to spare. The Air Age had arrived.

Lindbergh's epochal solo flight evoked a mass adulation unique in all history. It was as if he personified the release of man's bondage to earth — to the French who mobbed him that night at Le Bourget, to the English who lifted him high in London, to the Brussels burgomaster who said, "In your glory is glory for all men." But it is as an American hero that Lindbergh must be seen. Americans made him their hero in 1927, his biographer Walter Ross wrote, "because they were innocent enough or brave enough to commit themselves to an ideal."

Lindbergh embodied that ideal. Young, tall, slim, tousled, he was a small-town boy from the Middle West with an accent to match. Daring, outward modesty, and mastery over machinery were his all-American traits. He was single. He seemed to love his mother. He was Anglo-Saxon.

The feat he had accomplished, all by himself, was an authentic First, a breathtaking leap along the onward and upward path that Americans believed led to the future. Chief Justice Hughes hailed Lindbergh as "a youth who lifted us into freer and upper air . . . turning a vision into reality." "Lucky" was the word headline writers often used, and he was lucky in the rarest sense of charisma and grace. The daring of his life-risking gamble, transformed by self-possession into courage, exemplified what Americans thought best in themselves. The *Washington Daily News* called him "the fair-haired boy that every man would like to have been."

Best of all, he had achieved success by his own efforts. Was that not the essence of the American dream?

On the night of May 20, 1927, as the *Spirit of St. Louis* headed out over the Atlantic, the forty thousand fans massed at Yankee Stadium for the heavyweight fight between Jack Sharkey and Tom Maloney fell quiet. "Ladies and gentlemen," shouted Announcer Joe Humphrey, "I want you to rise to your feet and think about a boy up there tonight who is carrying the hopes of all true-blooded Americans. Say a little prayer for Charles Lindbergh."

Those still alive can tell you where they were the next afternoon when they heard the news: "Lindbergh has landed." It lives as one of those rare unforgettable fragments of nationally shared memory — like remembering where you were when you heard of Pearl Harbor, Roosevelt's death, Kennedy's assassination. In Minneapolis word came over the loudspeaker at Nicollet Field during the seventh inning, and the pandemonium in the stands and dugouts stopped the ballgame between the Millers and the Milwaukee Brewers for ten minutes. In Washington men slapped each other on the back and shook hands when they heard. Said the *U.S. Air Service Magazine*, "People who had been afraid to cross the street talked about flying the Atlantic." Airline pilot Paul Carpenter said later, "I caught the madness." Outside Chicago's LaSalle Street Station he saw the extras' black headlines and turned away from boarding the train to his market-research job in La Porte, Indiana. "Then and there I decided, I'm getting into the flying business." The news came to Wellwood Beall at his University of Colorado fraternity house in Boulder. Said Beall, who built thousands of Boeing bombers in World War II, "It touched off aviation. It touched off me. I decided to be an aeronautical engineer."

President Coolidge dispatched a cruiser to bring Lindbergh home. Every plane the Air Force could muster was over Washington on the day of his return, along with the dirigible U.S.S. *Los Angeles*. When his mother was brought dockside to meet him, "a new burst of cheering went up. Men wept, they knew not why." At the Washington Monument, where Coolidge presented him with a reserve officer's commission before a vast throng, he was Colonel Lindbergh. In New York, where three million rained ticker-tape on him and Mayor Jimmy Walker said, "The city is yours — you won it," he was Lindy. In St. Louis, where he dined with his backers and flew acrobatics for an ecstatic crowd, he was Slim.

Before receiving the Congressional Medal of Honor, Lindbergh was introduced by President Coolidge to Dwight Morrow. Morrow soon had Lindbergh visit his New Jersey mansion. At Captain Harry Guggenheim's Long Island mansion Lindbergh met more of the masters of business. When asked about his future, Lindbergh said he was interested only "in a dignified proposal that will enable [me] to continue to fly while helping bring about successful commercial aeronautics — passenger-carrying on a large scale — in the United States." That was just what the magnates wanted to hear.

Guggenheim suggested that his Foundation for Aeronautical Research should back Lindbergh on an air tour of the country in his *Spirit of St. Louis* to publicize air travel. For three months Lindbergh flew to all forty-eight states, covered 22,350 miles, led parades in eighty-two different cities, and was seen by perhaps half the population of the United States. To prove to the adoring throngs that the Air Age had truly arrived, he always arrived on time, no matter what the weather (he missed only in Portland, Maine). "Since Lindbergh's flight everything that happens in aviation appears to the public to be extraordinary," wrote the wondering editor of *Aviation*. United Airlines Vice President Robert Johnson said later, "Who else could have conducted himself as perfectly as he did, then and afterwards? He was able to ignite the interest of the public in flying."

Next, Morrow, who had become ambassador to Mexico, arranged for Lindbergh to fly on a goodwill tour to Mexico, and then to all the countries of Latin America. Back in New York it was announced that Lindbergh had joined Clement Keys's Transcontinental Air Transport (TAT), Inc., the most lavishly promoted of the new airlines. Such was the public adulation of the unassuming aviator that it was taken for granted that he lived independent of money, like a god. Actually, besides a salary of $10,-000 a year, he was given 25,000 shares in TAT at ten dollars a share, half their market price before he joined the firm and a quarter of the price afterward — his name added that much prestige to the company. The letter of confirmation from TAT Chairman Keys said, "I suggest that you do not put very much of this stock in your own name because when you sell it — and I hope you will sell part of it on the first favorable opportunity — either the delivery of the stock in your own name, or the transfer of it on the books would excite a lot of attention, which is quite unnecessary." Lindbergh did sell some shares. Trippe also signed him on as a technical adviser to Pan Am at $10,000 a year and gave him stock options. On October 3, 1928, Lindbergh sent, unsolicited, a telegram to Herbert Hoover endorsing his presidential candidacy. On February 12, 1929, the U.S. embassy in Mexico City issued this statement: "Ambassador and Mrs. Morrow have announced the engagement of their daughter Anne Spencer Morrow to Col. Charles A. Lindbergh." The marriage took place on May 27, 1929.

Lindbergh's flight brought a flood of money into aviation. *Aviation* reported in August 1927 that on Wall Street "the effect was instantaneous. Aeronautics securities that had been dormant became active." Aviation holding companies sprang up. Two bold young investment bankers, Averell Harriman and Robert Lehman, formed the Aviation Corporation, raised $30 million by issuing shares, and started buying up all sorts of aviation properties. The Boeing-Rentschler group's United Aircraft combine, its shares touted by the National City Bank, rampaged to the top of the stock-market surge. Keys, the biggest promoter of all, merged Curtiss and Wright and loaded up with so many airline companies that the United group, by a sudden proxy raid, was able to pry away one of the best. This was National Air Transport, which held the mail contract for the eastern half of the main Post Office route — New York–Chicago. United already held the San Francisco–Chicago "main line" segment and, by this sudden corporate swoop, turned itself into the nation's first transcontinental airline. It was a loss Keys could stand because he had already launched TAT as a transcontinental, and he still had forty-five other aviation properties. Forming the third big holding company, North American Aviation, he floated yet another whopping issue of stock. All told, the public poured nearly a billion dollars into aviation shares before the bubble burst in October 1929.

7

The Birth of Pan Am

On the morning of October 19, 1927, a tiny Fairchild floatplane alighted on the coral green shallows off Key West and taxied up to the old naval base. Cy Caldwell, an itinerant pilot on his way to a job in Haiti, was stopping off to refuel. Down the ramp rushed Jack Whitbeck, beard streaming in the wind. Whitbeck, former Chicago chief of the Post Office airmail pilots and now Key West manager for the newly formed Pan American Airways, had been tearing his hair over how to fly the company's first mail trip that day from Key West across the Florida Strait to Havana. The airline's one and only plane, a Fokker trimotor, had not yet arrived from the factory, and it was the last day to complete the flight as the contract required.

On promise of a $175 fee, Caldwell obliged, and thus on borrowed barnstormer wings, Pan American first lifted into the skies. No crowds waved, no bands tootled, no press agents chorused mimeographed hallelujahs for this aerial premiere. Carrying the mail sacks on his lap, the substitute pilot flew the eighty miles to Havana in one uneventful hour. The day's only hitch was occasioned by the steamboat-age communications that prevailed. Caldwell got to Havana before the wire reporting his departure. Clearance and inspection therefore took four hours at the Cuban end, and Whitbeck was anxiously awaiting word when Caldwell landed on the return flight — an hour ahead of the cable announcing his plane had left Havana.

No further flights occurred until October 28, and work went feverishly forward to clear coral outcrops on the muddy stretch that was to be Pan American's runway. Then the wood-and-linen Fokker plane, the first of three ordered by the company, arrived at last from New Jersey, and Chief Pilot Hughie Wells, with a copilot at his side and a radio operator behind him, commenced daily service on the international route. On January 2, 1928, when the second Fokker was placed in service, Pan Am carried its first passengers — four each way, all paying the published fare, $100 round trip.

* * *

How, it may be asked, from such picayune beginnings, did the world's biggest airline emerge in just three years? The answer is that the right people with the right resources met up with the right opportunity in the right place at the right time. A whirl of chaos — and out of nowhere the pieces flew together and fit. Such was the logic, inevitable only in retrospect, by which Pan Am, between 1927 and 1930, became this country's one and only international airline.

First, the opportunity. In the early 1920s, when foreign affairs for Americans were still largely hemispheric affairs, U.S. business plunged into Latin America. Corporations and banks, hitherto taken up with home markets, went into Cuba in a big way after World War I and then spread southward into South America. Between 1921 and 1927 U.S. private investments in Latin America multiplied fortyfold; $1.5 billion worth of Latin American government bonds were floated in the Wall Street boom of those years. Such volumes of business cried out to be expedited by air.

Second, the place. South America was uniquely accessible to the United States by air. The planes of the day could island-hop across the New World's mediterranean waters, and once aircraft arrived at the South American shore, they had enormous advantages over other forms of transportation. Because of towering mountain ranges and vast tropical swamps, the land routes of that continent were poorly developed. A large part of South America was a kind of archipelago of scattered inhabited areas between which the airplane could provide not only the safest but also the most efficient means of transportation. In many places the airplane was the *only* means of penetrating an area. Latin American governments therefore were eager to leapfrog right past the railroad-building stage of transport development and grant favorable franchises to daring promoters who thought to fly airplanes into such places.

Third, the time. French and German airlines were already flying up and down South America. As far back as 1919, Dr. Peter Paul von Bauer, an Austrian airman, had organized some of his old war buddies into a company called Sociedad Colombo-Aleman de Transportes Aereos, better known as SCADTA. Incorporated under Colombian laws, SCADTA was one of the world's first airlines, and the first to demonstrate what aviation could do for an underdeveloped country with a harsh geography.

Between SCADTA's base at Barranquilla, the country's main port, and Bogotá, the capital locked in behind several Andean ranges, the distance was 670 miles. To negotiate this journey by land took a week when conditions were favorable, more than a month in the wet season. SCADTA's single-engine Junkers monoplanes, covering the distance in seven hours, were a huge success. Although fares were high and crashes not infrequent, SCADTA became an integral part of the national economy. It carried passengers, cargo, and mail. The airline operated its own post offices, set its own rates, and kept all the revenues from the mail.

In the 1920s, when the U.S. government was bent on opening every possible door to business expansion, von Bauer came out with a plan: let SCADTA, building on its success in Colombia, expand its routes northward into a great air transportation network linking the two western continents. Twice he visited Washington, once in 1925 and again in 1926, asking for mail contracts and the right to land in the United States. Because the military saw potential danger in allowing a foreign airline to operate so close to the Panama Canal, the government turned him down. Young Captain Hap Arnold, future chief of the Air Force, proposed that an American company be formed to link the continents. The idea appealed to the government, which was already giving out airmail contracts right and left within the country. Clearly the time was ripe for a U.S. airmail service to push the nation's business against European competition in its own backyard.

Fourth, the people. What American promoters wanted to fly the mail to Cuba in 1927? In number, they were three. First was a company formed early that year called, as Hap Arnold had proposed, Pan American, Incorporated. A flying friend of Arnold's named Johnny Montgomery, and Grant Mason, a young, Iowa-born businessman, were the organizers. Montgomery was a bluff South Carolina freebooter looking for a chance to prove the possibilities of aircraft. His partner Mason had caught a glimmer of von Bauer's vision and had gone so far as to win an airmail contract from the Cuban government. Financial backing for Pan American came through another former Navy airman, Richard Bevier, whose father-in-law was Lewis Pierson of Wall Street's Irving Trust Company.

The second applicant for the short route to Havana was Florida Airways, started by World War I aces Eddie Rickenbacker and Reed Chambers. Their financial angel was Richard F. Hoyt of Hayden, Stone and Company, the Wall Street investment house that backed so many early aviation ventures. When their planes met with mishaps and they failed to land the Atlanta–Miami mail contract, the airmen dropped out, Rickenbacker to take over Eastern Airlines later and Chambers to head the country's biggest aviation insurance combine. But Hoyt, egotistical and hard-driving, was one of those Wall Street wheeler-dealers who, as they said in the 1920s, "made the market," stayed with the venture, and even brought in other fat cats.

The third promoter who had seen the possibilities of the Key West–Havana mail contract was Trippe. This was the same Juan Trippe who at twenty-six had founded and managed Colonial Air Transport. After his break with CAT, Trippe had been casting about everywhere, even Alaska, for a flying future of large size. By 1927 he knew where it was. In that year Tony Fokker flew to Havana to introduce his fancy new eight-passenger trimotor, the last word in commercial air transport, to the Cubans. Trippe and his friend John Hambleton of the Baltimore banking family went along. They met President Machado and his ministers. What Trippe saw in Havana, and even more what he heard when he checked afterward in Washington, convinced him that his future lay to the south.

Such were the men, the place, the time, the opportunity. But there was small chance of breaking out beyond the footling Key West–Havana beginnings without the fifth element — resources. Here Trippe had an advantage. The friends who backed him when he took on the New York–Boston airline went with him when he broke away. They were rich, young, light-hearted "sportsmen" who loved flying; and commercial flying, especially as presented by persuasive Juan Trippe, seemed an alluring lark in the untaxed Twenties. These friends, mostly Yale classmates with names like Vanderbilt and Rockefeller, Whitney and Hambleton, invested $300,000 in an enterprise to which Trippe gave the grandiose name of Aviation Corporation of the Americas.

In the future, however, more capital would be needed. So Trippe's group and the two others talked about getting together. The first proposal was that the three parties would take equal shares in the new route and put up equal capital. Pan American could raise only 20 percent as its share, but insisted that its Cuban mail contract was worth the difference. Trippe then showed his steel. He slipped away for a flying visit to Havana from which he returned with an exclusive concession for Cuban landing rights. That made Mason's mail contract of little value, and the Pan American group angrily sold out.

The haggling kept on; Trippe proved a tenacious conference-table arm wrestler. Not until August, after the Key West–Havana contract had been landed, were the partners able to give firm orders for the planes to fly it. Not until the near-fiasco of the first day at Key West did Trippe, Hoyt, and their various supporters get the corporate lines in order. By then a third group had decided to put money in the enterprise. These people, whose interest was almost exclusively in the return on their investment, were led by Robert Lehman of the Wall Street banking house of Lehman Brothers.

In the shakedown Hoyt, the heavy hitter from Wall Street who was director of fifty companies, took office as chairman of the Aviation Corporation of the Americas, which became the holding company. Polo-playing Cornelius Vanderbilt "Sonny" Whitney, heaviest of Trippe's sportsmen-backers, was named its president. Trippe became president of Pan American Airways, its wholly owned subsidiary. Since Pan American held the contract and flew the planes, Trippe was very much in charge of operations. But he was surrounded by his Yale friends, who were keen about flying and owned a lot more stock than he did. And he had to clear everything with Hoyt.

At first they all got into the act. Hambleton went after a concession to fly to the Bahamas; Sonny Whitney reported overtures to Mexico; and Grant Mason, the only Pan American originator still aboard, dickered for more rights from the Cubans. Right away, fired up by a government decision in November 1927 to authorize Post Office mail contracts beyond Havana, the directors called for "joint operation of key routes to South America and the Canal Zone" with SCADTA and talked of flying even longer distances.

Back in 1925 the Huff-Daland Company, which pioneered the use of planes for crop-dusting in the boll weevil country of Louisiana and Mississippi, got the idea of using their planes year-round by shipping them across the equator to dust cotton fields in Peru during the North American winter. As soon as the Huff-Daland men found out how long it took to order spare parts from Peru, they decided to start an airline to fly the mail. Their Peruvian Airlines flew only between Lima and Talara, but it quickly came to Pan American's attention because Dick Hoyt looked after Huff-Daland's finances. One founder of Peruvian Airways was C. E. Woolman, who later turned Huff-Daland's Louisiana crop-dusting operation into Delta Airlines. The other was a former Army test pilot, Harold Harris. Traveling home via Argentina and Brazil in 1927, Harris drew up a map of what he thought a future South American airline's routes should look like. In New York he showed the map to Hoyt, who immediately picked up the phone. "Trippe," he barked, "come on over. There's someone here who's two years ahead of us."

Before Pan American could fly to places like Colombia and Peru, it had to cross the Caribbean. Here Trippe's knack of knowing the right people helped. An early acquisition of a small West Indian airline brought more influential backers to the fold. By sale of treasury stock, Trippe gained the presence of Fred Rentschler of Pratt & Whitney on the board. The fact that a sizable percentage of the financial interest in aviation was aligned behind his enterprise commended Trippe to Washington. It helped him gain entrée to offices where decisions were made. Postmaster General New, who wanted solid organizations to compete with foreign airlines on overseas routes, was ready and willing to aid such a promoter.

An early Trippe move, taking advantage of a provision of the law already exploited by the domestic airlines, was to get the Postmaster General to grant an "extension" of Pan Am's contract route from Key West to Miami. The move to Miami gave the airline a much stronger home base with good connections to the north; it also doubled the mail revenues on each day's trip. At once, Trippe set about building an airport at Miami.

Meanwhile his employees (there were twenty at the end of 1927) had to keep the airline shuttling to Cuba and to be ready for the next moves, all of which involved overseas flying. Directly in charge of Pan Am flight operations was a little man with a thick Dutch accent and eyes that never seemed to smile. Andre Priester in some ways resembled Admiral Hyman Rickover, post-World War II boss of the U.S. nuclear submarine program. Intense and focused, he was a perfectionist. Priester came to the United States about the same time as his brother-in-law, Tony Fokker. Former Amsterdam manager for KLM, he worked for a time building all-metal planes for Henry Ford, and in 1926 he set up a model airline to Washington for the Philadelphia Sesquicentennial. Then, at Fokker's suggestion, Trippe took him on. Priester was responsible for imparting a certain order to Pan Am that was its distinguishing mark for years thereafter. "Vat iss your eggsperience?" he would ask the pilots he hired, fixing them with a long, disconcerting stare. He insisted on multi-engine planes from the start for the margin of safety that would assure dependable performance and attract passengers. On his office wall he pasted these words by an early English flying man:

Aviation is not in itself inherently dangerous. But to an even greater extent than the sea, it is terribly unforgiving of any carelessness, incapacity or neglect.

Though the first aircraft acquired were Fokker's landplanes, the trips they performed were all across water. Priester therefore recommended that as an added safety factor Pan Am planes should carry radios as did the passenger planes of KLM and other airlines of Europe. Trippe turned to another new company, the Radio Corporation of America, which sent a young engineer named Hugo Leuteritz down to Key West to outfit the planes. The first Fokkers staggered into the air with two-way radios as big as upright pianos and weighing almost as much. That was bad enough, but according to Chief Pilot Wells, they did not work. The signals were too weak, and the static was something fierce. Leuteritz adapted swiftly. Out went the bulky, balky voice radios; in went tiny sets with telegraphic keys. Then when the radio operator aboard the Pan Am Fokker started sending his dots and dashes, Leuteritz's men at two shore stations turned small loop antennas toward the plane's signal. Each shore station reported the direction from which the signal came, and the two bearings were quickly plotted on a chart. Where the two lines intersected was the location of the Pan Am plane.

This was essential information for the airport manager. But Leuteritz's system did more than that. His ground operators flashed the latitude and longitude back to the plane in Morse. The plane's radio operator, translating the coded message, could then tell the pilot where he was. It was primitive — static and corrosion on the antennas were still problems — but it worked. The system, rapidly improved by Leuteritz, became the means by which Pan Am could launch flights across hundreds of miles of water and hit their destinations right on the nose. When Leuteritz in 1928 asked for $25,000 to spend on research, Trippe "nearly fainted." But direction-finding radio became so important as Pan Am flew on that Trippe willingly forked over a million dollars in the next few years to sharpen signals and lengthen range.

When Trippe shifted Pan Am's base from Key West to Miami, he was looking for the chance to fill the mail plane's eight seats with revenue-producing passengers. That proved anything but easy. Pan Am's Miami traffic manager stood outside the little office on Biscayne Bay eagerly barking for trade among the thirsty and well-heeled crowd of winter tourists. His pitch: "Fly with us to Havana, and you can bathe in Bacardi rum four hours from now." But booze was plentiful in Prohibition-era Miami and the round-trip fare was $100. Still, it may not have been the price that held back the customers so much as the thought that if they rose to the bait they might never behold their loved ones again. On August 15, 1928, one of Pan Am's three Fokkers disappeared in the Gulf of Mexico. In all of that first year, Pan Am toted only 1,184 passengers.

One of them, nervous and scowling, was Al Capone, on an urgent errand of rum-running business, no doubt. "Better see it's a safe plane — if anything happens to us, remember, it won't be so healthy for you," growled one of the four bodyguards as he dropped a thousand-dollar bill on the ticket agent's counter. In Havana, Vic Chenea was Pan Am's pioneer passenger agent. On slow days he would go into the Sevilla Bar, take drinks with the American tourists there, and challenge them to fly back to Miami. Now and then, Chenea recalled years later when he was traffic manager for the whole airline, some benighted character would rise from his cups, beat his breast,' and buy a ticket. Chenea would then rush the man by car to the field at Camp Columbia, sometimes hugging him tight to keep him from leaping out. "It was a serious business," he said. "We had to eat — and sometimes the only cash we saw was from passengers picked up at the bar."

But from the start the new airline's operations were organized on a big scale, and Trippe was scheming for much longer-range activity. Priester signed up more pilots, especially those with experience in overseas flying; Captain Eddie Musick, who had flown the Caribbean for the old Aeromarine Airways, was promoted to chief pilot. To see about getting more aircraft Priester camped out at the small Bridgeport, Connecticut, plant at which the Russian-born designer Igor Sikorsky was building his first flying boats, a samovar simmering beside his drafting table. "We had some of that Russian tea every afternoon," recalled Priester's assistant Sanford Kauffman. Lack of airports being an obstacle to rapid expansion in the islands, Priester favored seaplanes. "Flying boats carry their own airports on their bottoms," he said. But Trippe wanted something that could alight on both land and sea. So the first planes Priester got from the

Russian tea room at Bridgeport were amphibians. With such planes, he figured, you could land on the ground where there were airports, and on water where there were not.

With the first of these, a twin-engine S-38 leased from the manufacturer, Musick began to make survey flights beyond Havana. These flights, ordered and arranged by Trippe, were carefully planned to traverse routes that, partly at Trippe's prompting, had been selected for consideration for U.S. Post Office Department airmail contracts. One of these led eastward from Havana through Haiti and the Dominican Republic to Puerto Rico — a thousand miles without a single railroad track to follow. The other, even longer, took off westward from Havana in a daring overwater jump to islands off the southeastern Mexican coast and swung down along the string of Central American banana republics to Cristobal at the Panama Canal.

Weeks before others knew of these routes and months before the Post Office advertised them for bids, Pan Am had its advance man in Central America collecting landing rights and airmail concessions. He was Jack MacGregor, a former oilman who had learned the Latin American ropes working for some of Hoyt's interests in Mexico. He ran into a little difficulty in Guatemala and Honduras, reporting, "They are afraid that the octopus of the north is stretching forth its tentacles to devour them." But everything was put right when Trippe got the State Department to send word that the Germans in SCADTA might have been talking to the opposition. The U.S. minister called at the presidential palace to point out that Tegucigalpa could expect scheduled air service within months, and the green light flashed for Pan Am.

In Cuba Pan Am's man had the task of getting clearance for Musick to fly over the western tip of the island on the first leg of the flight to the Canal. Again a timely Trippe nudge brought a cable from Washington that sent the U.S. ambassador scurrying to the palace to get President Machado's authorizing signature. That freed Musick's shoe-shaped amphibian to refuel on Cuba's south coast that weekend and launch into a three-hundred mile whirl into the wild blue yonder. The destination was Cozumel Island, off the coast of Yucatan. Flying to the mainland would have hazardously overtaxed the range of the Sikorsky, and anyway there were no airports in that part of Mexico. No plane had ever been to the next stop, either. This was Belize, in British Honduras. At Belize, however, the British authorities had designated a sports park as a likely landing ground. When Musick arrived overhead, a cricket match was in progress. The match was halted, the Sikorsky rolled smartly up to the clubhouse, and the Pan American party took tea with the cricketers, one of whom was the colonial governor.

The flight bypassed Guatemala, owing to temporary obduracy in the National Assembly, which had resisted MacGregor's eloquence. In Hon-

Juan Trippe and Charles Lindbergh in front of a Fokker F-7 in Miami, just before takeoff on the first airmail flight to San Juan in 1929.

duras the welcome was warm enough, but the plane had to land at Tela on the coast because the capital without a railroad, Tegucigalpa, also lacked an airport. Flying to Nicaragua presented no such problems. That country was occupied by the U.S. Marines, who had not only built airfields but could also supply a copilot for Musick. This knowledgeable squadron commander showed Musick the best way across the country, across neighboring Costa Rica, and across Panama all the way to the Canal. Musick duly doubled back over the two thousand miles of turbulent skies and seas, alligator-infested mangrove swamps, and smoking volcanoes, to report in his laconic way that Pan Am could fly the route three times a week, day in and day out, as soon as the proper landing floats were moored in certain places and landing fields laid out in others.

Shortly afterward Postmaster General New awarded Pan Am the airmail contract for the route. He also gave Pan Am the contract for the West Indian route from Havana to San Juan, with the intriguing proviso that the route might be extended to Trinidad on the north rim of South America.

Since both contracts carried the maximum compensation of $2 for every pound of mail carried, Pan Am's annual mail income shot up from $160,000 to $2 million.

Promising as these gains might be, the Pan Am venture was still raw adventure. One day Trippe and Hambleton narrowly escaped injury when their Fokker was forced down among Florida mangroves. Another day a Pan Am trimotor, blown off course, ditched alongside an Esso tanker, with the loss of one of its two passengers. To bring some measure of safety along the Central American line Priester carved a few emergency fields out of the jungle. But when he air-dropped bags of flour to mark one site, his ground party had trouble finding it because local inhabitants scooped up his white markers and baked them into bread.

On February 4, 1929, none other than Charles A. Lindbergh took the controls on the first passenger flight over the route. Lindbergh, appointed Pan Am technical adviser after he followed up his Atlantic flight with a South American goodwill tour, was accompanied by his bride, Anne. Trippe went along, together with his new wife, Betty, daughter of Morgan partner Edward R. Stettinius. "We were so young, and it was such an adventure," Betty Trippe later recalled. The plane Lindbergh piloted was the first of twenty-five twin-engine S-38 flying boats that Sikorsky built for Pan Am. Weighing five tons, it could cruise three hundred miles at a hundred miles an hour. Blazing a new trail almost directly south from Cuba to a landfall on Nicaragua, Lindbergh cut the distance to the Canal by a third and thus was able to make the trip to Panama in a record three and a half days.

Shortly before this high-spirited outing the Post Office Department had advertised for bids on yet another foreign airmail route — from Brownsville, Texas, to Mexico City. Several bidders appeared, including one who offered to carry the mail for less than a dollar a mile. Trippe was on hand and bid two dollars a mile, the highest rate permitted by law. Pan Am was awarded the contract. When rivals screeched, the Postmaster General pointed out that Pan Am was the only company that could actually assure performance of the contract.

The Mexican government at the time permitted no foreign airline to carry mail in its territory; that could only be done by a small Mexican concern called Compañía Mexicana de Aviación, or CMA. In his fore-handed way Trippe had got in touch with George Rihl, CMA's founder, the year before and bought 100 percent control of the little CMA line. Rihl became a Pan Am vice president. Two years later the Post Office extended the route south from Mexico City through Central America to Panama. Together with the lines from Miami to Puerto Rico and from Miami to Panama, that gave Pan Am three fat trunk lines. This last one linked up with Texas and California and brought another $1 million in mail pay into the coffers.

Running off plays so fast and furiously, Trippe had need of able team-mates. Priester and Musick and Leuteritz were keeping the planes flying, but so many agents were chasing landing permits, running supplies through customs, and currying favor with a dozen foreign governments that the legend began to arise that Trippe ran his own independent diplomatic service. Nothing, of course, could have been further from the truth: Trippe was an entrepreneur who understood what the great French historian Fernand Braudel meant when he wrote that "capitalism triumphs only when it becomes identified with the state." One day Henry Friendly, then a young lawyer in Elihu Root's law firm, filled out a form at a colleague's request for a bid on a foreign airmail contract. It was a simple task, quickly done, and Friendly, handing back the form, asked casually, "Who was that big and rather swarthy looking young man sitting in your office a moment ago?" The colleague said, "Oh, that's Mr. Trippe; he's the president of the company." From that first brush Friendly was drawn into so much more work for Trippe that he joined Pan Am as counsel and soon, in the judgment of one of Pan Am's pioneer officers, "knew even more about the company's affairs than Trippe."

Landing the first three or four foreign airmail contracts brought timely revenues to Pan Am; they also gave the airline a leg up in the race for more concessions. One day Dick Hoyt got wind that W. R. Grace and Company, with big shipping, banking, and mercantile interests in Ecuador, Peru, and Chile, was preparing to start an airline down South America's west coast. In his most imperious wheeler-dealer manner Hoyt proposed that Grace and Pan Am join forces to bag a mail contract. He turned the details over to Trippe.

Trippe was already working such long hours that when he came home early one day his wife asked, "What's the matter, are you ill?" In the midst of getting the flying services started to Panama, San Juan, and Mexico, building a big flying-boat base at Miami, and working his lines to Washington, he closed the deal with Grace to start an airline, owned fifty-fifty, called Panagra. Harold Harris, the crop duster of two years before, was put in charge of flight operations; MacGregor, the former Pan Am advance man, ran the business side. And when Panagra's bid for the Post Office contract was not the lowest, Friendly drew up a list of pluses and minuses showing with all the force of mathematical proof that Pan Am, already flying routes from the north, and Grace, with offices all along the route, offered ideal qualifications. The Postmaster General accepted the argument, and Pan Am was on its way into South America.

One very ticklish piece of business remained to be done. Between Pan Am's line to Panama and the Panagra line on the west coast of South America there was a missing link. That gap could be closed only with the approval of the Colombian government, whose airspace lay between. Colombians had been suspicious of gringo tricks ever since Teddy Roosevelt "took" Panama from them tò build the Canal. They also knew that the chief of their airline, Dr. von Bauer, a naturalized Colombian and something of a hero in their country, had been denied landing rights in the United States. The Colombian government was in no hurry to grant favors to the airline favored by Washington over their own.

Trippe broke the blockade by a strange and secret deal. Von Bauer, accepting the fact that U.S. preponderance barred his way to expansion, sold 80 percent of SCADTA stock to Pan Am. That removed SCADTA's opposition. But von Bauer specified that no one, and above all not the Colombian public, was to know of the Pan Am purchase; the stock was placed in a sealed bank vault in New York and was not even voted by Pan Am. No Pan Am representative sat on the SCADTA board. So von Bauer, in fact as well as appearance, continued to control SCADTA, a situation that was to cause no end of embarrassment to the United States and Pan Am later when the Nazis began to muscle in on the airline.

Besides his directors, Trippe told only the State Department of the deal. Official U.S. policy had always been to avoid making any commitment about air rights to other nations and to let private aviation promoters do all their own negotiating for rights abroad. In this case it was realized that regularization of Panagra's entry into Colombia was unavoidable. In the spring of 1929, therefore, what is solemnly recorded in the history books as the Kellogg-Olaya Pact was rushed to signature by the foreign secretaries of the two governments. The pact, oldest bilateral aviation treaty in State Department files, was thought of at the time as purely a face-saving device for the Colombians. The chief authors were Assistant Secretary of State Francis White and Juan Trippe. It never occurred to these men, as they scribbled in equal concessions on all air rights to both nations, that the Colombians might one day operate their own international airline. When after World War II they so desired, they found themselves possessing the freedom to fly to almost any point in the United States, a freedom that the United States would grant to no others.

Even while Trippe was scrambling into Panama and leaping for his first foothold along the west coast of South America, a rival American airline was forming to fly to the far bigger and richer lands along the continent's east coast. This was a head-on challenge to Juan Trippe and his backers, who tried to head it off by spreading word in financial circles that "you can't operate an airline in the hurricane zone of the Caribbean and the perpetual rain belt of northern South America." The rival line, the New York, Rio and Buenos Aires Airways (NYRBA), had a name that sounded as if it belonged to a railroad. The fight that broke out, with raids into Wall Street, lobbies in Washington, and even on-site tussles over right of way, resembled the nineteenth-century battles in the West for railroad supremacy.

In 1928 Captain Ralph O'Neill, a flamboyant airman with a gift of gab who had spent five years as a director of training for the Mexican air force, toured South America with his fellow pilot Jimmy Doolittle on behalf of Boeing and Pratt & Whitney. Their job was to sell military planes to South American governments. Instead, O'Neill got an idea for an airline — an airline that, bypassing the necessity for constructing airfields and taking advantage of the added safety of seaplanes over water, would operate flying boats over the entire distance. Boeing wasn't interested, but Major Reuben Fleet, who was building flying boats at his Consolidated Aircraft plant in Buffalo, was.

O'Neill found a backer in the multimillionaire James Rand, of Remington Rand. Fleet agreed to build a half-dozen nine-ton, twin-engine flying boats that could carry a three-thousand-pound load six hundred miles without refueling. With the help of the most influential lawyer in Buenos Aires, O'Neill won a concession to carry Argentine airmail to the United States and followed this up with similar agreements with Uruguay and Brazil. In January 1929 he announced that his airline, called NYRBA and pronounced to rhyme with "Minerva," would shortly begin flying the full seven thousand miles from New York to Buenos Aires once a week, making stops at thirty harbors along the way. With such highly connected Republicans as former Assistant Secretary of Commerce for Aeronautics William McCracken and Colonel William J. Donovan behind the line, NYRBA might well land U.S. mail contracts shortly. On October 2, 1929, Mrs. Herbert Hoover christened the airline's Commodore flying boat *Buenos Aires* in a festive ceremony in Washington.

From Pan Am's standpoint the extension of its Havana–San Juan airmail route to Trinidad now became pretty crucial. NYRBA's flying boats were already making their first flights over this route. It was already known, furthermore, that the new Postmaster General, Walter Brown, was going to advertise the Trinidad–Buenos Aires route for bids, and if NYRBA had hopes of winning that contract, the line from Trinidad north to San Juan must be an important link to them too.

Striking fast, Henry Friendly hustled down to Washington and persuaded the Postmaster General to extend Pan Am's airmail route from San Juan to Trinidad right away. Leaving nothing to chance, he got the papers in order with the department solicitor and caught a train back to New York. There he had the papers signed and the corporate seal stamped on them; then he got right back on the night train for Washington. By 9:00 next morning the last signature was affixed. Friendly was just leaving the department when he met Colonel Donovan, attorney for NYRBA, coming down the hall. "Colonel, what are you doing here?" said Friendly affably. Donovan said, "I'm going to persuade the Postmaster General to let us in on the Puerto Rico–to–Trinidad route." Friendly said, "Well, Colonel, I'm afraid you're a little late — we just got the contract for that route."

Trippe was now determined to get the Trinidad–Buenos Aires contract. He called his key staff people together at Havana. "I suppose you know about these plans," he said, gesturing at a scrap of paper on the table on which he had scribbled a few notes. "We're going to fly over there to the mainland." Trippe broke the news that they would have to start flying past Trinidad to Paramaribo on the north coast of South America at once, even though Pan Am had no mail contract for the run. The reactions of his operations people were explosive. "It iss impossible," exclaimed Priester. He said Pan Am had not even digested its airways across the Caribbean and Central America. "We must get there and stake out our claim before others," Trippe persisted. Toward midnight Priester said he was willing to go along "on the understanding that the job is impossible anyway." He pulled from his pocket a watch fob shaped like a pistol and put it to his temple. The aviation business, he said, "is too full of gas — what this company needs is blimps."

On the other side O'Neill was in a frenzied rush to get the world's longest airline into something like regular operation. He stationed local managers and fuel dumps along the way. He hired pilots, bought some Sikorskys, and dispatched them to Argentina to start a shuttle passenger service across the River Plate between Buenos Aires and Montevideo. He told aides, who included a couple of the disgruntled founders of the original Pan American, Incorporated, "The Argentines are likely to cancel our arrangements at a moment's notice. Pan Am is trying to sell them the idea we're a flash in the pan. The only thing to save us is to get some planes down there in a hurry."

For NYRBA there was a crisis every minute. Somehow the Sikorskys got to Buenos Aires and a four-a-day shuttle bridged the Plate estuary to

Montevideo. Somehow O'Neill's people also launched a Ford trimotor service across the Andes to Chile. To get the flying boat *Buenos Aires* to Washington in time for the christening, they had to cut more corners. In flight testing, cracks showed up in the boat's hull. The cracks were simply brushed over with hot tar.

Somehow the *Buenos Aires*, shipshape despite one landing and takeoff after another, flew its load of ten newsmen, photographers, and mechanics safely as far as Havana. There the ebullient O'Neill insisted on taking up twenty-five Cubans, including members of President Machado's family, for a goodwill flight. Once in the air, however, a storm came up. Unwilling to chance a landing in the crowded harbor in poor visibility, NYRBA's chief pilot, Bill Grooch, brought the big plane down in the swells just off Morro Castle. The impact opened the tarred-over seams. As water poured in, Grooch gunned the plane to shore. Up to their knees in water, the passengers screamed for help. Grooch yelled, "We're resting on the bottom — there's no danger." But by the time a launch took them off in blinding rain, some of President Machado's relatives were hysterical.

While O'Neill rushed off to stem other crises, Grooch flew the flying boat back to Pensacola where old Navy friends fastened staunch plates on the bottom. Back in Havana, he picked up his passengers and lumbered on. In Haiti they slept in the plane. Grooch flew to Trinidad, then on to Paramaribo, where crocodiles slid past in the river as he unloaded his passengers into dugout canoes. Next stop after that was Cayenne, where one passenger, a newsman, deserted to try for a story on the Devil's Island penal colony. In Belém at the mouth of the Amazon the party was delighted to be met by NYRBA's agent, who came out on the launch with the port officers and reported that another NYRBA Sikorsky had just flown through on its way south. The next day's flight brought them to a dockside mooring in São Luiz, where a wizened, unshaven American expatriate named Harry Izler, who sold Singer sewing machines, stumped up and said, "You can't stay there — you'll be high and dry." Thus warned against São Luiz's twenty-foot tide, Grooch shifted his mooring. "My house is better than the bedbugs over at the Maranhao Hotel," said Izler, and took them off to his thatched hideaway down the beach.

In this way the broad-beamed *Buenos Aires* and its complement of intrepid airmen and passengers made their way around the bulge of Brazil and south across the equator. Each stop was a new experience. Refueling at Vitória north of Rio, Grooch took off down-river, climbing through a narrow gorge. Suddenly high-tension wires loomed in his path. He said later:

I shoved the nose down and just cleared the wires. But the left wing hit telephone wires below, and slewed the plane around at right angles. I did a diving turn to the right with the engines wide open. Barely in time we leveled off, and landed.

Was I glad Shorty Clark was along. He folded a thin sheet of zinc over the gash in the wing and tacked it securely on.

Not much later the Commodore, biggest plane ever seen in South America, was circling Rio harbor while two Fox Movietone cameramen cranked scenes of Sugar Loaf, Corcovado, and the crescent beach of Copacabana. Three days later the Commodore thundered across the River Plate to alight safely at NYRBA's anchorage at the Buenos Aires yacht club. Wrote Ted Heath of the *Washington Star,* a passenger, "We just passed an American ocean liner that's been three weeks on the way — the airline will cut that to a week."

Flying seven thousand miles in seven days in hundred-mile-per-hour planes was a tall order. O'Neill, who seemed to think he had to fly the planes himself, took off northward with Grooch and another load of passengers. Approaching Santos, south of Rio, the weather turned stormy. Fearful that the plane was out of control, O'Neill rushed forward. Grooch said later:

I knocked his hand from the wheel, and made a rough landing. O'Neill kicked out a window, thinking the plane was capsizing. The passengers were in a frenzy. One drew a gun, shouted in Portuguese. The flight mechanic hit him over the head with a pyrene bottle. A huge comber dumped us on the beach. O'Neill was suddenly sick. Both lower wings were damaged. Mail bags were floating in water a foot deep in the cabin. O'Neill said, "I guess I'm Jonah. Every time I climb in a plane there's trouble."

Ashore, they phoned Joe Edwards, their agent in Rio, for a relief plane. Edwards replied that a small Brazilian airline had attached the plane, and everything else the company owned in Rio, in a lawsuit. O'Neill was more than equal to this crisis. He drove by car to Rio, and as Grooch related, "we beat the police launch across the harbor to our takeoff." Thus, in a second Commodore, after at least one more crash and various other adventures, O'Neill winged his way back to the United States.

Trippe's next move was to dispatch Vice President Rihl and his wife aboard a Pan Am Sikorsky for a goodwill visit to Brazil and Argentina. When the plane landed at NYRBA's Lake Montenegro facility in northern Brazil, O'Neill's employees, armed with monkey wrenches, drove off members of the Pan Am crew as they tried to berth at the NYRBA dock. After letting the Pan Am party sit for a couple of days, O'Neill cabled permission to gas up their plane, then sent Pan Am an itemized $1,500 bill, including charges for transporting the fuel from the United States. Trippe paid the bill, and Rihl went home.

In spite of the distances and scheduling foul-ups, NYRBA never lost a plane, a passenger, a mailbag. It maintained service month after month along its length and even added floats, barges, ramps, hangars, and other improvements. But while O'Neill was bouncing off breakwaters, making hairbreadth landings, and otherwise trying to get his new airline into service, the future of NYRBA was being settled in Washington and New York. The stock-market crash of 1929 ended all easy money for aviation ventures. Not even Jimmy Rand could pay the bill for the fourteen Commodores and ten Sikorskys O'Neill had ordered. NYRBA was hemorrhaging money at the rate of $50,000 a month.

By the summer of 1930 the airline began to send up distress signals. Had it won a U.S. airmail contract, it might have weathered the storm. But Postmaster General Walter Brown refused to heed the cries for help. Then the banks began to dig in for the long struggle. One by one the NYRBA promoters picked up their hats and departed quickly. Only Jimmy Rand stood fast. For a time he financed the airline out of his own pockets, hoping, apparently, to stave off utter collapse until negotiations could be completed for sale of the faltering company.

That was difficult because Pan Am was obviously the only buyer. Postmaster General Brown had convinced the promoters that no airmail contract would be handed out until the merger was effected. In the end he made it look as though he were doing a favor to NYRBA's backers. As George Rihl explained later, "Postmaster General Brown said to us, 'Buy them.'" On August 19, 1930, the sale of NYRBA to Pan Am was announced. NYRBA stockholders got about fifty cents on the dollar, which in the circumstances was generous. Pan Am, acquiring thirty-one Commodores and Sikorskys, almost doubled the size of its fleet. Just one day after the merger the Post Office advertised for bids on the South American east-coast mail route. Since Pan Am's only competitor had just been eliminated, the bid submitted was for the maximum rate.

By this decisive turn Pan Am (the holding company, Aviation Corporation of the Americas, was dissolved in 1930) emerged as the "chosen instrument" for carrying the U.S. flag through skies abroad. Chosen in this fashion by Postmaster General Brown, Pan Am found itself flying twenty thousand miles of routes in twenty countries, every one of them a U.S. Post Office contract route. All of a sudden, Pan Am was the world's biggest airline.

By the standards of American corporations, in 1930 Pan Am was still a very small company, not half the size, say, of the Wabash Railroad. In the preceding year it had carried only 6,824 passengers on its routes. Yet it was already the longest airline in the world, and its managers, with a $4-million annual income in Post Office payments alone, could shortly expect to be in the black.

8

Zeppelins

In the nineteenth century when the Prussian war machine dominated Europe, Count Ferdinand von Zeppelin was a highly unusual professional army officer. Behind his medals, epaulets, and walrus mustache, he was a dreamer and promoter. No Prussian at all, he was a Swabian from Friedrichshafen on the banks of Lake Constance. Having fought as a young volunteer Union cavalryman in the American Civil War, he tried to put the notions of individual soldierly initiative he brought back from America into practice in the Kaiser's army. Eased out of command, he proceeded at the age of fifty-two to invent the astonishing and romantic creation that bears his name.

The zeppelin — one reads about it now and wonders how it ever came to be. Yet the lighter-than-air dirigible offered promise of great range and load-carrying capacity long before the airplane amounted to anything. Passengers were riding about Germany in such airships before World War I, and after the war they crossed the Atlantic so regularly that when the Empire State Building was put up in 1930, it had a mooring mast atop it to accommodate expected airship traffic. Had the Germans won either the First or the Second World War it is conceivable that air travel might have evolved differently — spacious intercontinental zeppelins might still be providing the amenities of lounging, eating, and sleeping space prized above jet speeds by long-distance travelers.

From first to last, Germans seemed the only masters of this strange and wonderful creature of the skies. Airships call for big thinkers, and Count Zeppelin thought big when he got the basic ideas in 1890. Seizing upon the principle that the lifting power of such a craft would be proportional to the cube of its dimensions, he accepted the concept of enormous size and went on from there. He sketched out the idea of the unique metal frame that made rigid construction possible and added the further idea of many separate envelopes within the huge frame to seal in the gas needed to lift his whole grandiose contraption.

In bringing his giant airships into being, Count Zeppelin, of course, thought he was building superior new weapons of war. But that was not the only reason they captured the imagination of his countrymen. Nothing like them had been seen before. The airships built by the Count's organization between 1900 and 1914 embodied the world's highest technologies — German metallurgy, which took aluminum out of the university labs and fashioned it into framing girders; German chemistry, which purified, tamed, and packaged volatile hydrogen; German physics, which developed the original internal-combustion engine that gave the Count his needed lightweight propelling power. Only Germans, it seemed, could build such fabulous machines.

When the Count sailed his big ships over their cities, German people were filled with awe and pride. When calamities overtook them, when fire destroyed one airship on the ground at Kislegg in 1905 and winds drove another down and wrecked it at Echterdingen in 1907, the rich and the poor, the Kaiser and the schoolchildren, all flooded the Friedrichshafen headquarters with contributions that paid for bigger and better ones. Dr. Hugo Eckener, who had been writing critical articles in the *Frankfurter Zeitung,* shed his doubts. Following the Kislegg disaster, Count Zeppelin, in formal silk hat and yellow gloves, called on Eckener and won the journalist to his enterprise over dinner. Another fierce individualist with giant-sized dreams, Eckener had a great gift for popular persuasion. He became the greatest of all airship pilots.

At odds with the military, Count Zeppelin's company capitalized on public enthusiasm by starting an airline and flying passengers. It was this that made the zeppelin a national symbol throughout Germany. DELAG, as the airline was called, started flying November 16, 1909, from Frankfurt. Passengers bought tickets through the offices of the Hamburg-America steamship line, a big subscriber. The first ship was the 486-foot *Deutschland,* which had small elevator planes along the sides, multiple rudders at the tail, and eighteen hydrogen bags inside the hull. Overall the ship weighed 46,000 pounds, about the same as a four-engine airliner of the 1940s. Its structural weight, however, was 36,000 pounds, and only 2,600 to 3,500 pounds had to be added for fuel, so 7,000 pounds remained for useful load. Therefore the *Deutschland* could carry, besides water ballast, a crew of eight and twenty-four passengers. However, the best cruising speed of which the three Daimler 125-horsepower engines were capable was thirty-three miles per hour. Local joyriding was about all such a speed permitted.

After taking aboard twenty-five journalists for a twenty-hour cruise in 1910, the *Deutschland* was blown by strong winds and crashed in the Teutoburger Forest. The only casualty was a crewman's broken leg. More money was raised, and Eckener took command of the next ship. Despite mishaps, he made flights almost daily. Then he took the new *Schwaben* out of the hangar at Friedrichshafen and piloted it to Frankfurt, to Düsseldorf, and finally to Berlin, where crowds went wild Patriots leaped to pay the $125 fare, ride in the mahogany-paneled gondola, sample the fine wines and elegant cold dishes served on high, and look down through the wide windows at the German countryside. Disasters occurred, but they never seemed to hurt anybody. In 1912 the *Schwaben* burned on the field at Düsseldorf when static electricity set fire to the rubberized fabric of the gas envelopes and the hydrogen went up with a whoosh. The company brought out the *Hansa* and the *Sachsen* and advertised service on a circular route from Friedrichshafen over all the country's principal cities. Between 1910 and 1914 DELAG's three ships made 1,588 flights and carried 10,197 passengers without injury to any of them. But Eckener was careful never to go when winds were unfavorable. Since the flights did not follow regular schedules, they could not yet be called regular airline operations.

Then came the war. Not only were DELAG's three airships taken over by the military but also the Count's company built more than a hundred zeppelins for the German Army and Navy in World War I. Scores were dispatched over Allied capitals in history's first strategic bombing raids. London caught the worst until airplanes finally defeated them. Zeppelins also shadowed the British Navy, whose dreadnoughts were powerless to keep the huge silver sausages from hanging silently in the sky out of range and reporting methodically their movements by radio.

After the war the victorious Allies bloodied their noses trying to fly the ships they took as booty as well as those they built on their own. One by one they gave up their lighter-than-air programs, even the Americans, who possess the world's only source of helium, a noninflammable substitute for hydrogen. All the sweeter for the Germans, therefore, were the developments that followed.

In 1921 the U.S. Navy, still hoping to turn the art of lighter-than-air flying to long-range scouting across its ocean approaches, approached Dr. Eckener about building an airship capable of flying the Atlantic. That was the chance the zeppelinists were waiting for. Reopening the old plant at Friedrichshafen, Eckener built the ZR-3 and in 1924 flew it five thousand miles nonstop to Lakehurst, New Jersey, where it was rebaptized *Los Angeles* and placed in a specially built hangar.

When the 1925 Treaty of Locarno removed the limits on the size of airships Germany could build, Eckener and the thousand men trained in zeppelins in World War I were ready. With the help of a popular subscription, the reconstituted company at Friedrichshafen proceeded to build the 775-foot *Graf Zeppelin*.

The *Graf*, ten times the size of the old *Deutschland*, was another marvel of engineering legerdemain. Dr. Ludwig Durr, the Count's old master builder, was again the presiding magician. Under his hand there came into being an airship of surpassing size and delicacy — twenty-eight lengthwise triangular-shaped girders of duralumin encircled by twenty immense duralumin rings, all artfully trussed and braced for stress so as to make seventeen spaces for the gas envelopes. These seventeen envelopes of cotton fabric had to be made absolutely gas-tight. This was accomplished by lining their insides with goldbeaters' skin — the delicate outer membrane covering a cow's upper intestine that was used by goldsmiths to separate the leaves of metal. But what prodigious amounts were needed! Each such skin measured at most thirty-nine by six inches, so that it took 850,000 skins in all to line the *Graf*'s lungs. The cost today of such a quantity, taking into account the care with which they had to be handled in the slaughterhouse and the skilled handwork needed to assemble them at the gas-cell factory, would be prohibitive, and airship makers, settling for synthetic substitutes, never repeated it.

The *Graf*'s hull was made of varnished sailcloth, and so were the huge, thick fins at the tail that gave lateral and up-and-down steerage to the ship. Filling the lower part of the space inside the hull was a new kind of fuel, not the old gasoline, but propanelike "blue gas," whose special virtue was that it weighed the same as air. Thus using up fuel would not, as with gasoline, lighten the ship and therefore would not require hydrogen to be wastefully valved off to maintain static equilibrium.

A gondola nearly one hundred feet long and twenty feet wide was built into the bottom of the *Graf*'s hull. At the center and rear were quarters for twenty passengers and forty crew. The largest space was occupied by the passengers' lounge and dining area, carpeted and curtained in burgundy red, and lined by outward-slanting windows on either side that permitted the passengers to look out as they dined. Immediately to the rear were five pairs of sleeping cabins, each with a sofa that hinged upward to form an upper berth, and each with windows that enabled the passengers to look out from their beds over the sea or land below. At the end of the sleeping area were lavatories and a bath, and beyond them were galley and crew quarters.

Slung in separate pods beneath the ship's hull were five engines — 550-horsepower Maybachs that could drive the ship ahead at speeds of seventy-three miles per hour for up to one hundred hours. The engines were so far from the living spaces that passengers felt no vibration and heard little noise. They were controlled from the command center at the forward end of the gondola. There the elevator man stood at his wheel; the rudder man stood at still another wheel; and the captain, with instruments, gas-cell indicators, fuel gauges, and navigating gear before him, conned the ship. That was the word for it because piloting an airship was more akin to sailing a ship or submarine than flying an airplane.

On October 11, 1928, the *Graf* was ready for its inaugural trip across the Atlantic. Two Americans, one the owner of a factory in Silesia and the other a financier living in Switzerland, were among the ten passengers who paid the $3,000 fare to fly to the United States. The other passengers were German government officials and journalists. Lady Drummond Hay, representing the *New York Times*, climbed in at the last minute and then said, "Oh for heaven's sake, I've forgotten my coat." Eckener gallantly dispatched a car to her hotel for it and then cast off at 7:55 P.M.

Crossing France, the *Graf* skirted the Spanish coast and passed through the Strait of Gibraltar under sunny skies. Eckener had devised a system of dead-reckoning navigation so accurate that star sights and radio bearings played secondary roles. The system included frequent checks of wind drift by dropping smoke bombs. Steering first with the wind and then changing course by forty-five degrees, the navigator took bearings, plotted them along with the air speed, and obtained vectors that showed both the desired heading and the speed over the ground. In this way the *Graf* hit Madeira right on the nose.

Beyond the Azores Eckener met "a blue-black wall of cloud of very threatening aspect advancing toward us at great speed from the northwest." Updrafts thrust the *Graf's* nose fifteen degrees skyward. Breakfasts clattered into passengers' laps. Pots crashed in the galley. A mechanic climbed down from the tail to report that the cotton cover had been ripped off the port horizontal stabilizer.

The flapping shreds of cloth could jam the elevators. Eckener slowed the engines. He also made the weighty decision to wireless the U.S. Navy for aid. Then he called for volunteers. Six, including his son Knut, climbed up on the tail and spread blankets across the hole; Chief Rigger Ludwig Knorr sewed them in place. Then, canceling his SOS, Eckener speeded up the engines and resumed his westward course, sure that his voyage got the headline attention he wanted. Off Bermuda the *Graf* ran into another front, but this time Eckener reduced engines to half speed going through. Again the ship bucked and pitched, while rain and hail beat on the outer cover; the going was slow. Newspapers reported each radio bulletin, and when the *Graf* cruised slowly over first Washington, then Baltimore, and finally New York, all work slowed, traffic stopped, and thousands crowded into the streets to watch its majestic passage.

On the trip back twenty-five passengers paid Thomas Cook and Sons $3,000 apiece for the nonstop run across the North Atlantic. When they landed, a crowd of thirty thousand sang "Deutschland über Alles," and President Hindenburg hailed "a superb German achievement." The following year, with publisher William Randolph Hearst putting up $100,-000, Eckener took the *Graf Zeppelin* on a trip round the world, starting at Lakehurst. One of the earliest to understand and use "pressure" systems in the Atlantic, Eckener dodged to the south side of a big one, picked up a tail wind, and crossed to Friedrichshafen in a record fifty-five hours. Taking aboard Russian and Japanese government representatives, he then flew nonstop across the Soviet Union to Tokyo. The next jump, which passengers found "dull" because so much of it was spent flying through fog and cloud, took the *Graf* across the Pacific, over San Francisco, and down to Los Angeles.

At Los Angeles the *Graf* met up with a phenomenon that can cause trouble for airships. This was a temperature inversion — at 1,600 feet the temperature was seventy-seven degrees, but the air below was sixty-six degrees. As with any airship, the *Graf*'s lift was influenced by the difference between the outside temperature and that of the gas inside, and any change in either had to be watched and compensated for by the man at the controls.

Approaching Mines Field, the *Graf* with its gas cells expanded by the warm temperatures above refused to descend into the cool bath below until Eckener valved off a lot of hydrogen. Taking off the next night, the problem reversed itself. The *Graf* refused to ascend until Eckener off-loaded fuel, ballast, and six crew members who had to be sent on to the East Coast by rail. Then, ringing up Full Speed on all his engines, Eckener aimed the *Graf*'s nose higher. Flying the big sausage like an airplane taking off, he eased the forepart of the dirigible over the high-tension wires at the airport fence. The moment the nose cleared the obstruction, he quickly ordered the elevators turned down, and the *Graf* being 775 feet long, there was just time and room to inch the stern over the wires too. It was a close call, but the *Graf* got away and soared on over Chicago and New York to complete the circuit at Lakehurst.

A Broadway ticker tape parade and a visit to President Hoover at the White House followed. Total flying time was twelve days, eleven minutes, considerably faster than the fifteen days and three hours the U.S. Army fliers needed in 1924 on the only round-the-world flight completed before this. Most significant of all, from sixteen to twenty-two passengers had been carried on each leg of the trip, in such safety and luxurious comfort that the pioneering journey was for them a pleasure cruise as well as an adventure.

Many of the flights the *Graf* made were essentially publicity trips, their cost defrayed mainly by the largesse of newspaper publishers and by the considerable sums stamp collectors contributed for special *Graf* flight covers. But Eckener always carried paying passengers; on one hop to both South and North America he collected a $6,500 fare from Don Alfonso de Orleans, cousin of the King of Spain. Because of lighter winds, Eckener favored the southerly crossing, and when the Brazilian government built an airship station near Rio, the *Graf* started regular commercial flights across the South Atlantic. In the next six years the *Graf* made some fifty advertised and scheduled trips between Friedrichshafen, Recife, Rio, and Buenos Aires. The same staying power that enabled the *Graf* to circle the world in four jumps was demonstrated in another way on the South Atlantic in 1935. When a revolution broke out in Brazil, skipper Fritz Lehmann simply drifted off the coast for three days until informed that the government had put down the uprising and secured the airdrome; then he brought his ship in for a normal landing.

In almost six hundred flights, the *Graf Zeppelin* transported more than thirteen thousand passengers without casualty. This was a notable achievement by any standards in the 1920s but a prodigious one in the view of the English, French, and Americans, who had met with fatal accidents in all their efforts at lighter-than-air flying. The *Graf*'s record put Germany ahead in long-range air transport.

When Hitler came to power in 1933, the zeppelins, so long a German symbol, had to be converted into a symbol of Nazism. Goering and Goebbels moved in on the company; the swastika was painted on the *Graf*'s tailfin. When Eckener refused to allow the huge hangar at Friedrichshafen to be used for a Nazi election rally, Hitler made sure that the old master would not be allowed to command the huge new dirigible whose construction got under way there.

This was the *Hindenburg*, the fir. airship to be built for commercial service on the North Atlantic. Although only thirty-three feet longer than the *Graf*, the *Hindenburg* had almost twice the gas capacity, engines with twice the horsepower, and space for fifty passengers. And what space! The luxury and spaciousness of the passenger quarters on the *Hindenburg* were the equal of those on any ocean liner and far surpassed anything seen to this day on any aircraft.

Passengers boarded the great ship on the lower of two decks. Ascending the wide staircase to A deck, they saw at the top a bust of President Hindenburg. Along the corridor beyond opened twenty-five staterooms, each with an upper and lower berth, its own toilet, and hot and cold running water in a washstand. During the day the upper berth could be folded into the wall, and the lower made into a sofa — much like Pullman compartments on a train. During the day each cabin was its own comfortable sitting room, decorated in pearl-gray linen. By the door was a button. "When you pushed it, a drink was produced," said a passenger who crossed on the first flight in 1936. "This saved walking perhaps fifty feet to the handiest of the three bars that were maintained by the seven stewards all day and most of the night."

Farther along was the dining room, occupying an area measuring fifteen by fifty feet. In all the luxury and refinement of a small restaurant were tables and chairs of lightweight tubular aluminum created for the *Hindenburg* by a leading German designer. The tables were laid with white linen, freshly cut flowers, fine silver, and the china service created for the *Hindenburg* with the gold Zeppelin crest. In the adjoining lounge was a 397-pound aluminum baby grand piano.

On the B deck below were galleys, baths, crew quarters — and the smoking room. For a hydrogen-filled ship this was a startling feature. But safety was assured by pressurizing the room and sealing it off with an air-lock door. The steward of the bar inside had to inspect each guest to make sure that he was not carelessly departing with a burning cigarette, cigar, or pipe. He was also the sole custodian of matches, and if a smoker needed a light, said one traveler, the steward might strike a match for him, "but he retains hold of said match with great tenacity, from ignition to charred cinder. For the passenger to do his own lighting is *streng verboten.*"

With Lehmann as captain, but with Eckener beside him in the command gondola, the *Hindenburg* flew one flight to the United States and seven to Rio in 1936. But when it took off for Lakehurst on its first flight of 1937, Eckener was not on board. For the thirty-six passengers and sixty-one crew members the eighty-hour crossing was uneventful. Toward the end, however, headwinds delayed progress, and when the big ship flew over Lakehurst a thunderstorm caused a further wait until 7:25 in the evening. Then the *Hindenburg* dropped its landing ropes to the men on the ground and hovered just in front of the mooring mast.

The arrival of any transatlantic liner brought out reporters in those days, and on the ground were three newspaper reporters as well as Herb Morrison, who had come to make a recording for the Chicago radio station WLS. Morrison's agonized voice, broadcast on network radio that night, told millions what happened next:

> Here it comes, ladies and gentlemen, and what a sight it is, a thrilling one, a marvelous sight.... The sun is striking the windows of the observation deck on the westward side and sparkling like glittering jewels on the background of black velvet ... Oh, oh, oh!
> It's burst into flames.... Get out of the way, please, oh my, this is terrible, oh, get out of the way, please! It is burning, bursting into flames and is falling.... Oh! This is one of the worst.... Oh! It's a terrific sight.... Oh! ... and all the humanity!

The airship had dropped its landing lines to the ground and was just coming up to the mooring mast when flames suddenly burst out atop the hull near the tail. The stern of the great ship, blazing fiercely, dropped and the framework broke in two amidships. The whole structure collapsed to the ground as the flames ate their way forward. Human beings, their clothes on fire, leaped or fell to the ground amid the blazing wreckage. Some got up and staggered off. Others rolled over and over to save themselves. Others lay still.

Amazingly, sixty-two of those who embarked survived the holocaust. But thirteen passengers died, along with twenty-two of the crew, including Captain Lehmann.

Both the German and American investigations that followed concluded that leaking hydrogen had been ignited by a static electricity charge. Sabotage has since been alleged, but hard evidence is totally lacking.

At the time, the *Hindenburg* disaster was so crushing to its owners that the *Graf Zeppelin* was also grounded, never to fly again. One year later the company completed the other large ship under construction, and Eckener took it up on its first flight at Friedrichshafen. But it never carried paying passengers, and in 1939 the Luftwaffe took it over. A year later at Goering's order it was broken up and its metal was used for warplane construction. Thus the airplane defeated the dirigible.

9

Air Travel in the 1920s

On the splendid July Sunday in 1919 when scheduled daily passenger service was first attempted between New York and Atlantic City, Mrs. John A. Hoagland of Manhattan was on hand for a ride. Mrs. Hoagland was a woman of spirit. She was also an air traveler of some experience, having made two ascents in Paris in a Farman six years before. When the Travelers Company's morning flight turned out to be fully booked, she arranged instead to fly in the afternoon. Appearing at the seaplane dock at Eighty-first Street on the Hudson River, she brought along a companion, Miss Ethel Hodges of Texas; each paid $100 for the round-trip flight.

Ethel Hodges pronounced the trip to Atlantic City, her first, "delightful." With Ralph Hewitt, one of the company's three owners, at the controls, the Aeromarine flying boat, a slightly remodeled Navy patrol plane, took off in a glory of spray. From the windows of their enclosed cabin the two women spotted the liner *Leviathan* lying at its dock in New Jersey. As the plane gained altitude they craned for a look at Ellis Island and the Statue of Liberty. From a height of three thousand feet they found the lighthouses, the big hotels, the weekend beach crowds along the Jersey shore an "enthralling spectacle."

But by the time the flying boat taxied out past the Atlantic City break-water for the trip back, it was 8:00 P.M. Twice on the northward journey pilot Hewitt had to land on the water to make minor engine adjustments. The women in their cabin took this in stride, but Hewitt in his open cockpit was deluged with spray each time the craft plowed through the water. Near the end of the trip, just as the sun was setting, the skies turned very black. Gusts of wind sent the plane careening from side to side. A thunderstorm! Without seat belts the women clung to their wicker chairs and to each other. They shut their windows. Lightning flashed close. Hewitt, crouching low, brought the plane down from four thousand feet to just above the water. Then, as if bursting through a curtain, the plane broke out of the storm and into calm air. They were over Sandy Hook, and it was night.

Hewitt put on his running lights and flew upriver. No flares shot up from the anchorage to guide their approach; all the help had gone home. But after the pilot had brought his plane down safely, a motor boat came along. The women climbed off. To a reporter Mrs. Hoagland said she thought she would not like to fly at night again. Ethel Hodges said that at home she had ridden bucking broncos without fear, but on the return flight she had "done all the praying." It must have been quite a wind; their trip, forced landings and all, had taken but seventy-eight minutes.

Though the Travelers Company went out of business shortly thereafter, flying boats remained popular with those venturesome people who wanted to go places by air. When few airfields yet existed, seaplanes could take customers wherever there was water, and it soon became evident that passengers preferred a plane that, in trouble, could come down on the water anywhere. This was the guiding notion of Grover Loening, one of the most notable aviation entrepreneurs of his day. He was a slim, starchy man, an engineering graduate of Columbia University who had learned to fly as Orville Wright's pupil and personal assistant. In the First World War Loening began to build planes, and shortly thereafter his company produced what was called the Loening Air Yacht, a five-seat flying boat with a single Hispano-Suiza engine.

In the fall of 1922 Loening took a load of passengers on a proving flight from Port Washington, Long Island, to Detroit. One was a woman who wrote up the trip for *Aviation* magazine. The seats they occupied, she said, were wood and canvas yacht chairs and afforded a splendid view out the side windows, which were kept open in fine weather. The "great, throbbing bird" (everybody had stuffed cotton in their ears) seemed to "annihilate distances." Flying one hundred feet above the Hudson River near West Point, the party passed "a New York Central train bound for Chicago — it couldn't keep up with our Loening." After stopping for fuel and sandwiches at Lake George, the flying boat headed north along Lake Champlain for Montreal before turning west. This was a bit of a jump and it seemed prudent to come down for fuel at St. Charles, just across the Canadian border. "Ours was the first ship to make a landing at this town, and it was unfortunate that none of us spoke the language." There were no such difficulties at Montreal, where the travelers spent the night before heading west along the St. Lawrence River. After another overnight stop, they flew along the south shore of Lake Ontario. "The air was so rough that after flying one hour and a half we landed at Sodus Point to wait for the wind to subside." Late in the afternoon it was deemed safe to go on to Buffalo.

The next day the weather was not conducive to safe flying, but in the afternoon the rain ceased. We left our mooring at 4 P.M. and flew along the north shore of Lake Erie. At Rondeau Harbor we had to land for gasoline. It had to be carried some distance so we were forced to stay all night. There was no hotel, but we were taken in by some very hospitable people. Next morning we took off at 9:04 and followed the lake shore to the Detroit River. When we sighted Grosse Isle we knew our wonderful journey was coming to an end. In perfect visibility we landed at Detroit Yacht Club at 10:25 A.M.

The trip had taken four days.

Scheduled flying with the Loening Fleetwing, a modified version of the Air Yacht, began next spring. Vincent Astor formed the New York–Newport Air Service Company and bought three Fleetwings. He opened a regular service between a float at Thirty-first Street on the East River in Manhattan and a borrowed Navy launch in Newport. Rich men kept the seats filled, at least on weekends. On July 20, 1923, Harold Fowler of the McCormick farm-machine family and Cary Morgan, nephew of J. Pierpont Morgan, were the passengers as the day's flight approached Newport on schedule. But the plane flew right into the water. In the smashup Morgan suffered a fractured leg; four days later he died. Loening insisted that flights cease until the cause of the accident could be traced. Fowler said his friend, riding just behind the pilot, had been asleep with his foot stuck out when the accident occurred. Placing himself where Morgan sat, Loening stuck out his foot beside the pilot's seat — his shoe caught in the plane's controls. That was it. The pilot, trying to level off for landing, could not pull back on the stick because Morgan's foot had blocked the way. It was the end of the New York–Newport Air Service Company. But Loening's shop superintendent, Leroy Grumman, later took up the business, renamed it Grumman Aircraft, and made a fortune as the primary builder of U.S. Navy planes.

Overland flying, the only kind possible in most of the country, was slow starting. But as early as 1919 something that looked very like a modern airliner took to the skies over the Middle West. This was the Lawson Airliner. Built in Milwaukee by a fast-talking promoter-dreamer named Alfred W. Lawson, founder of the magazine *Fly,* this plane had two wartime Liberty engines, a single wing with a ninety-six-foot span, and a fuselage wide and long enough to hold up to thirty passengers. This was as big as the Douglases and Boeings of two decades later. With its wings painted red and body green, with *Lawson Airlines* emblazoned in white along the side, it looked like them as well.

When Lawson flew the big plane east to tout his project for a whole fleet, plying between New York and San Francisco, journalists and others were invited to try its wings. Crossing the country, Lawson had trouble finding landing fields — in Toledo lumber from the stands for the Dempsey-Willard championship fight was still lying about the field and the plane had to come down in a nearby pasture. But with Carl Schory, his engine specialist, keeping the Libertys humming, the Lawson Airliner barged eastward at a sixty- to seventy-mile-an-hour clip. In Buffalo Glenn Curtiss climbed aboard, in Rochester George Eastman came out for a look, and in Syracuse Mr. and Mrs. Willis Carrier, of the home-heating fortune, were persuaded to ride along to New York. They soon began playing bridge. Katharine Brady, also from Syracuse, declared on getting off at Mineola that the ride had been "easier than a Pullman."

For the flight to Washington, Lawson picked up more passengers. One couple rode the whole way holding hands tightly; another passenger wrote out his will, then tore it up on arrival. Washington was unquestionably the high point of this pioneering safari. Somehow Lawson talked Secretary of War Newton D. Baker and Mrs. Baker, Senator Hoke Smith, and a dozen congressmen into going for a ride over the capital. They got safely back to the ground only when Schory frantically wobble-pumped the last drops of fuel to lift the plane past a fence and onto the College Park landing field. A man from the Post Office from whom Lawson sought an airmail contract pronounced the verdict: the plane would fly, but not fast enough — besides, there weren't enough flying fields yet.

Both shortcomings became evident when Lawson, ever hopeful, loaded up with passengers and winged back west. Climbing to ten thousand feet over the Alleghenies, the plane ran into head winds. When it was clear that the airliner would never make it to the intended first stop at Wheeling, Lawson's pilot headed down to look for a likely pasture. Flying over a ridge, the plane hit a downdraft that pitched everybody out of their seats. A young man who had sat motionless for three hours in the rear seat crawled forward to ask Schory, "Is everything all right?" Clutching the back of the pilot's seat, Schory said over his shoulder, "We're heading for a big crackup."

The plane overshot the pasture and came down in a cornfield beyond. A wing hooked on a corn shock, the plane spun around, the nose dug into the ground. "We were all right," said Schory. "But we had to disassemble the plane and ship it to Dayton, Ohio, for repairs. The passenger in the rear took off. I never saw him again." And the Lawson Airliner? "That ended up as a hamburger stand near Milwaukee."

Seat belts were just coming into use by the time of the Lawson Airliner. Benny Foulois, later head of the Air Force, fastened the first one in 1910. The Navy began to use them after Jack Towers, later head of naval aviation, was thrown out of his seat and survived only by dangling from a strut as the plane crash-landed itself in 1913. In 1925 the early birdman Marty Jensen and a passenger had a remarkable experience when flying with seat belts fastened. Jensen was teaching a young ensign to fly over his airfield at Dutch Flats in San Diego. The ensign, in the rear seat, was supposed to put the plane into a spin. Instead of easing the stick forward, the ensign pushed it full forward — the plane went into an outside loop with such force that both their seat belts snapped. The ensign was thrown clear. Jensen, hurled against the top of the fuselage, managed to cling to the control stick and pulled back on it. The ensign, falling earthward, landed kerplunk atop the fuselage just forward of the tail, causing the plane to zoom up again, but smashing just the right-sized hole to contain his rear end. In that posture he rode the rest of the way to a safe and soft landing. In 1976 when Jensen related this unlikely story on television, a stranger named S. L. Potter of Alpine, California, called Jensen in great excitement to say he had seen it all from the ground that day in 1925.

It was in just such open-cockpit airplanes that the private carriers began performing their airmail flights the following year. One day a few weeks after Western Air Express began flying the mail between Salt Lake City and Los Angeles, one of the pilots dropped into the Salt Lake City shop where Maude Campbell clerked. An air buff since childhood, Maude said, "Won't it be wonderful when the planes can carry passengers!" The pilot said, "Oh, we can take passengers now, we have two extra cockpits." Maude said, "No kidding," and bought herself a ticket — $200 round trip.

Western Air had already carried Ben Redman and J. A. Tomlinson, two Salt Lake City businessmen, riding on mail bags, on its first scheduled trip on May 23, 1926. But Maude Campbell was the line's first woman passenger. She arrived at planeside wearing plus fours and carrying a small toilet case. Pilot Al De Gormo lent her another pilot's coveralls and helmet and told her how to use the parachute: "If there's any trouble with the plane, jump out, count to three, and then pull the ripcord." De Gormo explained that the toilet facilities, if needed, were a tin can. Later she said, "I waited until we got to Las Vegas. It was the only thing to do."

Her trip, and the return a week later, were a great success. Along the way pilot De Gormo passed her scribbled notes about the weather, the towns they passed, and such sights as Zion Canyon. When she got off the plane at Los Angeles, she was windburned, red-nosed, and a bit deaf. But Western Air President Harris Hanshue was on hand to present her with an armful of gladiolas, and fifty years later Western was still proudly claiming Maude Campbell, now a Los Angeles matron, as the first woman passenger of America's oldest surviving airline.

Boeing Air Transport, having won the contract to carry airmail on the western part of the transcontinental route, made provisions to carry passengers from the beginning. In its 40B mail planes was a small cabin forward of and below the pilot's cockpit where two, and later four, passengers could ride. Often nobody showed up, and the pilots tossed mailbags into the empty space. But one day the same Al De Gormo who carried the first woman passenger for Western Air found himself flying for Boeing from Salt Lake City to San Francisco with a single male passenger aboard his biplane.

In the early 1920s no passenger who boarded an airplane could be sure of getting to his destination. In those days you could not even buy air insurance, and many policies were written to be void from the moment of takeoff and valid again only after you were safely on the ground once more. This particular passenger, De Gormo had noticed, climbed in with a bottle of whiskey for trip insurance. But the official Weather Bureau report forecast stormy skies, and De Gormo figured that the ride would be so bumpy his passenger would drink himself into oblivion. Actually, the turbulence and thick clouds kept the pilot so busy staying on course that he clean forgot about his passenger, until he felt a strange vibration and heaviness on the controls.

Just as De Gormo began to wonder about his passenger he saw him — out walking on the wing, clinging to a strut, and waving happily. Frantically Al beckoned to him and yelled. He kept on yelling until finally the man climbed back. Badly shaken, De Gormo brought the plane down on the emergency field at Carson City and gave his obstreperous passenger, a radio manufacturer's representative from Salt Lake City, a talking-to. The passenger meekly promised to be good, and they climbed back into the plane. Ten minutes after takeoff, the man was out of his seat, wing-walking again. Again, De Gormo shouted and shouted. Just before the plane started over the Sierra the passenger, chilled and possibly a bit more sober, crawled back to his seat. De Gormo said later that his voice was never as strong thereafter: "That was from shouting at him — to get him to come back."

In the late 1920s Bill Allen, Boeing's lawyer, was working on negotiations to transform the company's three airlines into United Airlines and establish its headquarters in Chicago. "I always allowed five days for the trip from Seattle," he said later. "It was all contact flying, principally in single-engine planes. Many's the time I'd have made it faster by the train.

Going home from Chicago once in mid-December I got as far as Salt Lake City. Then I got up at 5:00 A.M. to get on the flight for Seattle. It went via Boise to Pasco, Washington, then down the Columbia gorge to Portland. The morning was not promising. There was weather to the north. We started out in a little plane, and were in a blizzard as soon as we hit the north end of Salt Lake. But the weather closed in behind. The pilot — his name was Congdon — had to get down. He did, and finally located his emergency field called Locomotive Springs. An elderly couple acted as caretakers and radio operators. There was nobody else around for miles.

Obviously it was a widespread storm. By way of conversation I asked, "What do you do for amusement?"

The old man said, "Hunt ducks."

"You've got sloughs here?"

"Come on."

"I've got no clothes for it."

Congdon said, "I'll give you my boots and leather jacket."

The old fellow said, "I have an extra shotgun."

We started out across the salt flats. The old fellow stopped. "Now we'll act like critters."

"What?"

"I'll bend down, and you lean over me with your hands and gun on my back."

We stumbled toward the slough, probably making the poorest excuse for a cow ever seen by a mallard. But it worked. We had duck for dinner that night.

Next day was 20 degrees and clear as a bell. But Congdon couldn't get the engine started. They sent up another plane from Salt Lake City, and I got home to Seattle late that evening.

Contact flying usually meant following the railroad. Sometimes, in bad weather, this meant following the railroad very closely. Duke Ledbetter, a famous American Airlines pilot, followed the tracks one foggy day so closely that he came to grief. The tracks disappeared abruptly into a tunnel. There was no time for Duke to pull up. He pancaked his plane against the side of the hill and climbed out unhurt. Along the Union Pacific across Nebraska, where the tracks stretched straight mile after mile, the pilots learned to keep to the right to avoid smacking head-on into one another.

In those days forced landings happened all the time. Once two Boeing officials were flying across Nebraska through a blizzard when the plane's engine began revving down and giving off black smoke. Pilot Rube Wagner picked out a wheatfield he knew south of Sidney and made a landing. There was just time, as he figured, to get his two passengers and his 1,100 pounds of mail to the westbound train. Stamping their numbed feet and wiping the rime from their glasses, the two executives mounted the steps of the Overland Limited at the lonely whistle stop.

Obviously only a tiny proportion of the public went in for air travel in the 1920s. Even at that early date the operators had identified "fear and fare" as the two main reasons why more people did not do so. No question about it, the majority thought flying was dangerous. When newspapers reported a fatal crash on the London–Paris service in 1922, the pioneering Aeromarine Airways felt the tremors five thousand miles away the next day when its flights from Miami to Havana flew empty. *Aerial Age* asked, "When will newsboys cease crying 'Terrible Accident' every time a pilot runs out of gas and volplanes into a cornfield?" So many thought planes lethal that Professor Edward P. Warner of MIT asked whether it would not have been better for the public interest in aviation if planes had not been used at all in the First World War. As for fares, they were so high that few could afford them. The rate charged by National Air Transport between Chicago and New York in 1927 was a prohibitive $150. Everywhere else rates were stiff, well above first-class rail fares, and double what they would be a decade or two later.

Why *did* people fly in the open-cockpit days? Ham Lee, the "flyingest pilot" of the early airlines, said, "Actors did it — for fast transportation and publicity. I carried Mary Pickford, Richard Barthelmess, Doug Fairbanks." Two stars who became celebrated for their early flying were Bebe

Daniels and Will Rogers. Rogers flew everywhere, knew all the airmen, and wrote about his experiences in his daily column: "To locate an airport all you have to do is to follow a high-tension line and it will lead you right to the runway." In the summer of 1928 Rogers did not hesitate, though several of his pilot friends excused themselves, to take passage on Western Air's first Lockheed Air Express out of Los Angeles. On landing at Las Vegas, the plane, the fastest ever placed in commercial service to that date, hit a concrete directional *T* alongside the runway and flipped over. Rogers crawled out unhurt and caught the next flight to the Republican National Convention at Kansas City. Seven years later, Rogers died in an Alaskan crash while traveling with the famous round-the-world flier Wiley Post.

As soon as the contract mail operators began introducing bigger planes designed for both passengers and mail, businessmen began to fly and were, in fact, the chief early customers. For one thing, they did not have to foot the bill. One survey in 1929 showed that some eighty leading companies allowed their employees to put air fares on their expense accounts. Somebody at Standard Oil of Indiana figured out that executives were worth at least one hundred dollars a day to the company. So if four of them flew the 1,100 miles from Chicago to field headquarters at Casper, Wyoming, a distance that took thirty-eight hours by rail but only fourteen by air, they could add two days to their working week and thus save the company eight hundred dollars.

But resistance was great. When Richard Hoyt of Curtiss-Wright and Pan Am addressed the National Chamber of Commerce in 1929, he told the leaders of industry, "You have given us help and money, but you have not said to yourselves, 'Now I am going to fly.'" As Paul Henderson of National Air Transport acknowledged, "It is one thing for a man to mail his letter and something else for him to put himself in an airplane."

One of the first of the airmail contractors to go after passengers was Colonial Air Transport. The line set up booking offices at the Statler Hotel in Boston and at the Pennsylvania Hotel in New York; "special buses" would pick up customers at these hotels and take them to the airport — across the Charles River on Commonwealth Flats in East Boston, and at Hadley Field, near New Brunswick, New Jersey, at the New York end. One day James Reddig, an Eastman Kodak executive, telephoned Colonial's Boston telephone number to reserve a seat on the next morning's flight to New York. But when he then said that he would not need transportation to the field on Colonial's special bus, "their operator hung up on me. I later learned that it had become fashionable to telephone for reservations implying intention to ride in an airplane for the sole purpose of impressing those overhearing the call."

Reddig persisted. He caught the subway to East Boston the next morn-

ing, changed to a trolley, and finally made his way to the airport on foot.
The Boston Airport, which had only been dredged out of harbor muck a
year or two before, was not much to brag about in 1926. Reddig said:

As I approached on foot through the spring gumbo mud, I was hailed by name
from a shanty that leaned against a National Guard hangar.

Colonial's man slogged out to meet me, shook my hand and seized my bag.
Then he led me back through the thick mud to the shanty, which reeked of the
stench of a coal-oil space heater. He shoved a mass of dirty engine parts along a
bench to make room for me, and retrieved a battered typewriter from the floor.
After wiping his hands on his coat, he rolled a Colonial letterhead, carbon and
tissue into the machine and carefully finger-picked my "ticket," following a sample tacked on the wall. Pulling out his misspelled handiwork with a grin, he
signed it with a flourish, took my $25, and carried my bag to the airplane.

That was how a passenger got to the plane in 1926. The factotum who
carried the passenger's bag, typed out the tickets, and in between worked
on engine parts was Sumner Sewall, the airline's "district manager," later
president of American Export Airlines and governor of Maine.

One of Colonial's stunts to attract business travelers was to sell blocks
of tickets at a discount. By the spring of 1929 a single fare between Boston
and New York had climbed to $34.85. Said Leonard Raymond, an early
sales-promotion specialist, "All we did was take a ten-trip commuter
ticket on the New Haven railroad and copy it." Ten-trip air tickets sold
for $27.88 — 20 percent off, but fifty-trip tickets could be had for $17.43 a
trip, or half price. Jordan Marsh and Filene's department stores bought
the first ones; Sewall happily reported a "deluge" of requests from other
firms.

Soon, on the strength of Raymond's survey of the first year's 4,016
riders, the airline's brochures could inform passengers on the new trimotors just who they were:

the businessman journeying to a meeting in another city; the salesman; seasoned
travelers and tourists; college students going home for the holidays; experts in
professions like architecture and engineering journeying to help solve some difficult situation; first-nighters going to a theatre opening, a political convention, a
social or sporting event; and those who "want to see what it is like."

For all the elaborate informality of this list, the last category must have
made up a pretty high proportion of the passengers. On the back of the
same brochure was a "certificate of flight" to be torn off and signed by the
pilot — something like a diploma from King Neptune for crossing the
equator. The Eastman Kodak man on his 1926 trip had gotten no farther
than Framingham when the door at the front of the cabin opened and the
pilot came back to take one of the empty wicker seats. He opened a
shouting conversation with the nearest passenger above the din of the engines. Reddig anxiously peeked forward and was slightly relieved to see

the overalled back of a mechanic at the controls, logging prized flight time. Reddig passed his ticket to the pilot to sign — H. Ponton D'Arce was the name, obviously quite a fellow. On his way to a party in New York, pilot D'Arce was attired in a tuxedo, white silk scarf, chesterfield, and derby. To Reddig's surprise, the pilot handed the ticket on to others for their signature. That was when he learned that all others on board were freeloading journalists riding along to write stories. "I was the first paying passenger to appear in 11 days."

To judge by Raymond's records, Colonial's passenger business had rather jagged ups and downs. On one weekend in August 1929, the airline flew extra sections to carry the 512 customers who wanted to travel between Boston and New York; some were even turned away. But when the snow fell in December, only a third as many passengers showed up in the whole month. Wives were believed to exert especially strong pressure on executive husbands to stick to the ground on their winter trips. Inevitably an airline traffic manager got the idea of offering free tickets to wives accompanying their husbands. The results were sufficiently gratifying for the airline to send out letters of appreciation to the wives, with a word of hope that they liked their flight well enough to try it again. Back came angry replies that indicated by no means all of the traveling companions had been wives. The traffic manager called off his campaign.

On the afternoon of July 8, 1929, the special bus brought out a full load of passengers for Colonial's morning flight from Boston. The trimotor Keystone Patrician, largest passenger plane then existing, had just gone into service, and every one of its fourteen seats was taken. Right behind pilot Hughie Wells sat Sumner Sewall and his friend Leonard Raymond. Taking off from the cinder runway, the plane climbed no more than twenty-five feet into the air before one of the engines cut out. Pilot Wells swung the ponderous plane around for an emergency landing back on the field. But a horse and wagon were crossing the runway. Wells sideslipped and still kept the plane level, but he could not avoid hitting a hangar under construction. Wooden scaffolding broke the impact as the plane plunged on into the excavation pit. After the crash there was a moment's silence. Through the open door Sewall could see liquid dripping down over the upper wing — gasoline. He could also see that the passenger behind him, a large, fat man, held a lighted cigar in his mouth. "Put that out," yelled Sewall. "What shall I do with it?" asked the man. "Swallow it!" roared Sewall.

But Wells had cut the ignition before the crash and there was no fire. Although the nose and fuselage were battered, none of the plane's windows had broken; nobody on board had suffered so much as a scratch. Eleven of the fourteen passengers promptly got on another plane and flew off to New York. "Not one passenger lost his nerve," reported the Boston Globe, explaining that the substitute plane had room for only eleven. According to the *Globe,* one Dr. Riggs of New York asked if such delightful thrills were provided on every trip.

When there were forced landings — Colonial had 116 in the first six months — the airlines had to put their passengers on the train. Sometimes, when the weather was too bad for flying, both passengers and the mail had to be put on the train — "trained," as the airline people said. On the most elaborate of all the airline operations of the 1920s, Transcontinental Air Transport, a train ride was a scheduled part of going by air.

TAT — pilots derisively called it "take a train" — was started by Clement Keys, who liked to say that nine-tenths of the job of air transport was accomplished on the ground. He and his associates did not believe that planes could then safely carry passengers across the Allegheny Mountains. But Keys, on the crest of the stock-market boom, thought the time had come to offer passengers fast transportation all the way from coast-to-coast. To that end, he teamed up with two of the nation's biggest railroads to carry passengers by train by night and by plane by day. The whole fancy setup — by overnight Pennsy Pullman through the hazardous Alleghenies; by Ford trimotor from Columbus, Ohio, till nightfall at Waynoka, Oklahoma; by Santa Fe sleeper to Clovis, New Mexico; and in another Ford trimotor from there to journey's end at Los Angeles — was designed to permit the well-to-do traveler to cross the country in forty-eight hours, and thereby save a day over the fastest train trip. This, General W. W. Atterbury, the Pennsylvania's president, told his stockholders, was an enormous advance on the road of progress.

There was to be nothing of the rubber-band and baling-wire flying by oil-smeared barnstormers about this airline. Millions were laid out in preparations, on buildings ($1.5 million for ground facilities alone), on surveying the route (done by Lindbergh, the line's "chief technical adviser"). More than a year was spent making sure the service would satisfy the luxury-loving Pullman patrons and puzzling over such questions as, "If a rich family travels by air, what do you do with the colored maid?"

Joseph Edgerton, representing the *Washington Star* on one of the dress-rehearsal flights in June 1929, duly noted TAT's attentions to personal service. As the travelers trooped to their plane at Columbus, the pilot and his assistant (not yet called "captain" and "copilot"), a mechanic, and a courier stood at stiff attention "in snappy blue uniforms similar to Navy blues." The courier, named Canfield, took charge of the passengers' baggage, handed them their tickets and seat assignments, and asked them to go aboard. Western Air and Pan Am had hired their first stewards the year before (the first of all stewards, one Albert Hofe, had joined German Lufthansa in early 1928); on TAT they were called couriers because the Santa Fe used that name for the young men who led parties of rail passengers on side trips to the Grand Canyon and such places.

The pilots, looking "just like the senior officers of a great ocean liner,"

Passengers board a Transcontinental Air Transport Ford trimotor in 1928.

made Edgerton "feel instinctively that they know what they are about."
As soon as they had got the plane into the air, courier Canfield "comes
down the aisle and tells us we can remove our light seat belts. The belts,
he explains, are used more to keep passengers in their seats for landings
and takeoffs than for safety purposes. He hands each of us a small oiled-
paper envelope containing cotton for the ears and a package of chewing
gum." There was nothing specially luxurious about the cold lunch
(chicken salad, cheese-and-egg-salad sandwiches) that Canfield served on
aluminum trays; several passengers were too airsick to want any. There
were ten stops before Waynoka was reached at dusk. The corrugated
metal walls of the Ford were relentless transmitters of sound. As Edward
Evans, president of Detroit Aircraft Corporation admitted, "When the
day was over, my bones ached, and my whole nervous system was wearied
from the noise, the constant droning of the propellers and exhaust in my
ears."

The man of all work at Los Angeles when TAT's superservice finally
got going was Oz Cocke, the local passenger agent (and later senior

vice president of TWA). Early in the morning he personally drove to the house of Walter Furie, a Long Beach dentist, to fetch him to the airport as one of the first paying passengers. Cocke's regular morning task was to make the rounds for passengers in something called an Aerocar. Specially designed for TAT, this was a sort of forerunner of the automobile trailer. There was one at every TAT terminal. The front part was a Studebaker coupe with beefed-up springs; behind this rolled a three-wheeled vehicle, a tubular steel affair covered with fabric, with blue-and-silver upholstered seats inside — a sort of "mobile living room." Passengers at Los Angeles usually boarded at the Biltmore or Ambassador Hotel. Inside Cocke kept a fish scale, and on the way to the airport he would weigh everybody's baggage and tag it; he would also check tickets. In this way there was no checking in at Glendale: when the Aerocar arrived the passengers were ready to walk on the plane.

On the inaugural day, twenty thousand persons turned out at Glendale to watch Mary Pickford break a bottle of champagne over the sharklike nose of the *City of Los Angeles,* first of the two Ford trimotors drawn up to launch the eastbound transcontinental service. Colonel Lindbergh was at the controls. At Columbus, where two other trimotors were poised to start flying westward, it was raining. But at the appointed hour TAT officials got word that the weather was better to the west and, with a crowd of three thousand watching from under umbrellas, the speeches ended and the two planes soared off carrying a total of nineteen passengers, all but two of them freeloaders. Among the *City of Los Angeles* passengers were Mrs. Lindbergh, two local newsmen, two TAT officials, three TAT stockholders, and Dr. Furie. The Lindberghs and the two reporters, traveling only as far as Winslow, Arizona, the first stop, were flying for business reasons; so were the TAT people. Dr. Furie also had compelling reasons to take to the air. He was on his way to Chicago to elope with a girl friend.

TAT was the railroads' biggest foray into flying. They were responsible for its leisurely start. In other respects it bore their iron stamp. When passengers climbed off TAT's trimotors at their western transfer points, they were shepherded into Fred Harvey Restaurants for meals, just as if they were passengers stepping off the Santa Fe. When Oz Cocke in his office at the Pacific Mutual Building sold a Los Angeles oilman the first air ticket clean across the United States (total tab: $351.44), it was as complicated as only the railroads could make it. Because part of the ride was by train, Cocke had to give his customer an exchange order for the railway segments. Before he could claim his passage to New York, the passenger had to go along to the Santa Fe office one floor above and wait for the rail portion of his ticket to be written.

In some ways TAT was a railroader's dream. General Atterbury took delight in planning the style of service and personally roped David Sarnoff of RCA into helping equip the planes with the latest in radio communications. Sons of rail executives and their friends eagerly sought jobs in an enterprise that seemed, both technically and geographically, out on the romantic frontier of transport; the passenger agent at Barstow, California, was Dwight Morrow's nephew. Aboard the planes the food service, after the Pennsylvania's chief steward was called in, was patterned after that of ocean liners — bouillon in the morning with little bread-and-butter sandwiches, tea and toast in the afternoon. Lunch in 1929 was served on Dirigold plates, with silverware and starched lavender napkins, and it was hot — broiled chicken kept hot in thermos jugs that Oz Cocke and others handed up to the couriers just before takeoff.

One of the most luxurious appointments of TATs airliners was personal radio service for passengers. On October 24, 1929, the Black Friday when the New York stock market had its great fall, "many passengers aboard TAT planes sent messages to their brokers, and received replies before landing."

The nature of these replies — $30 billion in stock values evaporated that day — drastically affected the airline, which had been operating only four months and, without a mail contract, depended entirely on passengers for income. In the Ford trimotors there were seats for only ten, and even before Black Friday there were seldom that many passengers. Six or seven was a very strong turnout, according to TAT pilot Otis Bryan, and more often than not there were only two or three. The Wall Street crash put a swift end to the railroads' venture into aviation. For the railroads the crash and the Great Depression that followed spelled much more — the end of the age in which they monopolized transport, dominated commerce, and maintained their special style of luxury service. To point up what happened to General Atterbury's rather heavy-footed leap into the air, it is worth noting that New York's Waldorf-Astoria Hotel, under construction at about the same time, was built with a siding for private railroad cars — which then went out of style. The Dirigold plates, the Aerocars, the pink-cheeked eastern college boys staffing Arizona airstrips, did not last much longer either. The airline lost $2.7 million in eighteen months.

In 1930 the nearly bankrupt TAT was forcibly joined with Western Air Express over Harris Hanshue's bitter opposition. Hanshue served a reluctant year or so as president of the merged enterprise. Executives of TAT, which was affiliated with the North American–Curtiss combine, hoped that the consolidated airline would switch to the twin-engine Condor biplane being produced by Curtiss. But advance word on the Condor said it was big and slow, barely able to top the Alleghenies, and pilots were offering bets that it would never clear the Rockies. Someone asked Pop how long it would be before he placed orders for Condors for western flying.

"Oh, not for a long, long time," mumbled Hanshue.

"But, Mr. Hanshue, they'll be out of the factory in six months," the man said.

"I wasn't thinking of that," said Pop. "I was thinking of what a helluva long time it's going to take to tunnel through those goddam mountains."

Those who persisted in flying on TAT to the end continued to see quite a lot of the trains. When summer faded and the winter months of rain and snow came on, more and more trimotors were grounded, more and more passengers were trained. The courier on the plane would ride along with them. The airline kept track of them as they rode from one city to the next and tried to get them back on an airplane. When bad weather caused cancellations, TAT would simply take the grounded plane, turn it around, and send it back with passengers in the direction it came from. Often, however, TAT officials had to meet the passengers coming off the train and give them the bad news that flights were canceled again and the airline would have to re-train them. Once a passenger at Kansas City said to Oz Cocke, "Will you take me down to the airport and let me just see an airplane? I've been all the way to New York from Los Angeles and back and haven't seen an airplane yet."

On one occasion the passengers got to the airfield at Kansas City, but at midnight they were still waiting for the plane to take off. One of those stranded was Harpo Marx. Merta Mary Parkinson, another passenger, sat at a nearby table with her dog, which had settled at Harpo's feet. A young woman came in and began gushing over the dog, patting it and oh-ing and ah-ing. Finally Harpo spoke up, "Take it as a gift. It was getting airsick anyway." The young woman had gathered up the poodle in her arms when Mrs. Parkinson yelled, "Hey, that's my dog."

When the trimotors came to California in the 1920s they still lacked the instruments that would have enabled their pilots to fly in any weather. Their cabins also lacked the pressurization that would have allowed the pilots to fly over it. Not that they didn't try. Tommy Tomlinson, octogenarian dean of the country's airline pilots, recalled that when he flew for Maddux Airlines in the early 1920s, and even later when he piloted Ford trimotors for TAT, he took few chances of slamming into the mountains east of Los Angeles. He said, "I would get a Ford or Fokker up to eighteen thousand feet, even twenty-one or twenty-two thousand feet, and fly reasonably well. Not for long, though. The passengers passed out."

If that sounds far-fetched, think of what happened to a load of passengers on Western Air's model airway between Los Angeles and San Francisco. The whole idea of this interesting experiment in passenger flying was to introduce some science into weather forecasting along a fixed route and thereby promote more reliable air travel. The trouble was that the data from the thirty-eight weather stations had to be funneled into the central office first, and there was no way for the pilot to get word of weather changes while in flight. As a consequence, passengers flying on Western Air's model airway sometimes came in for surprises. Here is the teletype exchange between Western Air's manager at Los Angeles, Bill Cole, and his opposite number at Oakland, Max Cromwell, on September 4, 1928:

COLE: Northbound ship unable to get through. Stand by.... For anyone's information that's expecting to come south, our pilot came back after going to 20,-000 feet altitude and flying up and down the Tehachapi ridge and could find no opening. Was too low on gas to go up coast which evidently is not bad.

CROMWELL: Did you say 20,000 feet?

COLE: Yes, he just about froze everyone on the ship. They all came back shaking to beat hell. Notify Mr. G. Murphy at airfield that Jean Murphy and J. A. Murphy will not be arriving but are coming by Sunset Limited.

Western Air's Fokkers had great climbing capacity, but their relatively light weight also made them roll like steers every time the air grew bumpy. By this time Western Air furnished "burp" containers at every seat, but these happened to be too small for relief when mal de l'air overcame passengers. At that moment the airsick passengers were grateful for the Fokker's much advertised "windows which can be opened or shut at pleasure." Many simply leaned out and upchucked. There is no record, however, that any of their fellow passengers experienced the indignity that befell Hudson Fysh, the old Qantas pilot, when he was flying as a passenger in Australia in 1930. Fysh had "taken the back seat in the cabin as safest in a crash." Somewhere north of Sydney, when he was leaning back with the window open and his eyes closed:

a passenger in the front seat was very sick into the usual container — which he then tried to throw out his window. I got the full blast in the back seat.

At that, airsickness on the Fokkers was so prevalent, according to Western's historian Bob Serling, "that planes frequently had to be hosed out after landing." Even on the fair day in 1929, when Joseph Edgerton crossed the country on TAT, he noticed fellow passengers overcome by airsickness and the courier "bringing a round oiled cardboard box and a towel to relieve the situation." It was years before the technology of assuaging the airsick improved very much, and it was not helped when American Airlines bought several hundred thousand ice-cream containers for the purpose and somehow never noticed that on the bottom of each one were imprinted the words: "Thank you — come again."

It would be a good many years before flying ceased to be both uncomfortable and hazardous. It was a profoundly new experience. At least a dozen passengers lost their lives before 1930 because they walked into a whirling propeller. The act of leaving the ground in defiance of the law of gravity was unquestionably unsettling to the nerves for many. This, quite as much as the subtle jostling of the balancing mechanisms of the inner ear, was the cause of airsickness. It was all very well for the facile admen of Colonial Air Transport to write in 1928:

You look out the window at the great wings, firm as though on concrete roads. There is the air, packed as hard as a flow of lava, iron or steel under those great wings. You cannot see it — but if you will open your window a little and let your fingers poke impudently out into the slip stream that sizzles along the side of a plane, you will realize the forces holding that wing to its cleaving way.

Despite all such assurances, flying was chancy in the 1920s. Knowledge that the engine might conk out, that the plane might have to come down in some hayfield, upset some passengers' stomachs. In 1928 the Eastman Kodak man from Boston knew only that flying to New York was an exhilarating adventure. He started out with the hand-typed ticket and got it autographed by a derby-hatted pilot on his way to a party, but he found that the airline people seemed utterly unprepared for his safe arrival at Hadley Field. Back on the ground:

... getting to New York became a most casual affair. Finally it was arranged that I ride in a Post Office mail truck to the Pennsylvania railroad station in New Brunswick, where I was assured a train came through from Washington at least every hour. Nobody told me that only a few bothered to stop at New Brunswick. It took longer to get from Hadley to Penn Station than from Boston to Hadley.

10

Harbors of the Air

When Secretary of Commerce Herbert Hoover testified before the Morrow board in 1925, he drew on maritime law to explain how the nation should think about its airports. The parallel between air commerce and waterborne commerce, Hoover said, was "complete." For more than a century the federal government had been providing certain types of service to water navigation. Among these were to light, mark, and chart channels; to inspect ships and those who manned them; and to aid in the development and improvement of ports and waterways. Without these services, Hoover told the board, waterborne navigation would instantly cease. Getting it going again would be impossible except "in the most primitive fashion." For commercial aviation it was exactly the same. Without such federal aid organized flying would "only develop in a primitive way."

Pursuing this analogy, Hoover went on to speak of airports. To make his point, he used the new name rather than the more familiar term "landing field." Airports were like the docks of waterborne commerce, he said, and in the management of waterborne commerce, docks were traditionally provided by private enterprise or municipalities. That was the way it was on the waterfronts of New York, San Francisco, New Orleans. So it should be with airports; they should be established at all important cities, but by the local authorities, not the federal government.

On December 1, 1922, the *Boston Transcript*'s front page headline was, "Seven Days for Boston's Archaic Postal Service to Get Parcel Post to New York." The charge, leveled by young Leonard Raymond of the Boston Chamber of Commerce and backed up with pictures of cancellations and insured-mail receipts, was another big blast in the campaign to get Boston big-city mail service.

Outraged that the Post Office was carrying mail between New York, Chicago, and San Francisco by air, Bostonians were determined not to fall behind. When Washington said it would help if Boston had a flying field, a committee decided that the handiest site would be a patch of tidal mud known as Commonwealth Flats at Jeffries Point in East Boston. Presumably it was land that nobody much wanted; only "thirty small residences" would have to be cleared. The U.S. Army was persuaded to say it would lease the field for ten years if it were suitably leveled and marked. After a sharp fight, the committee managed to push an authorizing bill through the Commonwealth's legislature. But the appropriation fell short by $15,000. Thereupon the Minute Men of the Boston Chamber of Commerce, joined by the Aero Club of Massachusetts and the Massachusetts Air National Guard, personally solicited businessmen at their desks and obtained five hundred checks of from $1 to $1,000 that made up the difference.

After its success (as the *Transcript* said) in "moulding public sentiment," the Chamber of Commerce pumped enough new mud into Commonwealth Flats to proclaim the opening of Boston Airport on September 8, 1923. It was a splendid day, and of course there was an air meet, attended by fifty thousand spectators. According to a contemporary account, "in going from one air meet to another, cross-country flights have never been attempted. Planes are taken down, shipped by rail, and then re-assembled. For Boston Airport's inaugural meet, no fewer than 25 planes arrived by air." Landings were made on a surface of packed cinders. In fact, a special prize was awarded for landings. R. P. Hewitt, piloting a tiny Farman, "with engine shut off glided to the target mark on the ground and stopped just on top of it, a test of skill and judgment as airplanes do not have brakes."

Sure enough, the Post Office's Air Mail Route No. 1 was duly awarded for extending airmail service between New York and Boston. The new airport, it was noted, was "probably the best in the country in its close proximity to the business section." The Chamber of Commerce report added that the field was "capable of almost indefinite expansion, without in any way approaching any crowded section," an observation that could hardly be made fifty years later when the encroachment of jet takeoffs from the same airport had generated an intensity of popular feeling in Boston second only to that stirred up by the school integration issue.

So landing fields became airports, the harbors of air transportation, and scores of other municipalities soon followed Boston's example. This was the era of *Babbitt*, and the chambers of commerce, Rotary clubs, and civic leagues did their work. Big cities dared not hold back for fear of loss of trade. Smaller ones looked for an advantage. Most of the early fields were only one or two stages removed from cow pastures. Most lacked lights. Few were properly drained. Facilities such as hangars — that was another word, like fuselage and aileron, empannage and longeron, that migrated into the language during the years between the Wright brothers' flight and World War I when the French seemed to have taken over aviation — were often lacking. Even at Boston Airport, Sumner Sewall of Colonial Air Transport worked in a tiny hut surrounded by mud and soggy cinders. The only maintenance gear at some fields was a lawn mower and a few posts to which visiting aircraft could be tethered.

Major Roy Alexander, a Missouri National Guard officer who often flew cross-country with Lindbergh, has said that right up to the Second World War most of the airfields in the country were grass. Some of the major fields, like Bowman at Louisville, were grass-surfaced long after Douglas airliners regularly landed and took off there. In 1930 a student of such matters reported that in almost all cities of less than 500,000 people the "all-way field," meaning grass, "is accepted." The leading article in a 1929 magazine for airport managers was entitled "Selection of Seeds for Airport Turf," and the article itself carefully warned that it took at least a year to build a turf firm enough for Ford trimotors to land on.

After the war many an airman became the operator of a field at which barnstorming pilots could land, shoot the breeze, buy fuel, ask for repairs, and perhaps trade for another plane — much like driving up to a small-town garage. Those early entities of commercial aviation, called fixed-base operations, often had their own airfields with hangars, windsocks, and other facilities. Every aircraft factory had to have a field to test and send off its products, and some of these were used by the first airlines. As late as 1929 private interests were completing a $3 million Grand Central Terminal at Glendale to serve Los Angeles air travelers, and Pittsburgh and Washington were served solely by privately owned airports.

Virtually every large city had from two to ten airports — the New York area had over thirty. Most were small. Only eleven airports in the whole country spread out over the 500 acres the planners thought advisable. Boston's was 165 acres, San Francisco's Mills Field, 170 acres. Often the determining factor in the site of a municipal airport was "the word of the local political boss whose henchmen happened to own practically worthless land they were anxious to sell to the city at a high figure." The chief criterion at Sacramento was freedom from fog. But proximity to downtown was important from the start. Lindbergh, who laid out the fields along TAT's cross-country route, said in 1929, "I think distance from the city is of primary importance." Of 181 airports surveyed the following year, 150 relied for their link to town upon bus or trolley lines. "Only 15 have regular field cars to carry passengers to their journey's end in comfort."

Under the Air Commerce Act of 1926 the Secretary of Commerce was empowered to examine and rate airports. But by 1930 only two airports — Denver and Pontiac, Michigan — had been put through inspection. Neither received the highest rating, which was, and is, A1A. At that time windsocks, drainage, repair facilities, lights, twenty-four hour service, and an area "at least 2,500 feet in all directions" were required. Nothing at all was said about runways.

Actually, by 1929 half of the 181 fields surveyed had "prepared" runways, varying from a light surface oil treatment to heavy concrete pavement. The airport that Henry Ford built at Dearborn in 1925, "the finest in the world," had a concrete runway, and so did the new Detroit Municipal Airport. Chicago had the most runways — eight in all, in a double Union Jack pattern — and a railroad track running right down the middle of the field. All eight were cinder, like those at Buffalo and Milwaukee.

As soon as the Air Commerce Act passed in 1925, the Aeronautics Committee of New York City's Merchant Association announced that "the next step in commercial aviation is dependent upon establishment of convenient landing and starting fields in the large cities," and that such "commercial airports should be provided at public expense." Said the committee's chairman, the ubiquitous Richard Hoyt, "If a private company should be strong enough to acquire a central terminal, then that company might exercise undue power in controlling service to the city."

Hoyt's committee called upon Postmaster General New to join in pressing the Army to release Governors Island for New York City's airport. It is interesting to speculate how differently the city, its harbor, and even its trade patterns might have developed had this suggestion been adopted. But the Army wanted no part of the proposition, and the earliest scheduled flights in and out of New York were conducted in New Jersey — at first at the Air Mail Service's base at Hadley Field, near New Brunswick, and later at Newark. The Air Mail Service had used the old Heller Field at Newark earlier but had moved away because it was narrow and hedged in on one side by a line of hills and on the other by factory chimneys. During the big municipal airport building boom of the 1920s, however, the city of Newark filled and cleared a site five miles from the old Heller Field, and before long this became the world's busiest airport.

The Cleveland Airport was also famous in the early days. The Post Office planes landed for a while at Glenn Martin's factory field. The pilots would chat with Martin and his talented young assistants, Donald Douglas, Dutch Kindelberger, and Larry Bell, while downing a cup of coffee. Then Martin moved his plant to Baltimore, and the young assistants went off to win fame and fortune building planes in Santa Monica, Long Beach, and Buffalo. Clevelanders felt a cold chill in 1924 when the Post Office started night flying and bypassed them because there was no adequate field.

The city's response was to call in World War I hero Eddie Rickenbacker, and on his advice, to take an option on seventeen tracts totaling a thousand acres twelve miles southwest of town. The Post Office Department promised to move its base there, prompting some in the city council to assert this was subsidizing the Air Mail Service. But City Manager Hopkins, invoking the example of the pony express, carried the day. He said:

Two things make cities, highways and the means for people to meet and exchange their ideas, their questions, or anything else. There never was a great city except upon great highways; there never was a great city that did not provide ample space for people to come and go and do whatever they wanted to do. We have now in the air the greatest of highways. The only limit upon it is the possibility of a place for the airplane to land.

With such high visions the city soon extracted authorization from the Ohio legislature to take title to the land for this, as for any other, public need. The council voted $1,125,000 to take over some two hundred acres of the site by eminent domain. Knolls were removed, hollows filled. Under the direction of Airport Manager Jack Berry, one of the first of his breed, crews scraped and graded the site, and then poured cinders over the entire field.

Thus Cleveland had an all-way airfield — one huge circular mat on which planes could land and roll for two thousand feet in any direction. Lights were installed, and the same night in 1926 that three thousand people sat down to a City Hall banquet, night flying began. The following day Ford Air Transport began flying daily schedules from Detroit, and Cleveland staged its first air show. Those were the days of races in which fliers tore around pylons a few feet off the ground; before spellbound throngs in Cleveland, stars like Jimmy Doolittle and Al Williams set their speed records. In the first year four thousand planes landed or took off from Cleveland's big round mat. But in later years, when fliers from all round the world flocked to compete in the annual Cleveland National Air Show, more than four thousand touched down during the September week of the races alone.

In fine, when air travelers voyaged through the air ocean in the 1920s, the terrestrial ports at which they ascended and descended were locally and municipally maintained. The day had long passed when the Post Office Department could brush off Alfred Lawson's airliner by saying there simply weren't fields enough for scheduled cross-country flight. New paths were in the making that would reshape the economy and polity of the United States as surely as did the canals and railways before them. But there was no national system of air routes yet; the time was still far distant when the airport would become the hub of commerce, the concourse of travel, the popular site for conventions, the podium for almost all presidential campaign speeches.

Part III

Takeoff
(The 1930s)

11

On the Beam

So great was the splurge that the tiny American airline industry made after Lindbergh's Atlantic flight that within three years U.S. airlines were flying more passengers than all the rest of the world's airlines put together.

Few of these new outfits made money, but all hoped to. All sorts of lines started up, most of them without mail contracts from the Post Office Department. They began by carrying a few passengers, and hoped for a chance to bag a mail contract later. Some were quite literally "poor boy" operators. Lines like Hunter Airlines, advertising scheduled flights between Little Rock and Tulsa, or Reed Airlines, offering service between Wichita Falls and Oklahoma City, consisted simply of a couple of single-engine Travelair monoplanes. Another early upstart was Century Pacific, started by the Chicago promoter and automobile manufacturer, Errel Loban Cord. It tried to break into the already crowded Los Angeles–San Francisco market. Century Pacific's Stinsons flew at a low level, and when the clouds rolled in from the Pacific, it was very easy to turn into the wrong canyon and wind up at a dead end. It was anything but conducive to passenger comfort and safety. Often the pilot, to complete his schedule, flew against his better judgment, especially when another pilot had succeeded in getting through. Often they violated regulations by flying passengers at less than 500-foot altitudes.

Department of Commerce regulation was only getting started. The Air Commerce Bureau was concentrating on developing airways, and it had yet to complete an inspection of its first airport. But a handful of inspectors were flying the airlines by 1930, checking on their performance and enforcing compliance with Air Commerce's regulations for good safety practices. One of these was Jack Jaynes, a World War I pilot and a Texan. He had plenty of chances to observe airline performance while flying as a passenger on his way to check lighting, emergency fields, and other features of the new government-built airways.

"Welcome aboard," said Ham Lee. It was March 1, 1932, and snow was falling thickly as Jaynes shook hands with the United pilot at Aksarben Field in Omaha. Bundled up in helmet, goggles, and heavy flying suit, Lee told his passenger he thought they'd make it all right to Cheyenne and stamped his feet while Jaynes climbed into the cabin of his Boeing 40B biplane.

The takeoff, Jaynes said later, was "a lulu." No sooner had the plane gained flying speed than Lee rolled over on one wing and took off down an Omaha street, flying in and out between buildings. Jaynes thought for sure they were already lost as the plane twisted and turned. Gluing his face to the little cabin window, he realized what his pilot was doing. Lee was banking around haystacks, barns, houses, windmills, trees. There was practically no forward visibility, and Lee was down on the bottom shelf hunting for a still lower level. He found it at the Platte River; the frozen surface was some ten to twenty feet below the banks, and his ceiling increased by that amount.

Ham Lee and Jack Jaynes flew all the way to Wyoming in this manner, making stops at Lincoln and North Platte. As they neared Cheyenne, the ceiling and visibility improved a little, but the temperature when they landed was twenty-seven below zero. Lee had been sitting outside and fighting the snowy fury for five hours, but he only snorted when Jaynes asked him about it. Said Jaynes later, "It was just another flight to Ham Lee, one of the greats. He had to be to survive 30 years carrying Uncle Sam's mails and then passengers like me."

At Salt Lake City Jaynes boarded a Western Air Express Fokker trimotor to fly to Los Angeles. Again he was the only passenger. His pilot was Fred Kelly, the handsome Olympic high-hurdles champion who was the first man hired to fly the line. After taking off from Salt Lake City, Kelly had to double back when weather walled off the way. But he still thought he might get through if he flew between cloud layers. Heading back, he set the plane on a true compass course for Las Vegas. Then, veering around some mountain peaks sticking up through the thick layer below, he noticed that the compass did not move. It was frozen in a fixed position.

The only thing to do was to get above the overcast, and luck was with him. The clouds above turned out to be only two or three thousand feet thick and, although there was another, higher layer, it did not look like trouble to Jaynes. Now the problem was to find Las Vegas without compass, without radio navigation aids, and with mountain ranges of twelve and thirteen thousand feet lurking everywhere below. With enough fuel to fly for only three and a half hours, there was not much time for fooling around. Kelly told his copilot to remove the compass, take it back to the heater in the passenger cabin, and see if he could thaw it out. Within thirty minutes the compass thawed and was put back in place. Joining the pilots up forward, Jaynes was struck by how thoroughly the veteran Kelly knew his landmarks. Said Jaynes later:

After some 2½ hours Kelly gave a smile. He had spotted a range to the southwest just visible through the undercast. He obtained the Las Vegas weather by radio — broken cloud and no snow. He found a hole, spiralled down until he could see the horizon, and in 30 more minutes we had landed safely.

Early in the following month Jaynes flew east from Oakland one morning in a Boeing trimotor with pilot John Guglielmetti, copilot Jack Favorite, and seven other passengers. Landing to refuel at Sacramento, the pilot learned that a blizzard was raging in the High Sierra ahead; Reno was reporting strong northwest winds with blowing dust and snow. Taking off, he found it impossible to fly directly over the Donner Pass to Reno. So he tried one canyon after another. Flying up the Yuba River, Jaynes saw treetops higher than the plane on the canyon wall alongside. Said Jaynes: "Johnny knew the landmarks so well that even when he couldn't see he knew how much room he had to turn around in." Going back to Sacramento, Guglielmetti studied the reports and decided to try the northern way. Threading a low pass near Quincy, he turned southeast toward Reno against a strong northwest wind that bounced his passengers around severely. When they landed at Reno, having flown more than twice the normal distance on the second try, they put seven passengers on a train for Salt Lake City. "It was almost dark," Jaynes said later,

when we took off with only the crew and myself. The wind was really howling; the dust and snow made visibility less than one half mile. After a short while we were down to 500 feet or less. The strong crosswinds made the Boeing buck like a wild Brahma steer.

In snow and darkness Guglielmetti tried to fly from one beacon to the next. Sometimes he had to turn back, circle the beacon he had just passed, and peer again for the next one ten miles ahead. The low storm clouds moved fast. At one moment he would see a flash ahead and start for it, only to lose sight as more clouds rolled in. The route led over one mountain range after another.

One such range had a pass with a peak in it known as Mollie's Tit. On the tit was a beacon. Other beacons marked the west and east sides of the pass. We made several attempts. Johnny would get a flash from the Tit beacon and would pour on full power for the climb over it. Each time storm clouds would cover it before we reached it. Then Johnny would do a fast vertical turn, dive back to the western beacon and circle it awaiting another chance. Finally Johnny determined he would make it on the next try.

Again we lost sight of the beacon. But as we passed we got a dim glow of its flash. Without pausing, Johnny dived down to pick up the beacon on the eastern side. The dive had to be perfectly timed because there was another ridge sticking up only 15 miles further east.

The next beacon came into sight, and the next. In this way, snaking around from beacon to beacon along the crooked airway, the Boeing tri-motor finally sneaked into Elko, Nevada.

Visibility was only about one half mile. A strong cross wind was blowing. To make matters worse, the throttles had frozen in cruising position. Johnny had to make his descent by blimping (turning off and on) the master engine switch. The swirling snow and dust made it necessary to leave the plane's landing lights off. Johnny had to feel his way down, holding the right wing down to counteract drift. The only lights he had to help him at the tiny field were the old fashioned boundary lights — no flood lights or approach lights at all.

After freeing the throttles and refueling, Guglielmetti pushed on. It was about eleven at night, and the storm seemed to be keeping pace with the plane. If anything the snow got thicker. The plane now hugged each beacon at five hundred feet before attempting the dash for the next one. But when the Boeing came to the narrow pass in the Ruby Mountains, there was no getting through. After several attempts, Guglielmetti had to fly back to Elko in the teeth of the storm. Again he had to resort to blimping the master switch in order to get down. The ground crew pulled the ice-covered plane to the hangar line. After sixteen hours of hard flying, they had advanced only four hundred miles. Said Jaynes later, "I happened to have my will in my briefcase. I thought of mailing it to my bank from Elko."

Thirty years and thirty thousand flying hours later, retired United Captain Johnny Guglielmetti did not even recall that particular trip, though his log book confirmed it, forced landings and all. In his diary Jaynes wrote in 1932:

One of the most thrilling trips of my career, and the worst as expressed to me by those pilots — fast moving storms cutting us off both forward and rear. Trip usually requiring 5½ hours took over 28. The next schedule out of Oakland after us failed to get through and cracked up at the base of the Sierra east of Sacramento.

Newspaper accounts attest that pilot Harry Huking, caught by a blinding snowstorm in the same Yuba Canyon that Guglielmetti had probed unavailingly the day before, brought his plane down in the only open place in the canyon — the middle of the river. The plane was smashed. Huking, another former Air Mail pilot, got a cut on the forehead. Shaken but unhurt, his seven passengers, two of them women, waded ashore. "We all owe our lives to the pilot," V. H. Oglesbee, a Chicago sales executive, said.

How many other passengers, how many other pilots, scraped over the western passes and through the eastern fogs in those years when only the flashing beacons lit their paths through the storms and the night? Collie

Collison lost his life when his plane struck a Wyoming peak; and Paul Wheatley and six passengers died in an Idaho crash on their way to Salt Lake City. Said Rube Wagner, who flew mostly across the middle of the country in those years:

We flew low, and when we had to we would land close to the beacon, and stay with the airplane, knowing that fog was always lowest at daylight. I often had breakfast with the farmer before taking off from his pasture. The Boeing trimotor could land almost as short as the De Havilland mail planes. We never lost a passenger in one.

Flying a Transcontinental and Western Air mail plane from Kansas City toward Amarillo on the night of April 30, 1931, Swede Golien found fog forming on the airway. But since he could see the glow of the beacon at the Canadian, Texas, emergency field, he, like Rube Wagner, decided to land and wait for clearing at dawn. It was an utterly blind landing but Swede, recalling that the beacon stood at the southwest corner of the field, just lined up on a northerly heading and eased down gingerly to the right of the light. He said later:

As soon as the plane touched down there were some alarming bumps and bounces. But when the plane jerked to a stop, there was nothing to do but curl up on the mail sacks and wait. In the morning the fog lifted, and I discovered I had not landed in the field at all but in a cornfield with deep furrows. The beacon I had thought was in the southwest corner of the emergency field was actually in .the northeast corner.

There was nobody at the field, which had been decommissioned. So Golien cranked up his Northrop Alpha, bounced away from the cornfield, and delivered his mail at Amarillo within the hour.

Up to this time pilots had survived by accumulating experience in seat-of-the-pants flying and using outside points of reference, such as beacons, to guide and control the airplane. Few had ventured to turn their eyes from the horizon to the instrument panel before them and deliberately, as the saying went, "fly blind." Some did, of course: those who flew long flights over oceans or overnight. Such pilots — Lindbergh, Doolittle, Balchen, Colonel Sanderson of the Marine Corps, Amelia Earhart — all had made remarkable flights and used their instruments well to do so, but not the general run of pilots. Said Walt Addems, one of the pioneer United pilots:

Some of the old pilots wanted no part of instrument flying. Hoppy Hopson had flown the Omaha–Chicago and Cleveland–New York runs for the Air Mail Service. He never believed in instruments. Neither did Elmer Garrison. They just kept going as before — one red flare told them to circle the airfield, two red flares meant: "Don't land." Both were killed.

All the chief airlines began to push instrument flying on their crews after Jimmy Doolittle made the first 100 percent instrument takeoff, flight, and landing at Roosevelt Field in 1929.

It wasn't until May 1932 that the Department of Commerce got around to making sure that airline pilots were capable of flying passengers with more than visual aids.

Six months earlier the airlines had taken steps on their own to instruct their pilots in instrument flying. In the fall of 1931 Bert Lott, United's chief pilot, made Walt Addems superintendent of flight operations and ordered him to start putting all the pilots through their paces. At the same time TWA hired a fellow named Jack Lynch to teach the rudiments of instrument flying to all its pilots assigned to carry the night mail. Lynch arrived in Kansas City in his Stearman. "Each of us," said Golien,

spent from four to six hours under the hood while Lynch sat in the rear cockpit. When first forced to fly without looking outside, a pilot usually experiences a form of vertigo: he feels disoriented and thinks he is flying upside down when he isn't. That was when we had to learn to control those erratic urges, and to do what the instruments told us to do.

Lynch went so far as to require us to throw the plane into a spin — we spun like the devil. A helluva sensation it was, cooped in, with the instruments going round and round, trying and finally succeeding in getting back to straight and level flight.

TWA bought its own Stearman and assigned Captain Hal Snead, Lynch's best pupil, to extend the program to all pilots, including those on daylight schedule.

Meanwhile Addems "gave many hours" of instruction time to United pilots. Addems began by teaching himself. At first he wore frosted glasses with pinholes in the middle to help keep his eyes inside the cockpit. Then he placed the hood over his head and relied on a check pilot in the rear seat for reassurance. Unlike Doolittle, Addems flew blind with only the bank-and-turn indicator besides the usual instruments — altimeter, airspeed indicator, rate-of-climb indicator, and compass. The artificial horizon was available, Addems said later, "but that would have made things too easy — better to learn blind flying with just the turn indicator."

In May 1932 the Department of Commerce announced the first Scheduled Air Transport Rating (SATR). Requirements: a proficiency in instrument flying and in navigating by radio. When Inspector Ernie Cutrell approached United to conduct the first tests, he found that four-course radio ranges had been set up, and Addems had worked out a routine drill for instrument training. Cutrell proposed that the pilot being tested would occupy the hooded front cockpit of his Bellanca. Cutrell would sit in the rear, take the plane aloft, and give the controls to the hooded pilot. He would then order the man in front to execute two or three turns. After that it would be up to the pilot to get back to straight and level flight, pick up the radio beacon, figure out from its signals on which side he was, and then, having done so, fly unaided until he had brought the plane over the beacon near the landing field. Addems, however, proposed that the pilots being tested take over the flying from the start, taking off and climbing as Cutrell directed. Cutrell said, "But it'll take an hour for the pilot to orient himself — we won't have time to qualify more than two or three a day that way."

Said Addems, who had trained his men that way, "Let me take the first check, and if you like it, all the rest of the boys can take the same."

Thus it was that Walt Addems set the standards by which airline pilots gained their ratings. Cutrell said later, "It took only 25 minutes. We ran 'em through — Bill Williams [another famous old pilot] was the number two boy — twelve United pilots in those first three days. Some flunked. Morry Mars had to have a second go-around on the fourth day." A week after checking out the United men in Cleveland and Chicago, Ernie Cutrell, flying blind for practice, arrived in Kansas City. ("You never thought about looking out for collisions then," he said in 1976. "The sky was wide open.") There he administered the same SATR tests to Snead and his pupils. Then he went to Newark, where Karl S. Day had been showing the ropes to American Airways' pilots. Day's book *Instrument and Radio Flying*, repeatedly revised, is still the U.S. military's guide to flying by instruments. Day, who enlisted in 1917 as a Marine pilot and died in 1972 as a three-star general, is also credited with developing the concept of airline dispatching. Dispatching assigns both the pilot and the ground operations manager responsibility for an airliner's flight: neither can overrule the other. Once the flight is airborne the captain is responsible, and the dispatcher is his eyes and ears on the ground, responsible for keeping the captain informed about anything that might remotely affect the safety of the flight. Dispatching, practically everyone agrees, is at least 50 percent responsible for the safety of airline operations. Day started it, and it spread all through the industry.

Instrument flying was a means of maintaining flight when the pilot could not see where he was going. By themselves, cockpit instruments were not yet sufficiently developed in the 1920s and 1930s to ensure safe and dependable flights across country. Somehow much more information would have to be furnished the pilot if he were to deliver passengers safely and dependably to their destinations.

There were essentially three methods by which the airman could find his way. Piloting by reference to landmarks, the only one available at first, was obviously not good enough. Commercial airlines operating in the uncertain skies of the North Temperate Zone could never hope to keep their schedules if everything depended on what the pilot saw. Although the elaborate network of lighted airways was hailed as a big advance when airliners were permitted to fly at night, flashing beacons were of no avail when clouds and fog swirled round the plane. The second alternative, equipping the airmen to emulate the sea captains and steer by the heavenly bodies, was out of the question in the 1920s. The flights of the day were too short, and too few were over water, for commercial airlines to school their men in taking star sights. In these straits, all those interested in developing organized air travel in the 1920s turned more and more to the third means of guiding planes across country. They said, "Give the pilot navigational information that he can hear when he cannot see." Their answer was radio.

Radio and airplanes grew up together. Two years before the Wright brothers made their first flights, Marconi sent his first message across the Atlantic on the air waves. As early as 1912 Lieutenant Hap Arnold took radio equipment aloft in a Signal Corps biplane. On the first radio-equipped airline planes the pilot, or someone else, had to dangle a long wire out the window as an aerial every time they wanted to transmit or receive. By the time that Hiscock of Boeing Air Transport got to work, radio engineers already knew that wire aerials on planes, whether trailed behind or fastened permanently to fuselage and tail, were almost totally ineffective. Static from the plane's engine was bad enough, but even after engineers learned to shield spark plugs and magnetos, noise interference on airplane radio receivers was deafening. It was especially bad when a pilot needed his radio most. Every time the plane flew into rain, snow, or even a fair-sized cloud, static blotted out radio reception.

It didn't take long to find out what was the matter. Hiscock got Bert Ball and a few other United pilots to fly day after day at Arcata, California, thought to be the foggiest place in the United States. Every tiny droplet of water in a cloud carries an electric charge, and when touched by a plane's wing, the static electricity is discharged. This is what causes the glow — the St. Elmo's fire observed by the ancients around ship's masts — that discharges in brushlike fiery jets around a plane's wings or propellers at night. This was what ruined the plane's radio reception. What to do?

It happened that David Little, a radio operator on Great Lakes ore-boats, had transferred to RCA's aviation office at Cleveland in 1926. Little remembered that the loop antennas used aboard ships for direction-finding equipment seemed to receive signals in foggy weather when his regular receiver, whose aerial was strung between the masts, did not. These loops were entirely covered by metal except for a paper-thin crack to admit the radio waves. A shielded loop, similar to the oreboat's but much smaller, was fitted to a TWA plane's antenna and, sure enough, static interference was reduced. But Bert Ball and others had to make many more flights at Arcata, and the phenomenon of static in clouds was not mastered at once. Eventually, by attaching bits of fine copper wire to the trailing edges of the wing, airline engineers succeeded in bleeding away the static electricity that built up on planes and gummed up radio reception. When very high frequencies were introduced in World War II, static was finally defeated. After that pilots could hear and be heard even in thunderstorms.

Radio, vital for talk between the companies and their planes, proved even more important in helping the airliners find their way through the murk. After twenty-five years, flying by eye had led the airlines, so to speak, up a blind alley. Just when it became clear that lighted airways led nowhere, radio came to the rescue, and enabled the pilot to fly by ear straight through clouds to his destination.

Radio navigation for planes had been long in the works. As early as 1908, the Germans discovered that radio waves could be directed. From an ordinary wire antenna, waves go out in all directions. But the Germans found that if the wire is shaped — in a loop, triangle, or square — the antenna will transmit or receive the greatest energy along the plane in which the wire is shaped. By the same token, it will send or receive the least energy in the direction perpendicular to the antenna. Exploiting this phenomenon, the Germans constructed a kind of "radio compass" to guide their zeppelins on World War I bombing raids over England. The first Telefunken compass was too unreliable to help the zeppelins at great distances from the German ground stations; other types of direction-finding by radio were tried. The U.S. Navy's Curtiss flying boats that first crossed the Atlantic in 1919 carried primitive radio directional equipment.

All this was as hush-hush as radar was in the Second World War. The U.S. Signal Corps asked the Bureau of Standards to see if this country could not improve on the German design. The Bureau's response was to bypass directional equipment for the planes (which required that the aircraft have a transmitter) and to concentrate on establishing a ground unit that would furnish directional signals to planes. They developed the aural beacon, or radio range as it was called. In this system, only a receiver would be needed on the plane; there would be no need for the pilot to speak up and ask for a transmission. All he would have to do was keep his head phones on, listen to the automatic signal sent out by the beacon, and steer his plane according to what these extremely simple signals told him.

Two Bureau physicists, F. H. Engel and Francis W. Dunsmore, rigged their first transmitter aboard the lighthouse tender *Mary* moored in the Washington Navy Yard. The key to their system was the way the antennas were placed. Antennas were made of two loops or coils fixed at right angles to each other (later at 135 degrees to each other). Each of these two-directional antennas transmitted its strongest signal along the plane of the wires and sent little or no signal at all in the direction perpendicular to this plane. This meant that one coil's coded signal was transmitted most strongly in two opposing directions, and the other coil's continuously repeated signal was sent in two opposing directions at right angles to the first. The effect was to create four distinct 90-degree quadrants around the transmitting beacon. In two of these quadrants one of the beacon's continuous signals could be heard; in two opposite quadrants the beacon's other signal blasted out with sharply, but uniformly graduated, degrees of loudness.

The two signals were "dit-dah," the letter A in Morse Code, and "dah-dit," the letter N in Morse Code. A pilot flying in either of the first two quadrants, or sectors, would hear nothing but dit-dahs — A's. If he flew in either of the other two quadrants, he would hear nothing but dah-dits — N's. But if he flew on toward the place where two quadrants came together, he would begin to hear *both* signals. That is, he would hear both A's and N's. Finally, when he reached the actual edge between the two quadrants, he could hear on his receiver that the two signals were of equal strength. Then, pointing the nose of his plane along the line where the signals continued to come in equally strong, steering the aircraft according to what his ears told him, the pilot could fly onward, the signals getting louder and louder, until he passed directly over the radio beacon.

In 1921 Engel and Dunsmore found out in their Washington experiments that the four zones in which A's and N's merged were about a mile wide thirty-five miles from the station. Then the Air Corps at McCook Field in Dayton, Ohio, took over and succeeded in timing the A and N transmissions so that in the equisignal zones the dots and dashes would form a continuous hum. They also narrowed the width of the equisignal zone from eight degrees to three.

By 1924 the government was ready to take the wraps off the new device. When the Air Commerce Act passed in 1926, the Department of Commerce announced that besides setting standards for planes and pilots and building lighted airways it would undertake to provide radio navigational aids. The Bureau of Standards took it from there. The men who built the first experimental beacons at College Park, Maryland, and Bellefonte, Pennsylvania, included Dunsmore, W. G. Kear and Harry Diamond — the same Harry Diamond who devised the proximity fuze in World War II. The Commerce Department pilots who flew most of the tests were Ernie Cutrell and his sidekick Maury Hobbs. Wesley Smith, the old Air Mail instrument pioneer now flying the same route for National Air Transport, was actively involved at Bellefonte.

There were all sorts of problems. The first antennas were loops, which made them subject to "night effect." That is, operating horizontally, they transmitted waves that traveled predictably enough by day, but at night the waves bounced erratically off layers of the ionosphere that formed at high altitudes when darkness fell. It was not until 1932 that vertical antennas were introduced which eliminated this source of wave disturbance and distortion.

By that time the Department of Commerce had planted a whole network of some eighty-two radio beacons across the country. Instead of merely homing on a single station in order to locate a field and make a landing, pilots began flying from one beacon to another — "as if," said Captain Mo Bowen, "the aircraft ran on rails." The men in the cockpit knew they had arrived at a beacon because the range was designed so that no energy at all was radiated immediately above a station. Pilots called this the "cone of silence," and flying into it, could know at that moment precisely where they were, no matter how bad the weather. For other points of importance along the airways, the Commerce Department designed aural signposts that became vital to successful airline operations. These were low-powered transmitters called marker beacons. They sent their beam straight up. A pilot could follow the beam until the signals fell silent over a beacon in a known position adjacent to an airfield. Then he could make a 90 degree turn, pick up the beam at right angles to the beam that had brought him in, and fly outward along this second beam until he heard the signal from a marker beam that told him he was positioned 500 feet short of the edge of the airfield itself.

With this kind of precise help in locating the airfield, the pilot could fly the rest of the way to a safe landing with the help of his instruments. That was exactly how Jimmy Doolittle, the Army flier who won so many air races in the 1920s and 1930s, accomplished the first all-instrument flight at Roosevelt Field in 1929. This truly historic feat was undertaken with the financial and technical support of the Guggenheim Fund for the Advancement of Aeronautics. Doolittle took temporary leave from the Air Corps to work on the project.

In the actual test, Doolittle in his Consolidated NY-2 trainer was under a hood, with a check pilot in the rear seat. He had compass, bank-and-turn indicator, sensitive altimeter, directional gyro, and the beautiful new instrument called the artificial horizon to guide him. He also had radio. As he took off and flew his circuit over the little field, he kept his eyes glued on his instruments and made all his turns according to what they told him. But the radio beacon was also transmitting information, and when Doolittle turned in for a landing, the marker beacon's signal at the edge of the field informed him that he was right where he was supposed to be and exactly how far he was from the end of the runway.

Before the airliner pilot got to this point, of course, he had to find his way across country, and with the four-course radio range that was not always so simple. A peculiarity of the system was that any beacon's antenna sent A's and N's in two opposing directions. Therefore, if the pilot were lost it was up to him to find out on which side of the beacon he was. He could be either north or south of the beacon and get the same signal. Of course the beacon's signals grew louder as he approached and fainter as he flew away, but valuable time was lost trying to make sure of directions, and there could be times when lives might be lost before a pilot could clear up the ambiguity of the beacon's message to him.

For the American airlines probably the biggest headache experienced in flying the beam was "mountain effect." This problem caused some bad accidents in the West, and many more frighteningly close calls. Just as radio waves could be knocked askew by bouncing off the lower ionosphere at night, so they could be deflected when they smacked into irregular terrain jutting up in their path. When radio range stations were set up in the Rocky Mountains, their waves ricocheted all over the place. Beams split, and the pilot could not know which he should follow. More than one airliner followed false beams straight into the side of a mountain. In some places the beam signal alternately waxed strong and then faded as reflected energy first joined and then bounced away from the direct signal. If the beam faded entirely, the pilot could mistake it for the cone of silence. The Commerce Department installed extra stations in the areas where mountain effect was worst. But relocating stations did little to eliminate multiple beams. Pilots had to learn to expect false beams in mountainous country when more than thirty miles from a beacon. To avoid being misled, airline captains had to compare their compass headings with the directional signals coming in over their headphones and to compute the plane's position from any other possible source, using every skill and resource they could summon. All this was done while staying ceaselessly wary of what the radio ranges were telling them.

Overall, however, radio was a tremendous navigational aid and served American airlines well in the 1930s. Overnight the phrase "on the beam" (and its equally vivid opposite "off the beam") entered the American language. Flying the radio ranges along the airways, simply making sure that they never wandered so far off line that they heard distinct A's or N's (the keenest pilots took care to fly on the edge of the equisignal zone where one letter could be heard slightly more than the other, rather than right on the beam because the beam was in fact a bit too wide for precise course-keeping), these men flew the same routes day after day. Pilots who had developed an uncanny knack for spotting small landmarks now discovered they could recognize the peculiarities of individual radio stations' signals. Such skills, almost unconscious on the part of the airline pilots, probably explain the sixth sense often attributed to experienced aviators; they extract useful information from evidence most people do not notice. And on the domestic routes, airline captains did not need much more than their radio beams, supplemented by these arcane skills, for their daily stint. These two fundamental ways of piloting, by radio and by landmark, served the domestic airlines well enough; their pilots had no need to choose routes like sea captains. But when aviation went overseas, a more sophisticated method of finding the way became essential for the airlines. This called into use the third method of air navigation — reference to the heavenly bodies.

12

Brown, Black, and Blue

M_{en} vote their interest," Aristotle said. In 1932 bread lines stretched long. Farms fell under the sheriff's hammer. The stock market plunged through the floor. Banks closed, many never to reopen. When 13 million workers were jobless, a quarter of city families were on relief, and uncounted thousands had lost their life savings in the collapse of banks and stocks, the voters of the United States rose in wrath. Who had led them from boom to bust? Who indeed? Who but Hoover — and the big business types allied with him?

Seldom has the nation seen such a political turnover as the Great Depression brought about. All through the 1920s, from Harding to Hoover, big business held sway. Now, after the Great Crash, Congress put business in the dock. Magnates who had been treated with deference, even awe, were haled to the stand and asked rude questions about how they had made free with other people's money. In one hearing room on Capitol Hill a lawyer named Ferdinand Pecora grilled the masters of banking and investment; J. Pierpont Morgan of Morgan's, Charles Mitchell of the National City Bank, and Richard Whitney of the New York Stock Exchange were only the most prominent forced to answer for their misjudgments — and misdeeds. In the next room a country boy from the hills of Alabama began a series of probes into monopoly practices, probes that clamped regulatory lids on major industry and catapulted him into the Supreme Court.

When Hugo La Fayette Black died in 1971, it was said, "Many believed he influenced American life more than any of his colleagues in modern times." Doubtless the *New York Times* was referring to the thirty-four years in which Justice Black's dissents turned into majority rulings in great Court cases. But before Hugo Black was a justice, he was a senator, the foremost inquisitor of America's hard times. In the three years before he went to the Court in 1937, Senator Black's investigative forays brought about acts of Congress that reshaped the economic arrangements of modern society.

If the characteristic form of nineteenth-century monopoly capitalism was the trust, in the twentieth century it was the holding company. It still is. As chairman of a series of Senate probes when an angry public demanded an accounting from irresponsible lords of business, Black riveted the nation's attention on holding companies that exploited government franchises. The climax of this campaign was the bitterly fought passage in 1937 of the law banning holding companies in the electric and gas industries. But it all began with the sharp, brutal episode in aviation history known as the airmail crisis. In that fight ten airline presidents were thrown over the side, Postmaster General James Farley was knocked into the scuppers, and on the quarterdeck President Roosevelt himself was socked squarely on the jaw. Before the row ended, Black had driven holding companies out of the airline business, and Congress set new terms by which airlines, clearly labeled as public utilities, would be supervised in the same way as railroads, broadcasters, and power companies.

As the public sized up Black in 1934, there was a deceptive mildness in his appearance — a placid, unlined face; soft, blue-gray eyes; an unhurried Alabama drawl. In certain ways he resembled another backbench senator who became unexpectedly famous. If Harry Truman had driven a gang plow pulled by four Missouri mules, Black had chopped cotton on Alabama's most impoverished, red-clay farm lands. If Truman had risen as a loyal straw boss in a corrupt Kansas City political machine, Black had climbed to the Senate by courting the Ku Klux Klan. But Populism colored Black's Baptist upbringing. As a Birmingham lawyer he mostly tried claims against corporations. In Washington he joined George Norris in the fight to produce public power at Muscle Shoals that ended in the creation of the Tennessee Valley Authority.

Given the assignment to investigate government mail contracts, Black saw at once that the setup handed the Postmaster General all responsibility for allocating among the airlines both their routes and their life-preserving mail subsidies. Under Hoover the Postmaster General had been Walter Folger Brown. An old-style Hamiltonian believer in strong central power, Brown had decisively used the government's mail payments to redraw the aviation map of the United States.

Never had an important sector of American economic life been more consciously and directly shaped by an appointed official of government. When the Interstate Commerce Commission had first subjected the railroads to supervision nearly half a century earlier, their routes and corporate structures were already well established. Radio, and later TV, got their networks, but these were never imposed by the Federal Communications Commission. The utilities of the United States were established in their present form before the Federal Power Commission began to regulate them.

Here, a Postmaster General, and a Republican at that, had stepped in when an industry had not yet hardened into form and, in effect, imposed a shape upon it. No industry was ever such a creature of Washington as the airlines. But in intervening to create three transcontinental airlines (TWA, American, and United), each owned by a Wall Street holding company, Brown overstepped himself. When Congress passed enabling legislation — the Watres Act — that kept the requirement for competitive bidding on the mail contracts, Brown simply choked off undesired competitive bids and forced the redistribution of routes.

Such high-handed actions were bound to give rise, at the least, to charges of favoritism. When the Democrats took over, independent airline operators rushed to tell how they had been frozen out by Republican Walter Brown. Then Senator Black met a young Hearst reporter named Fulton Lewis. Frustrated because his boss did not print the story, Lewis handed Black his report on how the Post Office had awarded the New York–Washington mail contract at a rate three times as high as a rejected bid by a smaller line. Scenting more than just favoritism, aides of the new Postmaster General, Jim Farley, were already searching for fraud and collusion in his predecessor's awards. So one day, at Black's direction, one hundred Interstate Commerce Commission agents fanned out through a dozen cities. Precisely at 9:13 A.M., their watches set by Western Union clocks, the investigators marched into one hundred aviation offices, simultaneously served warrants, and seized correspondence — there was never a chance for a warning to be flashed.

What emerged led Black later in life to theorize that in America "all great fortunes trace back to the government's treasury." It appeared that an infant industry had suddenly grown great, battening off government contracts to exploit the public. From military procurement contracts, the heads of the big aviation combines piled up salaries, bonuses, and earnings that seemed huge to Black and to the taxpayers of the hungry 1930s. To all this the promoters added speculative profits from stock they had floated in the expectation that their airline subsidiaries would capture lucrative Post Office contracts. Contracts, contracts, they all fed on contracts: for these early conglomerates of aviation, that was the name of the game. In highly questionable circumstances — avoidance of competitive bidding, covert destruction of official records, closed meetings with the combines beforehand — twenty-four of the twenty-seven Post Office airmail contracts had passed during Walter Brown's tenure into the hands of these three big outfits.

Comedy added spice to the revelations. On the first day a postal clerk testified that Brown had burned large quantities of files relating to mail contracts just before leaving office. Not true, cried Brown from his Toledo home. But on second thought, Brown brought Farley a suitcase stuffed with records, having "unexpectedly found them."

Remarking that papers were still missing, Black wheeled harshly to

other witnesses. The spectacular pyramiding by which financiers had inflated Pratt & Whitney Aircraft into United Aircraft, the biggest aviation
combine, was as sensational as any of the feats of Dan Drew, Jay Gould, or
other cynical waterers of nineteenth-century railroad stock. The paper
caper out of which Colonel Deeds's son Charles, the Rentschler brothers,
and other insiders collected fortunes, was exposed for all to see. Partner
Bill Boeing, whose $450,000 outlay had been far more substantial than
Fred Rentschler's $253, was questioned severely as he acknowledged he
had made $12 million from the stock flotations. "What about the public?"
asked Black. Boeing said, "I don't see where the public comes in on this."
When Black replied, "The public came in between $87 and $160," the
caucus room roared with laughter. The knock-down, drag-out session left
Boeing so enraged that he went home and quit aviation for good.

Reporter Lewis had spoken of "secret spoils conferences" at which the
big companies had carved up the air routes. Under questioning, D. W.
Schaeffer of Transcontinental and Western Air confirmed that he had attended the meetings in Brown's anteroom even though his company had
no contract. Paul Henderson of United confessed he had wondered
whether they were violating the Sherman Anti-Trust Act, but quoted the
reply of his lawyer Chester Cuthell:

If we were holding this meeting across the street in the Raleigh Hotel, it would be
an improper meeting, but because we are holding it at the invitation of a member
of the cabinet, and in the office of the Post Office Department, it is perfectly all
right.

It was confirmed that William P. MacCracken, former Assistant Secretary of Commerce for Air under Coolidge and Washington lawyer representing half a dozen airlines, had acted as chairman of the meetings. Out
to document collusion, Black demanded that MacCracken produce his
records. MacCracken refused, claiming the privilege of a lawyer safe-
guarding his client's privacy. Terming MacCracken a lobbyist, Black ordered him arrested for contempt of the Senate. When Black aides found
out that two airline executives had meanwhile extracted papers from
MacCracken's files, Black ordered these men arrested too. One hastily
handed over his papers, but the other, Vice President L. H. Brittin of
Northwest Airways, had torn his up. Black's men searched three hundred
bags of wastepaper, located the torn fragments, and pieced them together. Black had Brittin read a letter to a woman in Illinois that he wrote
during the meetings into the record:

Thanks for yours of the 29th. The airmail contractors are having a desperate time
in Washington. The Postmaster General was not able to get the necessary legislation in the Watres Bill to enable him to grant airmail contracts to the passenger-
carrying airlines without competitive bids. He has made up his mind to do this
anyway, and has hit upon a plan that is causing the operators no end of trouble.

He has conceived, probably in iniquity, a plan for three main transcontinental routes competitively operating, and several north-and-south lines as well.

To work things out he called the operators together, handed them this map and instructed them to settle among themselves the distribution of these routes. The operators have been meeting every day for two weeks and to date have arrived nowhere. The Postmaster General meets with them about once a week, stirs them up and keeps them going.

MacCracken and Brittin were found guilty and went to jail for ten days.

Nobody ever located the record of the meetings, but Black's questioning brought out the rest of the story. His memory "refreshed" by a copy of his memorandum to the directors of his company, Schaeffer confirmed that the operators, deadlocked, had handed the map back to Brown and "agreed to abide by his decision." But that was far from the end of the story.

Brown, when he took the stand, assumed full responsibility for making the awards. With the full support of President Hoover, he had made up his mind that there should be no such disorderly fragmenting of airlines as the nation had to endure with its railroads. Three transcontinental routes there should be, each held by a single outfit. Political paladin of the big business ascendancy that he was, Brown never doubted that these super-routes should go to the three supercombines that also dominated aircraft manufacture and military contracting. To Brown, big was beautiful. Liberty, in his lexicon, was freedom to combine.

He told how he forced the mergers. The San Francisco–New York route was already in United's hands — no problem. But when Western Air Express President Harris Hanshue balked at merging with the second group of Wall Streeters to take over the central transcontinental route, strong pressure had to be applied. "I did not force him," Brown said, but of course it was he who held the shotgun of the mail contracts over Hanshue that sent old Pop cursing and fuming to the altar. That merger, Brown said, was "the key to our whole plan of developing commercial aviation, and we naturally wanted to take as few chances as possible of disaster."

Then a series of transactions among the parties brought together by Brown allowed the third big combination to gain the southern cross-country route. When the winner, Fred Coburn of American Airways, took the stand, there was this exchange:

BLACK: There was an understanding as far as these three big companies were concerned?
COBURN: I did not fear any of these companies.
BLACK: Why?
COBURN: I understood they were not going to bid.
BLACK: That was an express understanding?
COBURN: Yes, sir.

The American Airways (Aviation Corporation) bid for the southern trans-continental route was unopposed, and the evidence before Black indicated that the award of the central route to the second combine was linked to it by a prearranged deal. That is, the one independent contender for the southern route, oilman Erle Halliburton of Oklahoma, was taken out of the play when American Airways bought up his Southern Air Fast Express for $1.4 million. But that sum, in turn, was exactly equal to the amount Transcontinental and Western Air made over to American for a hangar in Tulsa plus the value of American's stock in a unit absorbed by TWA. By such hug-and-slug tactics were the prize superroutes assembled and their mail contracts awarded to the supercompanies.

Black grudgingly accepted the necessity of mail contract subsidies if the air network were to be built up in the national interest. But the subsidies, he felt, needed the check of competitive bidding, and Brown had flouted this requirement. To channel awards to the combines, the Postmaster General had not only used his Watres Act power to "extend" routes arbitrarily but also had interpreted the requirement for "responsible" bidders to mean bidders with big-time financial backing. Though Brown insisted there had been no illegalities, Black asked, "What was the object of these conferences if they were not to arrange what companies should get what lines?" Black condemned the heaping of government favors on Wall Street kite flyers at the expense of the little guys who really flew the planes. He said:

The control of American aviation has been ruthlessly taken away from the men who could fly and bestowed upon bankers, brokers, promoters and politicians, sitting in their inner offices, allotting among themselves the taxpayers' money.

Late in January 1934 Black went to lunch at the White House. To President Roosevelt he said that the whole system of airmail contracts emanating from the clandestine 1930 carve-up was completely illegal, and he reminded Roosevelt that under an 1872 act the Postmaster General had power to cancel contracts obtained by fraud or conspiracy. In the Post Office Department, Solicitor Karl Crowley was also urging cancellation. Second Assistant Postmaster General Harllee Branch had sounded out General Benjamin Foulois, chief of the Army Air Corps. He asked, "If the President should cancel all contracts, do you think the Air Corps should carry the mail?" Benny Foulois had taken his lumps with Billy Mitchell in the losing fight for an independent air force. Now he saw visions of the acclaim and money his funds-starved service could win by performing the job again as it did when the Air Mail Service first started. He said, "Yes sir, if you want us to carry the mail, we'll do it."

The same day Farley carried the word to Roosevelt. "Furious" at the implications of contract collusion, the President wanted fast action. He asked only that Farley check first with Attorney General Hugh Cummings. Overnight Cummings read the Post Office Department's one-hundred-page indictment and said "undoubtedly" there had been wrongdoing. Without a word of discussion at the Cabinet meeting the next day, the bombshell was exploded. The White House announced that all airmail contracts had been annulled and the Air Corps had been directed to carry the mail in the emergency.

Any experienced pilot could have told the President what was going to happen. When the Air Corps took over, it was midwinter. Blizzards lashed the West; gales, sleet, fog, and cold gripped the Middle West and East. Army pilots were not trained for bad weather flying; many were not even trained for night flying. The Air Corps takeover was an instant fiasco. Second Lieutenant James Eastham, far off course in his little Boeing fighter, crashed in flames near Jerome, Idaho. Two other airmen, their Curtiss observation plane coated with ice, slammed into a mountain in Wyoming. An Army seaplane landed in freezing spray in high seas off Long Island and was soon smashed to pieces by the waves. At the end of the first week, five pilots were dead (three were killed in test operations before the Army officially took over) and six were injured; eight planes were wrecked. The country was appalled. Eddie Rickenbacker of Eastern Air Transport denounced it as "legalized murder."

The national shock suddenly gave the pent-up dissatisfaction with Roosevelt's New Deal a seemingly legitimate outlet. The airlines, ready to fight, wheeled their biggest gun into play. In 1934 the only American who could rival President Roosevelt in popularity was Charles Lindbergh. The shocking kidnapping of his baby son in 1932 had deepened Lindbergh's place in the nation's affections. Ever since his 1927 flight he had conducted a personal promotional campaign for aviation, first flying to all forty-eight states and sixteen Latin American countries to foster air-mindedness, later by identifying himself with the air industry. As technical consultant to Transcontinental Air Transport, he had pioneered what became the cross-country route of TWA — "the Lindbergh line." As an equally well paid adviser to Pan Am, he was constantly in the public eye making survey flights across the oceans. When the administration canceled the contracts, Lindbergh was a logical figure to express the industry's resentments.

He shot off a sharp telegram to the President: "Your present action does not discriminate between innocence and guilt, and places no premium on honest business. Your order of cancellation of all airmail contracts condemns the largest portion of our commercial aviation without just trial." Canceling the contracts without hearings was the weakest part of Roosevelt's case, and another of the nation's heroes protested its unfairness. Will Rogers wired the *Kansas City Star:* "It's like finding a crooked railroad president, and then stopping all the trains."

Without success, Secretary of War George Dern tried to get Lindbergh to join a committee to help the airmen. The collapse of the Air Corps had already signaled the outcome. After landing fifty miles off course and taxiing down the main street of Demopolis, Alabama, to a gas station, Lieutenant John R. Sutherland crashed his small biplane while landing at nearby Selma. Lieutenant H. L. Dietz, lost after his map blew out of the cockpit, groped through fog and finally came down in a Maryland tree.

"The wrath of the public descended on my head," said Jim Farley, who was left by the President to take the drubbing from outraged opinion in the clubs, smoking cars, and recently reopened bars. Then one day Roosevelt summoned Benny Foulois. "General, when are these killings going to stop?" he demanded. And for ten minutes Roosevelt gave Foulois what the general later described as "the worst tongue-lashing I ever received." But it was the President who had blundered, and by May, after fifty-seven accidents and twelve deaths in seventy-eight days, the commercial airlines were back carrying the mail.

Many years afterward C. R. Smith summed up the industry's memory of the near knockout blow:

I suppose the reason Roosevelt canceled the contracts was that a new administration had been inaugurated, and he wanted to show up the fellows that had been in before. On the question of collusion there was no real basis for what he claimed. But the airlines were in a rather peculiar position. They couldn't sue the Postmaster General; the Postmaster General had contracts to give out.

Actually, United Airlines did sue, and the U. S. Court of Claims finally ruled that the government was justified under the 1872 law in canceling the Brown contracts. But by then it was 1942, and there was a war on. President Roosevelt found time to read the decision and said he enjoyed it.

Before he threw in the towel, Roosevelt toyed with the idea of letting the Air Corps permanently fly the mail. It was a grim season for the airlines. None flew more than the skimpiest schedules, and all laid off help. When TWA President Richard Robbins furloughed all employees, Jack Frye went to Douglas and Wright Aeronautical, who were building planes and engines for TWA, and got them to promise they would hire TWA hands in the emergency. But in April Jim Farley, somewhat sheepishly, met with the operators at the Post Office Department, just like his predecessor. All the contracts were put up for bids, this time by competitive bidding. None of the companies whose contracts had been annulled were allowed to bid, but Farley was perfectly agreeable to the way they got around that prohibition. They simply changed their names. American Airways was rebaptized American Airlines, Eastern Air Transport became Eastern Airlines, and Transcontinental and Western Air simply added "Incorporated" to its name in its new incorporation papers. United Airlines's name was not changed at all because the original contracts had been awarded to United's subsidiaries. However, no contract could go to an airline that still employed anybody who had attended the 1930 "spoils conferences." A whole new lot of airline presidents therefore came forward at this time.

Since the new contracts were to last only until the Watres Act could be replaced with a new law, the operators bid well below Farley's new maximum rate of forty-five cents a mile. Their aim was to make sure they got the contracts, no matter what the pay, and then to hang on in hopes the new law would provide for higher payments. Because Farley's advertisements specified multi-engine equipment for the longer routes, the little guys championed by Black had small chance to take over the top contracts. The big lines generally regained their major holdings — United, TWA, and American their transcontinental routes, and Eastern its profitable routes along the eastern seaboard. But the independents made a comeback all the same. First was Tom Braniff, who had founded the Independent Scheduled Operators Association to fight back after he had been shut out entirely by Brown. Braniff took one of the fattest plums, the Chicago–Texas route, from United. Close behind came Delta Airlines, the little line headquartered in Black's home town of Birmingham, which had been forced to sell out to American in one of Brown's squeeze plays. Popping back after four years, Delta gained a contract to fly mail from Texas through Birmingham and Atlanta to Charleston on the Atlantic — the makings of a great airline to come. In all, twenty-nine airlines now held contracts, though six dropped out within a year and the Big Four hung on to the biggest and best.

By far the most important change was that put in motion by Hugo Black. This was the outlawing of the vertical holding companies by which big-business interests controlled both manufacture and transport. The dismantling of United Aircraft and Transport was first. The most powerful figure in American aviation, Fred Rentschler, left his Wall Street command center and went back to Hartford to build bigger and better engines, and Colonel Deeds, his second flyer in aviation over, returned to Ohio and the cash register company he started with. Under terms of the Black-McKellar Act of 1934, United split three ways: Boeing took over all manufacturing properties west of the Mississippi; Rentschler's United Aircraft Corporation received all holdings in the eastern part of the country; and United Airlines became a separate and independent entity, the largest air transport company in the United States, with W. A. "Pat" Patterson as its president. Late in 1934 control of the second big holding company, Aviation Corporation, passed from E. L. Cord to another big Wall Street operator, Victor Emanuel. Emanuel also chose to concentrate his efforts in manufacturing. He therefore sold off Avco's American Airlines stock, and C. R. Smith, newly elected president of American, found himself on his own and able to make the most of the line's independence as a new era dawned for U.S. air transport.

The third of the giant holding companies, North American Aviation, fought a rear-guard action until after Roosevelt's reelection in 1936. General Motors, which controlled North American by the time of the airmail crisis, sold off its 47.5 percent interest in TWA to John D. Hertz, the Yellow Cab magnate, and the Lehman Brothers banking house. But it was not for several years that Eddie Rickenbacker, with Laurance Rockefeller's backing, was able to buy the holding company's controlling interest in Eastern Airlines for $3.5 million. With that deal, North American retreated into manufacturing for good, and air transport in the United States was finally rid of its vertical combines.

In the splitoff of the airlines from the manufacturing companies, shares in the new air transport companies were distributed among individuals who held stock in the old holding companies. Accordingly, some of the old bosses' names popped up on the new airline boards, and New Dealers who had denounced interlocking relationships now talked darkly of "interlocking directorates" that limited competition. The most significantly "interlocking" of all airline boards in the 1930s was that of Pan American, whose very profitable overseas mail contracts had come through the airmail cancellation crisis unscathed.

Every bit as much as the three combines that gained control of the three transcontinental domestic routes, Pan Am was a creature of the big business–government alliance that prevailed before 1933. United, North American, Avco — all the big three aviation combines had their men on Pan Am's board. And Walter Brown as aviation's chief political officer actively supported the industry's "chosen instrument" abroad. When Black asked, "To give preference to Pan American over all other companies [overseas], was not that the acknowledged and accepted policy of the Post Office Department?" Brown said, "Well, perhaps it was. I would not dignify it with the term 'policy' but that is the practice we certainly followed."

Black, in his usual style, put corroborative correspondence on the record. Eastern's President Thomas Doe was shown complaining to Brown in 1931 that Pan Am was refraining from using his combine's Havana airport. To this Brown replied that Pan Am had turned over its New York–Atlantic City service to Doe's airline the year before because the company operating abroad should stay out of domestic business; by the same token Doe's company should lay off Pan Am and get out of Cuba.

Therefore, the routes that Brown had dictated survived the 1934 explosion to become basically the new trade routes that a resurrected and resurgent airline industry would before long impose upon the American continent. But another result of the convulsive outburst that almost grounded the airlines for good was to reduce the subsidies paid them from $19.4 million in 1933 to less than $8 million in 1934. No doubt Brown was neither so selfless and farseeing as he made himself out in his rebuttal before the committee, nor so wicked as Black proclaimed him to the press. But the slash in Post Office payments also speeded the result that Brown claimed to have sought — push the airlines from primarily mail-carrying to primarily passenger-carrying. Only airlines that could make money flying passengers could hope to make a go with such reduced mail payments. The new Douglas transports were just at this moment arriving, and with them came the capability, never before available, to accomplish the job.

For all the red ink spilled by the eruption in Washington, 1934 was not such a dreadful year for the airlines. And for the nation, the airmail crisis, by ending the unhealthy union of manufacturing with transport, was no bad thing. Hap Arnold later wrote that the Air Corps's ordeal had been a timely and "wonderful experience for combat flying," and "best of all, it made possible for us to get the latest navigational and night-flying instruments in our planes." For the passenger, informed that he could soon fly across country in the same swift twin-engine airliner in which Jack Frye and Eddie Rickenbacker spanned the country in thirteen hours as the airmail crisis began, 1934 looked like a year of real progress.

13

Arrival of the Modern Airliner

One morning in March 1931 Stew Baker was out feeding the cattle on his Kansas farm. The drizzle had stopped, but clouds still rolled overhead. Somewhere above he heard the engines of an airplane sputter. Then the motors stopped. Baker looked up just in time to see a big airliner flash down through the clouds. A wing tore loose and sailed like a falling leaf to land half a mile away. The plane, a Fokker trimotor, plunged into Stew Baker's pasture with such force that one of the motors buried itself completely in the ground. On the adjoining Seward ranch, fieldhand Clarence McCracken also saw the plane fall. The two men rushed to find all seven persons aboard the plane dead — one of them Knute Rockne, coach of Notre Dame's unbeatable football teams and one of the most famous sport personalities of his day.

Within hours, as the whole nation led by President Hoover mourned the death of the great man, Department of Commerce officials began an investigation of the crash. Almost to the last minute the plane, a Transcontinental and Western Air transport on its way from Kansas City to Los Angeles, had kept in radio communication with the Wichita field, where it was due a half hour later. As far as could be determined, Captain Robert Frye, an experienced and capable pilot, had been maintaining level flight above the cloud layer at about two thousand feet when trouble struck. The plane had been delayed forty-five minutes in Kansas City, but only because a connecting mail flight had been late arriving from the East.

Because Captain Frye had cut the switches before impact, there was no fire. Investigators quickly established that the engines seemed to have been running all right. The plane may have hit some bumpy air at the altitude it was flying, but a dive out of the sky midway through a routine morning flight? What the witnesses saw pointed to the cause: a structural failure. The Department of Commerce men found rot where the wooden members of the right wing joined with the fuselage. The rot had not been noticed on inspection. In fact, the way a Fokker was built, no inspector could get inside the wing to check the joint. Designer Tony Fokker, with his unrivaled experience, prestige, and Dutch stubbornness, had continued to insist on building wings of wood, not metal. Now it appeared that wood was the source of the wing's fatal weakness. Fokker's business never recovered from this blow.

The Rockne crash also had far-reaching consequences for Transcontinental and Western Air. After "Boss" Brown had tossed TWA one of the three rich cross-country mail routes, the newly merged company remained in disarray. The Pennsylvania Railroad, having lost heavily on the experimental air-and-rail passenger venture that blazed the New York–Los Angeles trail, dropped right out of the picture. The airline's big new backer appeared to be General Motors, which gained financial control through its ownership of the holding company, North American Aviation. But the Great Depression was nearing bottom, and the men in Detroit worried about the vanishing auto market.

It was not a progressive situation. TWA was the line whose heady promises back in 1928 had first set the air-minded public thinking about transcontinental plane travel. Yet the trail-blazing "Lindbergh Line" was still flying the same old trimotors it had begun with when the Rockne crash shook public confidence in March 1931.

Competition, Walter Brown insisted, was what he was after — competition between airlines operated by the three big combines of the aviation industry. He made much of the fact that until he redrew the aviation map of the United States, not a single commercial airline would schedule flights across the Allegheny Mountains. No passenger could fly between New York and Chicago; the one line flying the mail on that route would not sell tickets. So on October 25, 1930, when TWA started its first passenger service out of Newark to the West, Brown himself was on hand for the trip.

It was a blustery day. Lindbergh was wearing a thick leather jacket. Although he did not go along with Brown, Amelia Earhart, another famous TWA employee, did. The first scheduled stop was at Camden, sixty miles on. The wind was blowing fifty miles an hour, but the pilot came in too fast. Seeing he was going to overshoot, he gunned the trimotor. Barely clearing a hangar, the plane went round and landed safely on the second pass. A shaken man, Brown wondered if he should have gone along. Taking on more passengers, the plane flew on to Harrisburg, then headed for Pittsburgh. Again it was no routine landing; Bettis Field was a very small airport atop a round hill. But, as Brown said later, "we managed to hang onto it." Refueling, the trimotor flew on to Columbus where, having survived the first scheduled passenger flight ever attempted over the Alleghenies, the Postmaster General got off.

Determined to make his point that the day of carrying only mail was past, Brown boarded one of Eastern Air Transport's first passenger flights south from Washington. The plane was forced down by fog, and the Postmaster General had to proceed to Atlanta by train. But at Atlanta he obtained his reward. It was a telegram from Colonel Henderson, manager of the one airline that had been insisting its route was too hazardous for carrying passengers: "National Air Transport will start daylight passenger service New York to Chicago with trimotored airplanes on October 29."

NAT, last of the holdouts, finally got its service started December 3, 1930. Keeping up the pressure, Brown authorized TWA and American to fly into Chicago from the East. By the time he left office in 1933 all three transcontinentals were providing service between Chicago and New York.

Of the three big aviation combines favored by Brown, the most successful from the beginning was United Aircraft and Transport. Winning control over 40 percent of all aviation business, the Rentschler-Boeing combine began to exhibit some of the traits of a Standard Oil or a du Pont. Any plane that emerged from a Boeing — or Vought, or Sikorsky — factory was pretty sure to have Rentschler's Pratt & Whitney engines. Even the propellers were supplied by a United affiliate. And when the time came for the airlines to put on the bigger and faster planes that Boss Brown's new mail-pay system called for, the United command worked out a cozy arrangement. Boeing would make the new planes, Pratt & Whitney would supply the engines, and the orders placed by the four or five airlines belonging to the United system would not only be large enough to ensure a profitable market to both, but the orders would keep them so busy they could not possibly sell an airliner to outsiders for a couple of years. In short, United's new airliner would sew up the passenger market. To Rentschler this was the payoff for building his combine. To Columbia economics professor Edward Chamberlin it was a prize example of "monopolistic competition."

The monopoly's monoplane however did not spring Athena-like from Rentschler's head. Like all new planes, it was a creature of past experience, present art, and expectations of what the future market would bear. At Boeing Claire Egtvedt, the company's senior designer, had taken advantage of the power available in the new 575-horsepower Pratt & Whitney engines to design a sleek, all-metal, twin-engine bomber that promised speeds no fighter flying in 1930 could match. This was the XB-9, a low-wing monoplane that Egtvedt intended to carry retractable landing gear, something never tried before except on a few experimental planes. It would also have a new kind of wing, in which the polished outer surface bore a large proportion of the load.

Assisting in the sales department at that time was Fred Collins, a young man lately graduated from the Boeing School of Aeronautics in Oakland. Collins heard the excited shop talk about the breakaway bomber design that was to be submitted to the Air Corps against rival proposals from

Martin and Douglas. He kept asking questions. "If it's all you say it is, it will revolutionize aviation," he said. Why not widen the fuselage and revolutionize air transport as well? Collins drafted a memo on sales department paper. The day came when Boeing, Lockheed, Douglas, all of them, took the costly prototypes they designed for the military and turned them into commercial airliners. But in 1931 what Collins proposed had never before been tried. It happened that W. A. "Pat" Patterson, then a loan officer with San Francisco's Wells Fargo Bank, stopped by as young Collins was about to drop the memo into his boss's in-basket. Collins said, "Here's a copy, what do you think?" Patterson read it and said, "Man, I think you've got something here. I'm going to send this right off to Phil." Phil was P. G. Johnson, President of United Airlines in Chicago.

It turned out that the Air Corps was not ready to go ahead with Boeing's bomber. But a group was forming under George Mead, the combine's chief engineer, to develop a Hornet-powered airliner. Claire Egtvedt's bomber took shape as an all-metal, low-wing, twin-engine monoplane that would cruise at 155 miles per hour — 50 miles per hour faster than the trimotors then in service. "If the body's going to be metal, the easiest way is to make it perfectly round," he had said. But Boeing's chief engineer, C. N. "Monty" Monteith, was a conservative fellow who wanted to see all possible support for the wings. In the altered design, the wing passed through rather than under the body.

Meanwhile, Fred Collins had been assigned as copilot on the San Francisco–Chicago run to find out what the line pilots thought they wanted. The pilots were the most conservative of all. Many had flown for the old Post Office Air Mail Service. When Boeing built its trimotor, they had successfully demanded a design that kept the open cockpit. Now, led by Operations Manager D. B. Colyer, old-time Post Office airmail superintendent, they bridled at the idea of giving up the Wasp engine that had served them so well. They thought that a sixteen-thousand-pound aircraft would be too heavy to land safely, and Monteith himself held that the bigger you made an airplane, the more difficult became the problem of structural strength. Scaled down to a ten-passenger, Wasp-powered plane weighing twelve thousand pounds, the new Boeing 247 airliner was a creature of surpassing compactness, an eye-pleasing departure from the ungainly trimotors with their engines strung naked beneath the wings. Said Monteith on February 8, 1933, when the prototype first winged, silver-bright, over Puget Sound, "They'll never build 'em any bigger."

Although that turned out to be one of the most extravagantly mistaken prophecies in aviation history, the new plane made everything that had flown on the airlines before obsolete. As soon as it went into service in June, United's coast-to-coast flights needed only six refueling stops instead of fourteen. Through schedules were cut to nineteen and a half hours — twelve hours less than with the old trimotors. And to the eye, the sleek, silver-skinned 247 was without question the first modern airliner. The fuselage was indeed rounded, and aerodynamically and acoustically better. The wings, not quite so wide in span as a Ford's, merged into the body as if all were one. The engines were set into the wings in handsome nacelles modeled after a new National Advisory Committee on Aeronautics design that reduced drag enormously while still letting in plenty of air to cool the cylinders. With all this plus retractable landing gear — another airliner first — Boeing had achieved "full streamlining and freedom from parasite drag," according to the authoritative magazine *Flight*.

For the passengers there was heavy insulation in the cabin walls, a hot-water heating system, a double ventilation system to keep air constantly circulating, and well-upholstered seats. The height of the cabin was six feet, not quite enough for a tall man to stand upright. Especially awkward was a bulge in the cabin floor behind the first two pairs of seats. This was created by the main wing spar that ran right across the fuselage. Though padded, it jutted up some twelve inches, and anyone walking up or down the aisle had to step over it.

When United Airlines ordered sixty of these new transports, paying a record $4 million for their delivery, the man in charge of flight operations at TWA knew his airline's plight was desperate. This was Jack Frye, who had become the airline's operational vice president in 1930 at the age of twenty-six. Frye was a big friendly bear of a man with wavy brown hair and the florid face of a person who had spent much of his life in an open cockpit. Early in 1930 Western Air had taken over Frye's small Standard Airlines operation and extended its Los Angeles–Phoenix–El Paso service into Dallas as part of its maneuvering for the western chunk of whatever transcontinental route the Postmaster General might open into Los Angeles. Very shortly thereafter Western Air's ambitions were squelched when Walter Brown forced the company to surrender the Dallas–Los Angeles route to American Airways and become a mere component in the merged Transcontinental and Western Air, holder of the central transcontinental franchise. But Jack Frye and Paul Richter, skilled airmen and nimble managers whose abilities had caught the eye of Harris Hanshue, went along in the merger package. And so, in October 1930 Jack Frye became vice president in charge of operations for TWA.

Jack Frye knew airplanes, and particularly after the Rockne crash dealt a heavy blow to its Fokkers, he knew that his new airline would have to have something better. In the fall of 1932 TWA abolished its overnight stops at Kansas City, and its Ford trimotors began flying straight through between New York and Los Angeles on a twenty-six-and-three quarter-hour cross-country schedule. This was all very well, but the new Boeings that Frye knew were in the works would make TWA's fleet of slow-flying Fords look pretty shoddy. As for the Fokkers, a Frye lieutenant said later,

The all-metal low wing Boeing 247 introduced in 1933 was the first
of the modern airliners.

Left to right: Donald Douglas, E. F. Burton, and Arthur E. Raymond in 1946.

"We all saw we had come to the end of the wooden-wing airplane." Then Frye made inquiries at Boeing, and was told that United Airline's order for sixty of the new planes had tied up all the factory's output for two years. It looked as if you didn't have a chance if you didn't belong to the Boeing-Rentschler combine.

Frye therefore solicited the views of everybody at TWA on what they would like to see in a plane that could compete with and outperform the Boeing 247. Not surprisingly, since the trimotor had become something of a fixture on the airways, the sales department said that three engines would give the traveling public greatest assurance. The radio people drew up a list of conduit systems and other devices that would shield airborne sets from static. Others asked for such improvements as retractable landing gear, tailwheels instead of tailskids, and comfortable, heated soundproofed cabins for the passengers. As for the engineers, mechanics, and pilots, especially those who flew over the mountains, they demanded more powerful engines. Colonel Lindbergh, chief of the line's technical committee, went so far as to specify that the new airliner ought to be able to climb to a safe altitude after failure of one engine on takeoff from any point on the TWA route. In practice, this meant taking off with one engine out from the Winslow, Arizona field (4,934 feet) and flying over the Continental Divide (9,500 feet).

On August 2, 1932, Frye addressed the following letter to aircraft manufacturers:

Transcontinental and Western Air is interested in purchasing ten or more trimotored transport planes. I am attaching our general performance specifications, covering this equipment, and would appreciate your advising whether your Company is interested in this manufacturing job. If so, how long would it take to turn out the first plane for service tests?

The attached specifications left considerable leeway — "all metal trimotored monoplane preferred but combination structure or biplane would be considered: main internal structure must be metal." Weight was set at 14,200 pounds, and cruising speed at sea level was 160 miles per hour, only slightly more than the Boeing. At the end was this sentence: "This plane, fully loaded, must make satisfactory takeoffs under good control at any TWA airport on any combination of two engines." As might have been expected General Aviation, a member of the General Motors combine that controlled TWA, agreed to go to work on such a trimotor.

Two other manufacturers replied to Frye's letter. One was Douglas Aircraft Company, a small outfit in Santa Monica, California, that had hitherto built only military and mail planes. The company was a member of none of the three big aviation combines. It had been started in 1921 by Donald Douglas, a soft-spoken Naval Academy dropout who had taken up aeronautical engineering at MIT and been chief engineer at Glenn Mar-

tin's plant in Cleveland. Deciding that California had more favorable flying conditions, an easier climate, and lower labor costs, Douglas had
pulled away from the icebound eastern winters and set up his own plane-
building shop in an old movie studio on Wilshire Boulevard in Santa
Monica. As soon as the Morrow board put through the nation's aircraft
procurement policies in 1925, Douglas landed his first substantial military
order — seventy-five observation planes. When Frye's letter arrived, the
company was still small. Douglas's office was a little cubbyhole off the engineering bullpen where he often turned to his own drafting board. Outside in the bullpen sat a dozen shirt-sleeved young designers, including
such future greats of the aviation industry as Art Raymond, Jack
Northrop, and Ed Burton. Another was Donald Hall, a spindly health nut
who insisted all windows be opened when everybody else wanted them
closed and who often stood up at his desk, opened a bottle of castor oil,
and let the vile-smelling stuff gurgle down into his opened mouth. Gifted
stress analyst though he was, others were glad to see him leave — he
drifted off to San Diego, where soon afterward, with Charles Lindbergh
standing impatiently by his drafting board, he turned the Ryan monoplane into the *Spirit of St. Louis.*

When Hall left, somebody else had to take over his arcane specialty.
Douglas called Art Raymond into his cubicle and asked him to perform
stress analysis for him on the pontoon bracing for a Navy seaplane the
company was about to build. For two days and nights Raymond struggled
without finding the answer. On the third morning it suddenly came to him
that the truss had a redundant member. Solution followed quickly. Minutes later Douglas called Raymond in to ask if he had the answer. Coolly,
Raymond handed the boss his calculations. Douglas grunted satisfaction.
Raymond was to become Douglas's right-hand man, father of the most
successful plane in airline history. But, he said later, "If Doug had called
earlier, I would not have been around the next day."

Don Douglas thought that life on the job should be more enjoyable in
California. Everybody wore golf knickers to work, and at lunch went out
in an abandoned reservoir and played volleyball. In 1925 Dutch Kindelberger arrived from Cleveland, where he had held Douglas's old job with
Glenn Martin. He became chief engineer, and the drafting room began to
fill with such talents as Ed Heinemann, Frank Collbohm, Jerry Vultee,
and Ken Ebel, all of whom were to win fame in the expanding California
aircraft industry. Dutch was from Wheeling, where his father had been a
steel worker. He was a first-rate engineer and manager, friendly and gre-
garious — Jack Frye was his friend, and so were the top airmen at the
other lines. Douglas himself was convivial, and liked parties. But he was a
Scot, and slow with the buck. Once Raymond, newly married and earning
thirty dollars a week, hit him up for a five-dollar raise. Douglas finally
agreed, then, growing expansive, confided, "Why, if you stay with the

company I am almost sure that in time I'll be able to raise you to forty dollars a week."

In 1932 came Frye's letter. Douglas was interested at once. He thought the letter called for a trimotor only because everybody thought you would have to have three engines to cross the Divide with one engine out. Raymond said later:

We all thought, Doug, Dutch and I, that we had a chance to meet the requirement with two engines. The simplicity of that solution appealed to us. It would clean up the whole front of the airplane. You would not have the propeller in front of the pilot; you would not have the aerodynamic drag of the engine up there. With no engines in front, you'd have less noise and vibration in the cabin, and no gas lines or fumes in the fuselage. You'd have a simpler, less costly design. It was just a better solution.

But the question of three as opposed to two engines was crucial. Frye's requirement for crossing the Divide meant a twin-engine plane had to have a ceiling, with one engine out, of ten thousand feet. It happened that Raymond, teaching part-time at Cal Tech, had been working with an advanced class on formulating more exact methods of estimating airplane performance. In fact, he had set one of his best students, Bailey Oswald, to writing a thesis on precisely this problem. Raymond himself sat down to apply the fruits of the study to the three-engine vs. two-engine question. In just five days, the Douglas group, convinced they had the answer, whipped together a proposal for a twin-engine passenger plane, which they called the DC-1, or Douglas Commercial Number One.

Thereupon Art Raymond set out for New York, along with Harry Wetzel, Douglas's financial man, to try to sell their idea against the trimotors being proposed by the combine's General Aviation and a second outsider, Sikorsky. On the train east, Raymond finished his preliminary calculations for the proposed plane, using a circular slide rule and the analytical method of performance prescribed in Oswald's Cal Tech thesis (freshly published in a National Advisory Committee on Aeronautics report).

In New York, while Wetzel talked to TWA officers and directors, Raymond made his pitch to Lindbergh. Flashing his circular slide rule and going through his computations, Raymond demonstrated that it was 90 percent certain that Douglas's twin-engine design would fly at ten thousand feet on one engine. ("That other 10 percent is keeping me awake nights," he told Douglas on the telephone.) The case was strong enough to convince Lindbergh that, while the combine started building its new trimotor, Douglas should get a chance to try its twin-engine answer to TWA's need, too. Whenever Lindbergh made a technical recommendation to the TWA board, the directors simply nodded yes. Others claimed later that Lindbergh wasn't really backing Douglas, only making sure that the combine's trimotor got some competition. Be that as it may, Raymond received word that Douglas had a contract. Climbing on a TWA Ford trimotor at Newark, he flew to Kansas City to work out the specifications in much greater detail with Jack Frye, Paul Richter, and the rest of the airline's operations staff.

The airplane that Art Raymond wanted to build for TWA, so different from the trimotors that had been flying the nation's airlines for the past six or seven years, incorporated a wholly new form of construction. This was the monocoque form, in which all the loads are taken by the outer shell — the claw of the lobster is a good example in nature. The word "monocoque," in French, means "single shell." Before and during the First World War, European designers had experimented with this type of construction, lining the fuselage walls with glued sheets of plywood. In practice, they found that the shell needed inner support, especially in the wings. In 1919 Dr. Adolf Rohrbach devised a kind of compromise. He found that an outside skin of smooth metal could bear a high proportion of the load after he had designed box-shaped metal spars for interior reinforcement. The result was semimonocoque construction — almost the only truly monocoque plane ever built for serviceable use was the British high-altitude Mosquito bomber of the Second World War, whose skin was of extremely light curved wood panels that carried all the weight without a single support within.

In 1926 Rohrbach, on a visit to Los Angeles, gave a lecture on his way of putting all possible load on a smooth, minimally supported metal skin. To this semimonocoque method he gave the name "stressed-skin" construction. His ideas fired the imagination of Jack Northrop, the gifted young designer who had sat beside Raymond among the original dozen denizens of Donald Douglas's drafting room. Northrop, who by then had left Douglas, took his ideas to Lockheed. The stressed-skin fuselage of the Vega he designed for Lockheed made room for 35 percent more cabin space than in the old-style framework structure and reduced weight and aerodynamic drag as well. Wiley Post flew a Vega around the world in eight days in 1931. But the Vega was made of wood. Forming his own company, Northrop built for the Air Corps America's first metal plane of semimonocoque construction. In the Northrop Alpha the shiny metal skin of the fuselage was backed by quarter-inch duralumin "stiffeners" that ran around the circumference of the interior like so many inside barrel hoops. But the wing was the most arresting feature of Northrop's new plane. Such interior hoops would have been inadequate to withstand the pressure the wing had to bear. So to back up the smooth metal surfaces of the handsomely tapered wing Northrop created for the Alpha, he installed a cleverly compact and lightweight lattice inside of crisscrossing aluminum spars.

In flight, the wing's stressed skin carried the weight of the whole plane — its undersurface in tension, the upper surface under compression. The task performed was so visible that the pilot could look out from the cockpit of his Northrop Alpha as he flew and see the skin wrinkle slightly between the ribs because of the compression loads. But the results, in swift and economical flight, were eye-opening. After the Alpha, no one ever thought to build anything but a monoplane, and everybody turned to metal construction. Northrop's company became a Douglas subsidiary, and Northrop's Alphas were bought by Frye to carry the mail for TWA.

Inspired by Northrop's innovations, others turned to stressed-skin construction. When Claire Egtvedt in Seattle had remarked that to design a really efficient airplane body you ought to make it round, he was reflecting the Northrop ideas. Stressed-skin was precisely the technique embodied in the Boeing 247. Its sleek and smooth-skinned surface fully entitled the plane to be called the first modern airliner. But the Boeing engineers, when they came to construct the 247's wing, had not yet broken away from established practices. Even Northrop's wings had spanned no more than forty feet, and for a transport wing nearly twice as wide, the men in Seattle preferred to stick with bridge-type support. So in Boeing's plane a stout spar of aluminum extended from inside one wingtip of the airplane all of the way across to the other, right through the fuselage. The same feature, incidentally, cropped up in the trimotor that General Aviation was designing for TWA simultaneously with Douglas's DC-1. When Ernie Breech, the General Motors troubleshooter who later headed both Ford and TWA, first joined TWA's board in January 1933, he cast about for ways to staunch the heavy outflow of funds. His eye lit on this costly item. He said later:

The General Aviation trimotor was designed to be supported by a large aluminum spar — hogged out of a solid bar of aluminum, a very expensive proposition — extending from wingtip to wingtip through the center section.

Though General Aviation had already spent $800,000, Breech axed the order for its trimotor. The runway was cleared for the DC-1.

To Breech and his fellow directors, the Douglas plane appeared to be much lighter and more economical, even though the DC-1 that emerged weighed more and cost more than the original contract prescribed. The body was essentially monocoque but the wings were not. Raymond's proposal had stated, "The wings shall be of cellular, multi-web construction similar to that used in the Northrop Alpha." And so they were — built up like an egg crate in several sections. The wing's center section, in contrast with the Boeing 247's transverse spar, passed through the body beneath the flooring. Outer wings were fastened to the center section by the same heroically simple joint that Jack Northrop had devised for his smaller plane. The joint was flanged, held together by a large number of bolts, and a fairing of aluminum fitted smoothly over it and merged with the skin. One of the advantages of this type of construction, in contrast with the impenetrable innards of the Fokker, was that when anything was dented or broken it was easy to take apart and repair. And the four or five interior spars allowed room for four gas tanks close by the engines inside the wing's center section.

The engines that were to go on the DC-1 were Wright Cyclones. While Douglas was building the plane, 87-octane aviation gasoline, introduced in the U. S. military in 1930, became commercially available, and Wright succeeded in increasing the power of its engine to 710 horsepower. By an equally fortunate circumstance, the Hamilton-Standard Company came up with a two-speed propeller at just this moment. It was not yet the variable-pitch propeller that enabled the pilot to maintain his plane at constant speed by slight adjustments in the angle at which the propeller bit the air. That came a few years later. But the two-speed propeller that Frank Caldwell devised gave the pilot the means to set the propeller at the most favorable angle for the two most critical conditions of flight — takeoff and cruising. On takeoff he could set the blades to grip the air most firmly; later, after climbing to desired height, the blades could be moved nearer to the perpendicular so that the pilot could throttle back and use far less fuel for smooth and level flight.

It was a youthful and high-spirited group that brought forth this break-away airliner. The Douglases favored boating, the Raymonds camping. Elfin Eddie Allen, the test pilot who was also an engineer, chubby Bailey Oswald, and lanky Frank Collbohm were a trio who often swept through the engineering office, their shirttails streaming behind them in virtual airfoils, so eager were they to get to the next flight test, or look at the flight test data. "Such excitement helped your morale," said Maynard Pennell, who was later to design Boeing's first jetliner. "In the same rooming house with Eddie Allen lived Al Reed and Gene Root, fresh out of Cal Tech. They rode bicycles to work and arrived in a sweat. We all had some responsibility for structure — I did stress analysis on the engine nacelle."

That was the housing for the engines. Like Boeing's 247, the Douglas airliner was the beneficiary of a recent innovation. The enclosed cowl, developed by the National Advisory Committee on Aeronautics, evolved into an aerodynamic marvel on the DC-1 wing. The same nacelles enfolded the plane's landing gear after takeoff. The DC-1 also led the way with the first automatic pilot and the first efficient wing flaps ever installed on an airliner, and the wings wound up with a sweptback look that had not been present in the original design. This was because as the weight grew, the center of balance moved aft, and Raymond found it easier to make the outer wing panels sweep back than to move the entire wing.

The day came when the prototype was wheeled out to fly. It was in July

The Douglas DC-2, the first "stretched" version of the DC-1 monocoque design.

of 1933 at grassy little Clover Field, now smothered under 215 acres of concrete and called Santa Monica Airport. Carl Cover was the pilot, with Fred Herman of the design staff as copilot, and the first flight almost ended in disaster. When Cover tried to put the plane in a climb, the engines cut out. Somehow he managed to nurse the plane around and back onto the field for a safe landing. Mechanics located the trouble: the carburetors had been installed backward! The next day Eddie Allen, with TWA's check pilot Tommy Tomlinson alongside him, took the plane up. They put it through all kinds of tests — with two engines, with one engine. When they landed Tommy shot off a telegram to TWA President Richard Robbins in New York: "This is it."

But the contract specified that the plane would have to be flown over the Continental Divide on one engine before the airline would accept it. Came September, and with Eddie Allen at the controls and Frank Collbohm as copilot, the DC-1 flew to Winslow, Arizona, for the big test. Sandbags were loaded to bring the plane to the required weight. TWA posted observers every hundred feet along the runway. Tommy Tomlinson climbed up to ride standby in the aisle just behind Allen and Collbohm. Then the chocks were pulled; the plane raced across the grass.

Allen pulled back, the plane rose, and Collbohm pulled hard to retract the landing gear. At this moment Tomlinson reached past Allen and cut not the throttle but the switch to the left engine. This was not in the game plan; the Douglas men were taken by surprise. The plane settled; according to Tomlinson, it came within six inches of the ground. The TWA watchers gasped. Then the plane rose steadily and, on one engine, the first Douglas commercial airliner flew over the 6,500-foot hump east of Gallup and landed safely at Albuquerque.

There was only one DC-1, and it was only briefly put into service by TWA. But when the Roosevelt administration canceled all airmail contracts in February 1934, Jack Frye used the new plane to make a defiant gesture. On the night of February 13, with Eddie Rickenbacker (vice president of Eastern Air Transport) in the copilot's seat, he took off from Los Angeles and flew it across the country to Newark in the record time of thirteen hours and four minutes.

The DC-1 had a deficiency. There was room for only twelve passengers, and TWA had asked space for "at least" that number. Before the plane went into production, Douglas found a way to put in two more. As he was soon to discover, one of the most profitable features of his designs was that the planes could be "stretched." In a fashion never quite equaled by Boeing, Consolidated, or even Lockheed, Douglas planes grew and grew. Now, in the first modest bit of beanstalking, the men of Santa Monica lengthened the DC-1's fuselage by eighteen inches. By carving a further eighteen inches out of the front baggage compartment, they were able to create three feet more cabin space. Thus they made room for another seat on each side of the aisle — a total passenger capacity of fourteen. With a few other changes, this became the production model, designated Douglas Commercial Number Two.

Douglas's partner in the next advance was American Airlines, a makeshift outfit that had even further to go to catch up with the van of air progress than TWA. However, by this time the most admired figure ever to run an American airline had taken office as president. The collaboration of Cyrus Rowlett Smith with Donald Douglas proved to be one of the most fruitful events in the history of commercial aviation. The result was the DC-3, which became all the world's airliner. But in the first instance the deal with Douglas played a big part in the transformation of C. R. Smith's airline from a bunch of routes as random as a grasshopper's jumps into the nation's foremost air transport system. To take the measure of this transformation, and of the man who led it, you have to go back to their beginnings.

American Airlines, even more than TWA and United, was a child of the promotional impulses of Wall Street stock speculators. A group of such men formed a holding company, the Aviation Corporation. One of the leaders was Averell Harriman, then a young man remote from politics with no thought other than to make as great a fortune in aviation as his father had in railroads. The second was Robert Lehman, nephew of New York's lieutenant governor and driving force in the potent investment house of Lehman Brothers. Raising $38 million by selling stock, they opened an office and bought aviation properties right and left. They bought a plane-making firm, an engine plant, some airports. And they bought airlines — in all, eighty properties. They did not know a great deal about aviation. They hired some Army Air Corps people who did not seem to know much more. Their holdings were not consolidated in any way; some of the lines they bought did not even connect with others.

The first president of their combine was Graham Grosvenor, who gained his experience in air transport as vice president of the Otis Elevator Company. Grosvenor was succeeded by Charles Coburn, who came to the job from a firm of New York management engineers. Coburn tried to make sense of the melange of holdings. He was responsible by 1930 for pulling the various airlines together into a confederation called American Airways. And it was Coburn who was in office when the combine played footsie with Postmaster General Brown and made the series of bids, deals, and corridor arrangements whereby American gained the third, or southern, transcontinental route. But even after these favors, American Airways lost money month after month, and Coburn was ousted.

At this point Aviation Corporation, the parent holding company, was rocked by a proxy fight. The Wall Street plunger E. L. Cord had turned his acquisitive attention to the disorganized aviation business. Advertising for shares in the newspapers while selling Avco short in the market, he wrested control of the combine from Harriman and Lehman. Never in his life had Harriman absorbed such a beating; he got right out of Wall Street and took up a new career in the Roosevelt administration. Lehman retired to his less exposed position on the Pan Am board. After the victory, Cord's first move was to make Lester Seymour, a onetime Army engineer hired from United Airlines, president of American Airways. Seymour turned out to be a ridiculous choice — he didn't like to fly in airplanes. "He knew more people on the Pennsylvania Railroad than he did on his airline," growled one of his managers.

This was where C. R. Smith came in. Smith was a Texan, tall, taut, and leathery, who had been swept into the system along with one of the small airlines the Wall Streeters had picked up along their way. Smith had the clear blue eyes of an airline pilot. But he was no pilot; he was an accountant. His vision for numbers was better than twenty-twenty. When he looked into the wild blue yonder, it was dollar signs he saw.

"Cy" Smith, as Texans called him, was born in the little town of Minerva. One of six children in a family abandoned by their father, he worked his way through the university at Austin. His first employer, Peat, Mar-

INTERNATIONAL NEWS PHOTO, I[

Cyrus R. Smith in 1939

W. A. "Pat" Patterson

wick and Mitchell, assigned him to a Texas utility whose owner bought a small airline. The company needed a treasurer. Both Cy Smith and his boss thought this was the important job at Texas Air Transport, which had mail contracts to fly between Dallas, Fort Worth, San Antonio, Houston, and Brownsville. Smith said later, "He asked me to take over the operation, and I told him I'd take the job while he looked for a replacement." Smith took the job and fell in love with the business and the people in it. He traded bookkeeping tips for flying lessons.

From the first Cy Smith hustled to get passengers. All the Dallas hotel porters got a cut of the fares they could steer to his planes. Oilmen were the best customers: whenever there was a new oil strike — and in the 1920s there was one in Texas almost every month — Cy Smith's planes stopped to discharge and receive drillers and roustabouts in the nearest pasture. In the western part of the state his planes had to fly over the Guadalupe Mountains, the highest mountains in Texas. When the peaks clouded over, Cy Smith flew his passengers to the foothills, then bussed them through to the far side, where another plane was usually able to fly them the rest of the way into El Paso.

Smith's Dallas desk was in a big hangar amid tires and plane parts. He ran his operation with gusto, answering his own phone, typing his own letters. He took his bourbon neat and excelled in kicking out light bulbs on overhead cords; at the hangar cafeteria he heaped collards on guests' plates with the shout; "Eat it up, it's good, goddam it." His marriage fell apart, but the men in the operations huts swore by him. Obviously, Cy Smith also knew how to handle big stockholders like Amon Carter, the sombreroed Fort Worth publisher who cut a big figure in Austin and Washington. Growling at "those damn New York bankers," Smith drawled, "I haven't done business with any damn Yankee unless I had to."

When Texas Air Transport merged in Southern Air Transport, Smith ran the new airline. When Southern Air Transport became part of American Airways, C. R. moved to St. Louis and, as vice president for operations, helped the Yankees try to whip the unwieldy system into shape. He saw a lot of Ralph Damon, president of the Curtiss factory that built Condor airliners in St. Louis. The trouble with American's new transcontinental route, gruffed Smith, was that it was so much longer than either TWA's or United's. Flying American to California, a passenger saw both Canada (at Buffalo) and Mexico (at El Paso) on his way. You had to sell flying anyway, said Smith, but to sell flying cross-country on American you had to have something extra. So he worked out a scheme with Damon to convert Condors into sleeper planes for American. Then, for its longer ride, American could sell *Pullman* comfort.

But Smith fought Cord, who was cutting staff and threatening to hire pilots at bus drivers' pay. He also tangled with Cord's new president, Seymour, who sent him back to his old Dallas desk. Smith grew more and more disgusted with the way things were going. Among other things, Seymour decreed that pilots must not fly more than an hour above overcast — after sixty minutes, they must turn round and go back. One pilot asked Smith, "Suppose I've been in the air 45 minutes, and I can see clear skies 30 miles ahead — what then?" Smith pecked out a memo to Seymour: "Our people think that rule is foolish and dangerous."

Other divisions had other complaints. Morale, never high in an outfit that seemed to have been put together with no thought of coherence, sagged. Depression deepened; deficits got worse. At this point Cord called Smith to his Waldorf suite in New York. As Smith told it later, "I said to him, 'I think I've called you a sonofabitch on just about every street corner. I don't expect much; you better count me out.' Cord answered, 'Well, if you work as hard with me as you've worked against me, everything will be all right.'"

In September Cord made Smith president of the renamed American Airlines. "He was a superb businessman — my close friend," Smith said later. However, when Cord met trouble with the new Securities and Exchange Commission over his stock dealings and had to bow out, what Smith inherited was a congeries of airlines. Its scattered units harbored just about every kind of plane there was — Fords, Lockheeds, Stinsons, Bellancas, Vultees, and the lumbering Condors. "It was years before we got any order in the fleet," he said later. On the company's circuitous transcontinental route, "we had to fly night and day to keep up." The Condor sleepers went into service the same month as Frye's speedy DC-2s. Smith made a brave show of launching the first sleeper plane. The Condor was quiet and roomy (fifteen berths). But it was slow and notoriously underpowered for getting over mountains or weather. "One of the world's worst airplanes," Smith was to say later. He could hardly wait to get something better. When the DC-2 became available, he ordered some right away. But this too was awkward: passengers on the transcontinental flights had to change at Dallas from DC-2s to the big Condor sleepers, and the DC-2s carried no more than fourteen seats.

Smith began assembling a strong staff. His chief engineer, Bill Littlewood, was one of the best. Littlewood said to Smith, "We could easily make a better plane than the DC-2, and make a little money with it." He had made his calculations, and Smith was just the man to translate them into ledger figures. The proposal was irresistible: a "wide body" plane. It was what airline operators were demanding in the 1960s when they called for jumbo jets, and it was what Smith and Littlewood sought in 1935, and for the same reason. With a wider body on the DC-2, the plane could be turned into a sleeper. But Smith's dollar-sign vision saw further. With a wider fuselage on the Douglas, another row of seats could be added, and the airliner's daytime capacity could be increased from fourteen to twenty-one seats.

Smith said later, "I took Littlewood and Damon with me, and we went out to Santa Monica to see Donald Douglas. I said, 'We have a very firm idea that you could make a much better airplane than we have now. You could widen the fuselage and make more seats.'"

Douglas was not enthusiastic. He said, "We can't even keep up with orders for the DC-2s." Smith said he would buy twenty of the bigger planes sight unseen. Douglas finally said yes, he would order a design study. Smith said, "There's no use, sweetheart, drawing up any damn contract." And Douglas remained dubious. Afterward he said to Art Raymond, "So they want to buy twenty? We'll be lucky if we break even. Who the hell is going to buy a sleeper plane? Night flying is about as popular as silent movies."

As it turned out, the Douglas Transport Sleeper (DST) occupied the same sort of fleeting place in the evolution of Douglas airliners as the original DC-1. The DST was duly built and put into service by American Airlines, but long before that everybody's attention, both at the Douglas plant and in the airline offices around the country, had turned to the version of the same plane that Douglas was building with twenty-one seats. This was the DC-3, the "plane that changed the world."

The first step in creating a plane is to make the design, the second is to build a miniature scale model, the third is to build a life-sized wooden mock-up of the plane you are going to build, and the final stage is to manufacture the machine. At the start of the work for American, Dutch Kindelberger left to run his own show as head of North American Aircraft; Art Raymond became Douglas's chief engineer. The design team for the new plane included Raymond, Ed Burton, and Lee Atwood, who shortly left to join Kindelberger. Raymond said later, "By the time the plane flew we literally knew every nut, bolt, and rivet — it was the greatest fun I ever had."

Douglas said:

The original plan called for a bigger fuselage but essentially the same old DC-1 and DC-2 wing. Of course we added five feet more to the wing on each side of the fuselage but in plan form and in airfoil it was the same wing. We soon found out however that just putting on more wing didn't give us the lift and stability we needed. It was really a case of redesigning the wing.

First to discover this were Bailey Oswald and Mage Klein, the Cal Tech professor who acted as a consultant on aerodynamics. When the two men put the miniature model of the new plane in the wind tunnel at Cal Tech, they were compelled to call for modifications right away. After repeated changes, the model still showed such drastic instability that some thought Douglas would have to call the whole thing off. On the eleventh model tested, Klein and Oswald altered the wing again. This time they narrowed the airfoil, thereby changing the center of balance of the plane. That moved the fulcrum on which the plane balanced in flight. Looking over the results, the two men thought the shift was just enough to make everything fall into equipoise. Bailey Oswald told Douglas, to everybody's relief, "I think we've hit on the answer."

They had. As Raymond said mildly, "Take a second try at something and you can do a better job." The new airplane turned out to be a mighty stable airplane. And the beauty of the final design was that structural loads were taken in a distributed rather than concentrated fashion, which made the DC-3 a durable airplane, one of the most durable of all time.

Like the DC-2, the new airplane was a truly low-wing monoplane. And inside the wings were the same neat Northrop egg crate spars — except the later model had benefited by improvements in detail. Between the spars, the top skin was now reinforced by corrugated sheeting whose wrinkles, running spanwise, enormously reinforced the wing's capacity to withstand compression. Better yet, the aluminum industry had now developed an aluminum alloy, known as 2024-T3, that was 25 percent stronger than the German-invented duralumin on which plane builders had previously relied. This aluminum of superior strength was introduced for the first time in the new plane's stressed-skin construction. It also helped stiffen the wing's outer sections, in the shape of artfully placed stringers that gave interior reinforcement. All these parts were bolted together in the Northrop manner, which made the plane easy to inspect and repair. Extreme simplicity of structural design was a big reason for the plane's success.

In other important respects, technological improvements came along just in time to be added to the new Douglas transport. The two engines were still Wright Cyclones, but they were more powerful — 1,000 horsepower instead of 710. Frank Caldwell, the propeller wizard at Hamilton-Standard, had finally perfected a gear in the hubcap that automatically changed the propeller's pitch as needed for takeoff, climb, and economical cruising. The landing gear of the DC-2 had been one of its worst features. Pilots had denounced it as "stiff-legged"; one copilot said the plane landed so hard that once his head went down between his knees. This primitive rig now gave way to a new, hydraulically operated system. Cushioned shock-absorbers of an improved type took up the jolt of bouncy landings, and the fancy new mechanism freed the copilot from his laborious task of hand-cranking the wheels up after takeoff and lowering them again for landings.

The cabin of the new Douglas, requiring the design of berths for sleeping and seats for daytime travel, was such a special problem that, as Art Raymond said later,

I let Bill Littlewood have almost a free hand with it. My main concern was with the changes that transformed the plane from the DC-2 to the new configuration. The DC-3 was the product of cooperation between the builder and the user, one reason why it was so successful.

The DC-3 flew fast enough to be practical and commercial.
Shown is the sleeper version.

Littlewood and his assistant Dan Beard applied themselves to the interior of the sleeper mock-up. Into the cabin proper, which was nineteen feet six inches long and seven feet eight inches wide, they fitted twelve berths. Upper berths were designed to fit into the ceiling when not in use; lowers folded away into the walls and seats by day in Pullman fashion. By adroit use of every other inch inside the body, they made room for two more berths. These were forward of the main cabin on the right in what American later grandly called the "Sky Room." Stewardesses later dubbed this the "honeymoon hut." There were separate dressing rooms for men and women and separate lavatories, hitherto unknown aloft, at the rear. All told, fourteen passengers could travel in the Douglas Sleeper Transport, with a pilot, copilot, and stewardess for their crew.

From the first, Raymond and Littlewood had settled on the number of berths and seats — fourteen passengers by night, twenty-one by day. But as Littlewood worked, he gave more and more attention to enhancing what he called the "human efficiency" of the passenger accommodations. To offset the racket of the plane's high-powered engines, he brought in an "acoustician" named Stephen Zand. Zand lined the walls with a sound-absorbing plastic material. He inserted a noise-deadening bulkhead between the engines up front and the passengers in the rear. He set every seat in rubber.

In the flying business everybody talked about the weather but nobody had yet done much about it for the air traveler. Moreover, the plane Douglas was creating for Littlewood's passengers was designed to fly through the weather rather than over it. Almost all who flew felt some uneasiness about taking to the air, and Littlewood was sure that this vague uneasiness had something to do with the frequent onset of airsickness. Having made studies of color, he found that certain shades of green worked on some passengers in a most unpleasant way. Furthermore, too much red induced excitement. Even patterns and designs in the plane's interior might aggravate airsickness, it seemed. Accordingly, Littlewood decreed that the decor of the new planes must be subdued, blended. The carpets and lower walls of the cabins were finished in darker tones — to give a feeling of strength and security underfoot. The upper walls and ceiling were tinted light blue and gray to lessen the feeling of confinement.

Long before this point, Douglas knew he had a winner. When the new plane was still in the mock-up stage, he began to see that the wider fuselage coupled with the more powerful engines now available would produce a plane that all the airlines would want. Well before it flew, he informed his stockholders that the new plane would carry a payload a third greater than that of any airliner in existence.

Originally the uncertainty had been so great that Smith had not even signed a contract. The two men had simply shaken hands on Smith's affirmation that he would buy twenty planes and pay for them on delivery. Since American Airlines was no longer the ward of a Wall Street combine, Smith had to go elsewhere for his financing. In the midst of the Great Depression, hard-pressed businessmen sometimes turned to Washington for emergency aid. Smith said later:

I got the money from the Reconstruction Finance Corporation. Jesse Jones (the chairman) was from Texas, and a good friend of mine. I went to see him and I said, "Mr. Jesse, this outfit of yours has been organized by the government to lend to those who can't borrow money anywhere else." I said I wanted money to buy 20 planes, and I got it.

The RFC advanced Smith $1,235,000 in return for 5 percent equipment trust certificates, and until the loan was retired four years later, American's passengers could find a small metal plate on the cabin wall of the new planes: Property of the Reconstruction Finance Corporation.

As soon as the prototype flew, American Airlines changed its order to eight sleeper planes and twelve of the daytime DC-3 version. Smith said later, "We paid for the planes as we got them." C. R. himself handed Donald Douglas the $100,000 check for the first DST, in which he then flew nonstop with Littlewood and Beard to Chicago on June 7, 1936.

At twelve noon on June 25, American's new airliner began service as a day plane on a nonstop flight from Chicago to New York. Captain Walter Braznell was in command. W. A. Miller flew as copilot. Flying time was three hours, fifty-eight minutes. Three months later, with suitable fanfare, American introduced the first DC sleeper on transcontinental flights. Shirley Temple, age eight, got the first ticket.

The "giant" new airliner was half again as large as the DC-2, and a big advance over its predecessors in just about every way. To the pilots, it seemed at first too big. Bert Lott of United called it the "flying whore" — because it had no visible means of support — and others noted with alarm that in flight the wings flapped like a bird's. The wingtips of a DC-3 really did move up and down (as much as five degrees in rough air) as innumerable passengers later saw with their own eyes. To all this, the reply was made that the wings weren't meant to be rigid, but to roll with the meteorological punches. And because the load was distributed so evenly over the bearing surfaces, the DC-3's wings could take some thunderous blows. Very shortly after they went into operation, American's veteran pilot Joe Hammer was flying one from Chicago to Detroit. Over Michigan he ran into a storm so violent that passenger seats were ripped from the floor. Safety belts snapped like rubber bands. Yet Hammer pulled out, leveled off, and flew to Detroit. Landing, he summoned mechanics who swarmed over the plane. Nary a loosened rivet did they find.

To the airline operators the DC-3 was, as C. R. Smith said, "the perfect airplane for its day." Not only did it have the autopilot and higher horsepower, it also had wing flaps, better brakes, and several new cockpit and radio aids. Simple in design, it was simple to maintain. Except for the control surfaces — rudder, elevators, ailerons, and flaps — the plane was all metal, hence easy to clean. The wing panels came off readily for inspection and repair; control cables were easy to get at. The nacelles that held the engines came with fittings into which fuel lines and electrical connections could be plugged, just like wall sockets. These conveniences, never before seen in an airliner, made it possible to change engines in two hours.

All these improvements meant that costs dropped. Costs fell even further when all the other airlines followed suit and ordered the DC-3. Standardizing on one passenger plane, the companies were obliged to stock fewer spare parts, and their mechanics needed to know only how to fix DC-3s. Douglas had said that to fly the plane nonstop from Chicago to New York cost roughly eight hundred dollars. This was a bit more than it cost to operate the DC-2 over the same distance. But the seven additional seats on the DC-3 meant that the cost per seat was 30 percent lower on the new plane. In the face of such figures, no competition could keep up with the DC-3.

And of course the DC-3s were faster. United pilot Weldon Rhoades said:

Oh, United tried to advertise that its Boeings were just as fast as the DC-3s but you knew it wasn't true when you saw them pass you up in the air. We said they were pushing their engines. It wasn't true. There was an honest 10 to 15 mph edge in speed — and that's 10 percent: they could publish a schedule that was significantly faster. We could try that, but we would be late — and there's nothing that makes passengers madder than arriving late all the time. So United got rid of its 247s and bought DC-3s too.

Being faster, the DC-3s cut flying time across the country to fifteen hours flying east and seventeen hours flying west. That meant further savings, because in commercial air transport speed pays. The airlines were just starting to find out the basic truth that operating costs are mainly related to flying time whereas operating revenues are mainly related to miles flown. If the experience of the 1920s and 1930s taught them anything, it was that their planes were too slow. They were too vulnerable to adverse winds, which compelled them to carry large fuel reserves and small payloads, and which frustrated efforts to schedule planes, crews, and overhauls with any sort of punctuality or regularity. A minimum capability of 175 miles per hour was needed if airlines were to put themselves in a position where they could utilize their men and machines. Thus it was no accident that the DC-3 was the first practical and commercial airliner. It was the first to fly fast enough.

One big result of the economy of speed was the astonishing reduction in the number of planes needed to carry the nation's air traffic. Before the DC-3 came along, domestic airlines had 460 planes in passenger service. Five years later they were operating only 358 passenger planes — 80 percent of them DC-3s — although the number of people carried had meantime quadrupled.

No airline gained more by this advance than the airline that had sponsored the DC-3. The year before American got its DC-3s, it lost $758,000 and C. R. Smith told a congressional committee, "Either raise the mail pay rate or the line will go out of business." Two years after the first twenty Douglases began flying for American, the airline made its first profit. The following year American carried more passengers than United, and twice as many as TWA. Its profit, although mail pay had been cut in half, was $213,261 in 1938 and $1.4 million in 1939. Truly, as Smith said, the DC-3 was "the first airplane that could make money just by hauling passengers."

14

Love and Fear of Flying

The majority of Americans still remained convinced in the 1930s that any man or woman who lifted an airplane to the clouds had bade farewell to common sense, caution, and the right to call anyone else audacious. Pilots, then, were a race apart; obviously they held this view themselves.

Flyers were not like other people. Qualities demanded by the military, which trained most of them, set them apart. Physically, they had to withstand the rigors of altitudes and the stress of G-forces better than other mortals. Physiologically, they had to be demonstrably quicker in reaction time, surer in depth perception, better able to tolerate rotary and confused motion.

When airmen still ran most airlines, they tended to share the view that every young pilot, if he was any good, set out to be an acrobatic pilot. For example, when Ham Lee, United's senior captain, joined the Air Mail Service back in 1919, he held the record for having looped-the-loop 105 times (he ran out of gas on the 106th try and had to land). And Jimmy Doolittle, the airman's airman, had flown upside down under more bridges than anybody; after his first instruction flight, it is said, he was confined for thirty days because he climbed out of his seat and sat on the axle as the plane landed.

Pilots were men who dared. For a long time the Navy asked flying cadets whether they favored Mom or Dad, believing those who hunted and fished with Dad were the boys to bet on. A standard Air Corps question — "Which handle of a motorcycle has the grip for the throttle?" — was framed just to find out who rode motorcycles, the youngsters the service wanted. A distinction of sorts was made as cadets left Kelly Field: the "quick" ones went off to be fighter pilots, the "slow thinkers" would be steered to observation and bomber planes. Fighter pilots ran the Air Force in World War II. And the fighting ace of World War I, Captain Eddie Rickenbacker, bossed Eastern Airlines.

For all their physical endowments, it was the emotional makeup of the men so singled out that was unique. In his famous essay on Leonardo,

Freud wrote that "the wish to be able to fly signifies in the dream nothing but the longing for the ability of sexual accomplishment." Even in the 1970s, when airline pilots were mainly monitors directing extremely complicated machinery, Dr. Ludwig Lederer, the old American Airlines aeromedic, liked to say, "The pilot's life is founded on three things — sex, seniority, and salary, in that order." Dr. Constantino of Pan Am said, "Their sex drive is strong. They're very physical, strong-willed, macho types." When the airlines expanded in the 1930s, Captain Dean Smith said later, "Of course sex reared its lovely head. After all, you had these young crews and young girls." Once Northwest had to fire a skipper when a hotel night clerk reported him for trying to break into a stewardess's room at the Billings layover.

Under the duress of World War II combat, flight surgeons found a fairly consistent pattern in the personality structures of successful fliers. At the profoundest level, the act of flying yields a distinct gratification, particularly to the pilot. And it appeared that this libidinal devotion served the airman as a powerful shield against the thought of failure or death. The characteristic that certainly set the early airmen apart was their daily closeness — a courted closeness — to death. They themselves, when they talked together, sometimes spoke as if some inexpressible bond existed among them, binding them in a sort of mystical unity. André Malraux, an airman as well as a novelist, said, "Aviation unites them as childbirth makes all women one."

If only the self-selected few took up flying in the earlier days, only the fittest of them survived. Otis Bryan, the tall, unflappable Air Corps graduate who hired hundreds in his years as chief pilot for TWA, said that as the 1930s began his airline was "losing 5 percent of its flying personnel a year." Danger, even as the airlines painfully lifted themselves out of the hit-or-miss style of flying and up to the rigors of scheduled performance, was never far away. The number of fatal accidents was nothing compared with all the near misses. Everybody had his close calls. One night United's Larry Nelson bailed out of his Boeing east of Cleveland. The plane was equipped with a radio. But come-apart cords had not arrived yet, and the wires to his earphones almost jerked Nelson's head off. He climbed back in, detached the jack, then jumped clear, to safety.

Bryan himself had a narrow escape while routinely flying the night mail in the winter of 1931. From Lindbergh's time, mail pilots carried parachute flares at their feet that they could release through the bottom of the fuselage to help see to land in an emergency. Somebody at TWA, thinking to make this easier, had removed the wire the pilot pulled first and put a little gunpowder inside that would blow the flare out the tube for him. But nobody thought what might happen if the gunpowder got wet, as it did on the clammy night that Bryan, running into fog, tried to drop a flare from four hundred feet before landing at the New Florence, Missouri, emergency field. When Bryan pulled the flare, the wet powder blew off and ignited the flare while it was still in the cockpit. His flying suit on fire, Bryan took his plane down fast, jumped out, and rolled in the snow till the fire was out. "That's how we learned," Bryan said later.

As the Douglas airliners came along and flying passengers in comfort and safety became an overriding consideration, the companies found themselves signing on a new breed of pilots. In the left-hand seats of the sleek new planes were the veteran captains, many of whom had grown up with aviation and had picked up their flying skills around county fairs. Some of them had learned to fly in planes they had built by themselves back in the days when there were few airplane factories. These were older men, fatherly, grizzled, stern — "Greyhound bus driver types," was how one early stewardess described them. The newcomers were smoother, slicker. They had been to college. Almost without exception they had got their flight training in the Air Corps or the Navy.

Inevitably there were strains when these young whippersnappers began reporting aboard. In the late 1930s when TWA had a hundred captains and a hundred copilots, one of these upstarts who took his place at the captain's right was Floyd Hall, later president of Eastern. Hall said later:

I was a fresh product of Air Corps training. I thought nobody in the world was better than I. At Kelly Field they took your blood, and put tiger milk in there instead. I was sure that I was the best flier in the Air Corps.

Well, I learned to fly *after* I got to TWA. I flew with Captain Fred Richardson for one whole year, and he taught me all I needed to know. He sure took the steam out of me fast. He worked me over. On the DC-2 the copilot had to pump up the landing gear manually after takeoff. He said, "I want that gear up before we're across that fence." He was the sort of man who would get up to go back to the head or to say hello to the folks in the cabin — and he would put his hat on his head, look at you and point to the trim-tab control and say, "That's what you keep it level with." He just took you to pieces.

The old pilots never accommodated. We copilots would tell 'em, "We're the next generation, there's twice as many of us." As copilot I learned I always carried the hostess's bag for her. The captain saw to that — and you didn't smart-aleck him. It was "Yes sir," "No sir," when you spoke to him. The copilot ranked slightly above the colonel's dog.

Ham Smith at American Airlines started out as the number two boy on the stately old Curtiss Condor — the Mighty Monarch of the Airways, as the ads called it. A biplane, the Condor was also the world's first sleeper transport, but in the East, where Smith flew, it was always used as a day plane with fourteen seats. Like all airplanes, the Condor had a personality. This the old pilots seemed to understand.

The old pilots who had flown the airmail knew how to live with it. They taught me the tricks. They would fly through blizzards. One of the tricks was to take coffee in a clear glass cup and place it on the glare shield just above the instrument panel. As long as the coffee in the cup stayed level, no matter how bad the weather was, the pilot was OK. But if the coffee tilted in the glass, it was not a well-coordinated turn.

Of course they took some liberties when they weren't flying passengers. On one mail trip Cy Bittner carried along a full bottle of bourbon and put whisky in the glass. Every time he got too cold, he took a nip. The cup was soon empty, and then empty again, and empty again, until he had no instrument.

He was flying low, contact, and saw a New York Central train bearing down on him. Its headlight was shining level right into his cockpit. Cy put his landing lights on, and maybe pulled up a little.

The engineer, with the blinding light suddenly in his eyes, threw on the train's emergency brakes. The Twentieth Century slammed to a halt. The dazed engineer said, "Those lights almost ran into me." His superintendent demanded, "What lights?" The plane had hopped over his train, and was gone. There was only one person in the plane, and he was flying mailbags. It made the papers; they could never explain why the train stopped.

Of course the copilots were resented by their elders — as tattletales, as threats to jobs in the depths of the Depression. And the on-the-job training the seniors dished out was not always what it might have been. Ernie Gann treasured the memory of his very first trip, flying copilot on an American DC-2 out of Newark. As they approached the first airport at Wilkes-Barre, Sam Ross, his captain, turned to him and said, "It's your landing." That was rare indeed. Far more likely, Gann said, were captains who "let you hold the wheel while they ate lunch."

Weldon Rhoades, one of the copilots hired for United's new fleet of twin-engine Boeing 247s, also got his start out of Newark. It was unforgettable for a very different reason:

There was no training at all. In fact it was the first time I had ever been in that type plane. We had a full load — ten big fat businessmen from New York. The captains had no time for us copilots. They were more or less impatient having us around, and saw us as a threat to their jobs. This captain, Clarence Hudson, started rolling while I was still back in the cabin seeing to passengers' seat belts and handing out newspapers.

That plane needed full power for takeoff, and at full power you needed to pump fuel. There was a wobble pump the copilot could work to provide the extra pressure. But it was also my job to pump up the landing gear.

I jumped into the copilot's seat. The plane was already moving fast. Near the end of the runway an engine quit, probably because of my not operating the wobble pump. I was just getting up the landing gear by hand when we hit. The wheel yoke on my side hit my face a good hard blow. Our captain had landed us in the middle of the swamp at the end of the runway.

Nobody saw us. The radio was dead. I got out of the plane. We were 150 feet beyond the end of the runway. I carried the passengers one by one to the end of the runway, all ten of 'em.

About this time it came to the attention of the ground people that there was an airplane in the swamp. They came out, got the plane out of the mud, washed it off a little. They started the engines up. There was some metal work to do, and the engines had to be changed for damage to the bearings. Some of the passengers went back to New York. Later we took off. This time I was in my seat before Hudson began to roll.

In the transition United got rid of a lot of old pilots. R. T. Freng, a go-getting Norseman who had flown the Salt Lake City–San Francisco route, was put in as director of flying. In the West especially, where weather was better, a number of pilots were low on instrument proficiency. Some did not like to fasten earphones and listen to the radio beam *A*'s and *N*'s, and simply had their copilots tell them if they deviated to right or left. A few would not learn the newfangled instrument tricks, and just got by. But when better instruments permitted flying with less and less visibility, they were not up to passing the airline's twice-yearly checks. Freng, who had flown with them and knew them all, decided to clean house. Perhaps ten or fifteen captains and a few copilots were furloughed, most of them to fly later for the Air Transport Command in World War II. The others were not told, but left to guess why they left. But Rhoades said later:

Some of those men had it coming to them. They would turn to me, the copilot, and say, "Should we try to make it or not?" That was pitiful. Such men shouldn't have been flying passengers.

Fresh from Kelly Field, Hal Blackburn noted a like change at TWA. He said later:

They were getting to drink on the layovers in Chicago, and they got canned. Even the union said it wouldn't stand by them.

Some of those old pilots, World War I types, had a mania for flying low: they had to look for familiar landmarks. One had vomit from one end of his plane to the other. He should have known better: when you flew low in summer you could see the heat rising — it was thermal heat, producing thermal turbulence. When I said that, the captain said I was smart-ass.

A captain who insisted on flying as if radio had never been invented was responsible for an accident that took the life of a famous movie star and shocked the nation in 1942. TWA Pilot Wayne Williams was a strapping old-timer who had logged thousands of hours in Fords and Fokkers and was known for taking advice from nobody. One clear, cold evening in January 1942 he took off late from Las Vegas with eighteen passengers on the last leg of a DC-3 transcontinental trip to Los Angeles. Had Williams simply followed the beam, he would have flown straight on a heading of 205 degrees through the pass to his destination. Sure that he knew his way, he paid no heed to the *A*'s and *N*'s. Flying the course of 215 degrees carelessly marked on his flight plan, he rammed his plane instead into the side of Potosi Mountain, killing all twenty-three aboard, including actress Carole Lombard on her way to join her husband, Clark Gable. The Civil

Aeronautics Board's (CAB) scathing verdict on the cause of the crash: "Failure of the captain to use the navigational facilities available to him."

The new breed of pilots the airlines sought, Dr. Lederer said later, were simply healthy, well-compensated, normal human beings; men who loved to fly, but who were also family men, well educated, respectful of discipline, with a feeling for other people. The companies actually preferred pilots trained in bombers because such men had had the experience of serving in multiple crews. Daring was out; getting along with people was in. To manage the larger and more complicated flying machines, an airline executive said, "we wanted unanimity in the cockpit." "We looked for a stoic type," Dr. Constantino said.

"Pilots aren't scared, they're busy," explained one of the new professionals. Jack Hulburd, schooled at Penn and Pensacola, flew thirty years for TWA. One morning shortly after World War II he was driving to La Guardia Airport for his day's flight. Just as he was approaching the turnoff for the field, a United DC-4, struggling to take off, passed so low over Grand Central Parkway that one of its landing wheels dented the top of Hulburd's car. A few hundred feet farther on, the DC-4 crashed and burned with the loss of forty-three lives. (The accident is described in Chapter 25.) Hulburd proceeded to the TWA operations office, wrote his flight plan, piloted his Constellation 1,127 miles to Chicago and Kansas City, walked from the plane into the Kansas City ready room — and collapsed. Later Hulburd sued United for the damage to his car.

Ernie Gann, another of the new stoics, wrote a book about fated flying, but did not mention the stroke that brought down the old instrument pioneer, Wesley Smith. From the day Donald Douglas designed his DC-1 to fly over the Continental Divide on just one engine, innumerable pilots faced and made the decision, when an engine failed, to press on to their destination. It was the fate of Wesley Smith to lose an engine one afternoon in 1936 when his DC-2 from New York was seventy-five miles east of Chicago and to lose the other one as he began his final approach to Midway Airport. By a fantastic bit of flying, he brought his machine down on a vacant corner lot without a scratch on his twelve passengers. But by the half-mile he fell short of his destination, Smith was deemed guilty of a palpable error of judgment and was dismissed. American's Paul Carpenter, on the other hand, never had an emergency in thirty thousand hours of flying ("My record is as clean as an angel's drawers," he said). But years after the event, he cited his many near misses and like other early birds upheld the prowess of the unloved and unlucky Smith. Perhaps the sense that fate was the hunter, stalking all and inscrutably sparing some, was the one bond that united these airmen, old and young.

15

Into the Stratosphere

P_{ilots} always called the DC-3 a "forgiving" airplane, allowing uncommon latitude for their frailties, but in one fundamental respect, the twin-engine airliners of the 1930s represented no particular advance over the Ford and Fokker trimotors of the 1920s. That is to say, the airliners flew at low altitudes. Back in the 1920s pilots had flown their Fokkers to fourteen and fifteen thousand feet trying to get above bad weather, but they had to knock it off when their passengers passed out. In the 1930s the airlines and the pilots accepted the necessity of flying their Douglases through the weather rather than over it. Most pilots flew their DC-3s through the thick of it at five or six thousand feet — or even lower, as Hal Blackburn had caustically observed.

One early airline pilot who had taken his Fokker as high as eighteen thousand feet and found the going better was Tommy Tomlinson. New York state–born and Annapolis-bred, Tommy Tomlinson was one of the great fliers of airline history. Surviving two hairbreadth crashes and courts-martial at San Diego, he founded and led the Navy's fabled stunt team, the Sea Hawks. In the mid-1920s he became chief pilot for Maddux, the ancestral line of TWA in California. Tommy was a member of the technical committee that checked out the first Douglas Commercial airliner, and when the DC-1 flew, it was Tommy who tested it and handed over the check for it to Donald Douglas. In sixty years he test-flew 225 different kinds of airplane and landed just about every place on earth from the decks of the Navy's aboriginal carrier *Langley* to a cemetery in Harrisburg. When he finally retired and took to fishing off the Mexican west coast, his restless intelligence noted startling changes in ocean temperatures. He took his observations to the Scripps Institute of Oceanography at La Jolla, where they were confirmed and subsequently identified as causally related to the drought that in the next years struck the U.S. Pacific Coast.

In the 1930s Tommy was TWA's experimental pilot and chief engineer. He said later, "Jack Frye was a pilot, and he and I understood each other — we knew we had to get up there." For a starter, they turned the original DC-1, the only one ever built, into a high-altitude test plane. They outfitted its Wright Cyclone engines with two-speed blowers and Curtiss electric, fully controllable props so that it could fly up to twenty-seven thousand feet. After piloting the DC-1 in and out of thunderstorms for most of 1935, Tommy said, "I knew we had to go higher."

It happened that Frye had ordered the Northrop Gamma, a very high-speed single-engine plane, to replace TWA's Alpha mail planes. When the airmail cancellation broke up this plan, one of the two Gammas delivered to TWA became available for Tomlinson's experiments. They borrowed a turbosupercharger from Air Corps experimenters at Wright Field, got a special engine from Wright Aeronautical, a special carburetor, and intercooler. When there still wasn't enough exhaust at thirty thousand feet to spin the turbine, they fitted on special pumps to beef up fuel pressure.

Once they got the bugs out, Tommy soared daily into the stratosphere. Between July 6, 1936, and January 24, 1937, he clocked more time (forty hours, ten minutes) above thirty thousand feet than all other pilots together up to that point. One day the turbowheel flew off the engine, smacked into the wing, and sent Tommy into a dive earthward from which he barely recovered. Dr. Sanford Moss, the General Electric engineer who had developed the supercharger that made high flying possible in the first place, rushed to Kansas City to make corrective repairs. Aeromedics also watched Tommy's outings with keen interest. Once, flying to New York through solid overcast, Tommy took the plane to its absolute ceiling of thirty-six thousand feet to top the clouds and pick up the Newark radio range's beam. Lack of oxygen so numbed his brain that he mistook the south quadrant's signal for the north's. His report:

The writer finally discovered the error when about 150 miles out on the Atlantic ocean at an altitude of 5000 feet. Just enough gas remained for the plane to arrive over Princeton, N.J., when the fuel supply failed. With a 300-foot ceiling, half-mile visibility and freezing rain, a safe forced landing was made on the side of a hill.

This kind of information was so new to aviation medicine that on December 15, 1936, Dr. H. A. Armstrong of Wright Field went along to find out more. Tomlinson reported:

At 34,000 feet writer noticed distinct pains in the knee joint, apparently under the kneecap. These are believed to result from so-called "bends," or freeing of nitrogen bubbles from the blood with the decrease of air pressure at such altitudes. Later in the evening experienced unusual fatigue in the legs, and upon going to bed occasional muscle spasms in leg before sleep.

Dr. Armstrong observed that the small expenditure of energy by himself and Mr. Hiestand [of TWA] brought them very close to limit with available oxygen supply. At one time both were on ragged edge of passing out. Both later noted a slight soreness in little finger and tendency for it to contract — manifestation of the so-called bends. Dr. Armstrong noted that at 30,000 feet and above he was unable to clear his throat of phlegm by coughing, because of thinness of air at that altitude having no force to dislodge the mucus and bring it up. This would be important for any person with a cold.

But perhaps the most astonishing discovery Tommy made was of the strong winds from the west-northwest that he bumped into above thirty thousand feet. When day after day he came down and reported winds of as much as 148 miles per hour in the stratosphere, TWA meteorologists Ed Minser and Parky Parkinson threw up their hands. They were at a loss for an explanation. As Tommy said later, "I was the first pilot to hit the jet stream." Today all the world knows about the tremendous currents of air that circle the globe at heights of thirty to fifty thousand feet, and all the airlines routinely program their planes to fly at altitudes that take best advantage of them. But in 1936 it was a big mystery.

In one experiment with high-altitude flying, the Douglas staff had tried filling the entire cabin of a DC-2 with oxygen-enriched air and had taken members of the board, their wives, and children up to twenty-five thousand feet. They wanted a cross-section of humanity for the test, and the company's doctor moved up and down the aisle checking everybody's pulse, blood pressure, and breathing. At the front of the cabin chief engineer Frank Collbohm kept releasing oxygen from a big tank. When the plane flew over Boulder Dam, it was necessary to scratch frost off the windows to see. Some members of the party lit cigarettes to see how large the flame was at twenty-five thousand feet. Long afterward Art Raymond recalled such casualness with horror. "Luckily we emerged unscathed," he said. "But how naive can one be?"

Back in 1921 Lieutenant Harold Harris, later manager of Panagra and president of American Overseas, had flown the first pressurized plane. On that occasion the object was to provide a camera, not people, with the regularized environment. Experimenters at McCook Field, out to obtain the highest possible precision in high-altitude photography, rigged up a pressurized compartment to house a camera which snapped pictures through the fuselage floor. The difficulties with a wood-and-linen plane were prodigious and Wiley Post's solution, when he set altitude records in 1931, was to design his own pressurized suit to withstand the height.

Tommy Tomlinson had been telling Frye that if the airline only had a plane that could fly up to twenty thousand feet, 80 percent of the weather problems would be eliminated. To this Frye replied, "You know we can't have passengers breathing oxygen." Once again the American aviation industry, with a lift from the military, crashed through with the innovative

The Boeing 307 Stratoliner, with the first pressurized cabin,
carried thirty-three passengers.

answer, just when the airlines had to have it. The answer was Boeing's 307
Stratoliner, first commercial air transport to be equipped with a pressurized cabin.

Frye knew that Boeing had developed a bomber that could fly high.
This was the B-17, the four-engine Flying Fortress that was to become the
standout bomber of World War II. The plane, burning the 100-octane
gasoline put into use by the military in 1936, had superchargers that enabled it to fly at twenty thousand feet and was fitted with oxygen tanks for
its crew. But the all-metal monocoque fuselages offered new possibilities,
and in 1937 a Lockheed-designed experimental bomber flew with a pressurized cabin. So when Frye went to Boeing that year, the company
agreed to redesign its big bomber for TWA with a pressurized cabin.
Wellwood Beall, the bluff and hearty production chief in Seattle, presided
over the design of the B-29 as well as the B-17 and probably turned out
more bombers than any man in history. But he had created the giant Boe-

ing 314 flying boats for Pan Am, and saw TWA's approach as a way to get back some business from Douglas. He said later:

Our design had a compressor, and you could regulate the pressure as desired. We decided the maximum pressure a passenger could tolerate would be 8000 feet and no higher — a pilot could set it lower if his flight plan permitted. Air Research of Los Angeles built the regulator for us.

Beall said later he got "awful scared" as his engineers worked to shape a cabin strong enough to withstand a pressure differential of two and a half pounds per square inch:

We crawled around inside the plane looking for windshield leaks with a paint-brush in one hand and a can of soap water in the other. Fast as we found them, we sealed them from inside with neoprene tape we got from du Pont. We beefed up the window frames; we tried double windows. I said to Bill Allen, "All we have to do is blow a window and we're in the soup."

I told Bill we needed to hold a symposium. We invited people from Air Research and Douglas and Lockheed to Seattle — also Randy Lovelace, the doctor who had been doing experiments at Mayo Clinic and at his clinic in New Mexico to get at the problem of oxygen want.

The Boeing 307 had the B-17's sturdy wing and the same supercharged engines. To provide the twenty sleeping compartments that TWA wanted to beat American and United in the transcontinental race, Beall designed a bigger body, and this proved costly. The luxurious interiors, designed by Raymond Loewy, were furnished by Marshall Field's. Pan Am ordered only three of the planes, for its Latin American service. And TWA, ordering five, found for neither the first nor last time in its stormy history that it had bitten off more than it could chew. The airline's early success when it introduced the DC-2 had been more than overcome by American's sponsorship of the spacious and much more economical DC-3. TWA was further handicapped by the fact that among all the big airlines, its route losses in the 1934 airmail contract cancellations had been the severest. For five years thereafter, the airline had to operate 40 percent of its system without benefit of contract subsidies.

These conditions disturbed some of the stockholders to whom financial control of TWA had passed after the split-up of the aviation combines. The chief of these were the Wall Street banking house of Lehman Brothers and John D. Hertz, a big market trader who by 1934 had become a Lehman partner. Still controlling Yellow Cab, Hertz was thought to have some interest in a rental car tie-up with the airline. But TWA had yet to pay a dividend, and Hertz took a dim view of what appeared to him to be extravagant outlays for new planes that were bigger and fancier than the line's competitors were willing to order. In December 1938 the TWA board voted to cancel Frye's orders for the high-flying Boeing 307s. A showdown appeared inevitable.

Jack Frye in 1935

At this point Jack Frye took the step of his life. Facing certain ouster over a decision he felt was vital to the airline's future, he flew to Los Angeles for a secret visit with Howard Hughes. As Frye said later, "One thing about Hughes was that he *did* have an understanding of the airplane." Frye did not have to spend much time explaining to Hughes, who had just completed a record-smashing, globe-circling flight through fog and rain in a Lockheed Electra airliner, why it was important to an airline to have a plane that could fly over weather. Hughes understood, and Hughes had money.

The upshot was that in the last week of December 1938, Frye returned to Kansas City, obtained from the corporation's secretary a list of stockholders, closed his office door, and began copying out in longhand the names and addresses of all who held three hundred or more shares of TWA stock. After a day or so of this, he asked his secretary to carry on and then to put the lists in envelopes and mail them to Hughes's post office box in Los Angeles. By early January 1939, Hughes had the names and was quietly buying up stock. By April Hughes had such a substantial number of shares that the Lehman-Hertz interests decided to sell out. Thus it was that Howard Hughes acquired ownership and control of one of the Big Four airlines.

For Hughes the acquisition of the TWA shares looked like one of the more logical leaps in a career that seemed to vault all over the American landscape. Like Lindbergh, like Boeing, like so many who took to aviation in the early years, he was a tinkerer, a technological "nut." And Hughes was born rich. He was recklessly successful, still young. He was tall, dark, and handsome. Newspapers said he had "the face of a poet." Hero he was, though no "lone eagle." He was a lone wolf.

Howard Hughes was born in 1904 in Houston, Texas, and grew up in the shadow of a dynamic, hearty, earthy father who was known as Big Howard. Big Howard's only son and namesake was called Little Howard, or Sonny, and later he said, "[My father] never suggested I do something; he just told me. He shoved things down my throat and I had to like it. But he had a hail-fellow-well-met quality that I never had. He was a terrifically loved man." Big Howard was the founder of the Hughes Tool Company, which rented its revolutionary oil-drilling bit — a device with 166 cutting edges that he had invented and patented — for $30,000 a well. Then, and later, this bit was the base of the Hughes fortune. It was a gold mine.

In his father's backyard shop, Sonny Hughes toyed with ham radio sets, bikes, old cars; once he took flying lessons from a barnstormer. When he was still in high school, his mother died. After a trip abroad, he was enrolled as an engineering freshman at Houston's Rice Institute. Then his father dropped dead, and from that traumatic moment everything changed.

He abandoned his classes, had himself declared an adult at nineteen, and bought out his relatives' shares in the Hughes Tool Company. To Dudley Sharp, who had traveled with him to Europe, he appeared to have "an absolute mania for proving himself. He didn't want to stay in Houston. I think he even disliked hearing his father's name. He wanted to get out and find something he wanted to do. He didn't know exactly what, but nothing was going to stop him." By Sharp's account he played sick and persuaded Dudley to get word to Ella Rice, daughter of a prominent family he did not even know, that he was in a coma and continually called her name. She came for a sickroom visit, and within months Hughes had married Ella Rice, younger sister of the Standard Oil president's wife. Off they went to Hollywood, where his father had liked to go for fun and where his uncle Rupert was writing and directing films. Howard Hughes, age twenty, set out to make a success in movies.

Gangly and awkward, he picked up film-making points while hiring a Los Angeles accountant named Noah Dietrich to watch over his finances. After bankrolling a couple of indifferent films, Hughes conceived the idea of producing a spectacular picture, no matter what the cost, of life in the British Flying Corps in World War I. Impatiently dismissing the director, he bore down on the smallest details. Scorning the conventional film-shooting fakery, he insisted on getting together a fleet of real wartime flying machines. He hired 137 pilots to do the work. Not content with their efforts (several crashed to their deaths), he learned to fly and performed one stunt himself. The picture, *Hell's Angels*, took shape as the all-time air epic of movie history.

In 1931 Hughes turned to aviation, and soon Noah Dietrich was saying, "Howard has three ambitions — to become the world's best golfer, the world's best flier, and the world's richest man." At this point, when Dietrich set up the Hughes Aircraft Company, Hughes had a two-stroke handicap in golf, was the holder of a limited transport pilot's license, and was one of a great many oil millionaires.

In his obsessive way, Hughes set out to learn the flying business from the grease pit up. On October 16, 1932, he got himself a $250-a-month job as a copilot with American Airways. For some months he toted passengers' bags between Los Angeles and Chicago. Then with Dick Palmer, a designer who had worked for Douglas and Lockheed, he plunged day and night — mostly night — into the planning and building of the H-1, the Hughes Racer. It was a rakish skyrocket of a machine with a big Hornet engine, the first flush-riveted skin ever seen on an airplane, and the first power-operated retractable landing-gear. Having pored over the Pratt & Whitney engine manuals, Hughes flight-tested the plane and made so many small changes he nearly drove the hangar staff crazy. Then on September 15, 1935, he took the H-1 up over Santa Ana and set a world's

Howard Hughes and the first plane he designed and built: the H-1.

speed record of 352 miles per hour, a record that stood for years. The following January he loaded the little plane with some of the first 87-octane gasoline released by the military and, sucking oxygen through a crude hose, flew high across the country to New York to set a new transcontinental speed record — seven hours and twenty-seven minutes. It was during this Los Angeles-to-New York dash, Hughes said later, that he decided some day to "build a plane that could carry a large number of passengers coast-to-coast in complete comfort, and like a bat out of hell."

Thereafter Hughes bought and flew almost every commercial airplane built in the United States. In January 1936 he flew a Northrop Gamma nonstop across the country; in May of that year he flew a Consolidated Fleetster from Miami to New York in a record four hours and twenty-one minutes; and in November he was in Kansas City to eat Thanksgiving dinner at the family table of TWA Vice President Paul Richter, one of his old *Hell's Angels* stunters, and to try out TWA's first DC-2. But it was not until July 1938 that Hughes was ready for his climactic flying feat. This was to take a stock-model airliner, load it up with 100-octane gasoline and last-word flight gadgetry, and fly it round the world faster than anyone had ever flown before.

The plane Hughes chose was the twin-engine Lockheed Electra, then in service on a half-dozen U.S. airlines. The route he plotted across the northern part of the globe was considerably shorter than that followed by Wiley Post when Post set his amazing solo globe-girdling record of seven days and eighteen hours in the single-engine Lockheed *Winnie Mae* in 1933. Hughes took along four passengers — two navigators and two radiomen. He also went to his usual extravagant lengths over preflight details (he tested fifteen kinds of bread for the trip). Then he was off.

It was another Hughes spectacular. He crossed from New York to Paris in half the time it took Lindbergh, rested for eight hours, and flew on across Germany to Moscow. The Russians turned on extra radio aids to help him land in fog, but wondered at the millionaire who wore patched pants. As he winged across Siberia each stop — Omsk, Yakutsk — became progressively shorter, but the automatic pilot helped. After Fairbanks Hughes paused only thirty minutes for gas at Minneapolis. Then he swept into Floyd Bennett Field at New York after a trip of only three and a half days — half the time it had taken Wiley Post. A crowd of twenty thousand, led by Mayor LaGuardia, cheered his arrival, and gala receptions followed at the New York World's Fair, on Broadway (1,800 tons of tape vs. Lindbergh's 1,600), and in Washington. Then he disappeared into Long Island with his latest flame, Katharine Hepburn. "All hats are off to Howard Hughes for a superb flight made to demonstrate the practicality of commercial transport," said the *New York Times*.

This was the man with whom Jack Frye struck his bargain later that year, and it is still a question exactly what Hughes, at thirty-four, wanted to do. Like almost everybody else interested in the aviation business at that time, he was actively working to create a new airplane that would land a big contract in the nation's rearmament program. And by night he was back in the movies planning a blockbuster of a film called *The Outlaw*.

So with the backing of Hughes's millions, and no real notion of how far and in what ways Hughes would take part in the affairs of the airline, Jack Frye closed the contract with Boeing for $1,600,000 worth of high-flying airliners. Because of the delay in clinching the order, TWA obtained delivery of its five Stratoliners only in 1940, scarcely a year before Pearl Harbor put an end to normal commercial air transport activity in the United States.

In that brief time, the Stratoliner made a strong showing. There were the usual crises at the start. Even before the first 307 left Seattle, one crashed. This was the second prototype. It was lost with all hands when a prospective buyer from KLM, testing the plane's stall characteristics, put the craft into a spin that broke off its outer wing and sent it plunging uncontrollably to earth.

After Boeing installed a slotted leading edge on the outer wing to rule out a recurrence, TWA's Chief Pilot Otis Bryan flew one of the first 307s on a proving flight from Kansas City to Albuquerque. Twenty thousand feet over southeastern Colorado, he bumped into a thunderstorm with snow. All four engines suddenly quit. Down through the overcast the big plane plunged. On the way down Bryan caught one fleeting glimpse — not the Rockies below but foothills:

I saw a canyon and hoping there would be a flat mesa-top alongside it, I came out at 500 feet. There was no canyon, and no mesa either. But I saw a level place and went down. The wheels were still retracted. But by luck the belly doors were open. They acted like a plow. They dug into the ground and shortened the slide. The plane came to a stop on its belly — about 200 feet short of one of those canyons.

Not one of the fourteen aboard was hurt. Neither was the plane. "We went and got a few pieces of angle iron, and a few days later I flew it out," Bryan said later. It was another case of ice forming in the carburetors and choking off the air supply to the engines — "like a car with its filter clogging, only in this case with ice," said Bryan. So even at twenty thousand feet, weather still created grave problems. The Stratoliner's system for channeling heat from the engine exhaust to warm the air intakes was beefed up.

The experience proved, Captain Swede Golien said later, "that with the Stratoliner we could get high enough to get into the middle of the roughest part of the storm." But the high-flying Boeing 307 brought much greater comfort and service to passengers. It was a hit, especially with Hughes's movie-colony friends who traveled a lot between Los Angeles and New York. The controlled pressure, temperature, and airflow in the cabin introduced a new class of air travel. You could go to bed after leaving Los Angeles and arrive in New York the next day after making only one stop, in Chicago. At their usual cruising altitude of fourteen thousand feet the Stratoliners flew at 200 miles per hour, crossing the country in thirteen hours and forty minutes, two hours less than the DC-3s. As sleepers, they had berths for sixteen; as day planes, they carried thirty-three passengers. The Stratoliners saw service too briefly to make much impact on the public or on the airline's profit-and-loss figures; but their value to TWA by other measurements was great. For in the war in which America shortly found itself, these were the nation's only commercial landplanes capable of transocean flight. TWA's five Stratoliners became the means by which the airline that Jack Frye ran, but Howard Hughes controlled, became a world airline.

*   *   *

Hughes was the most powerful new force to enter commercial aviation after the airlines cut loose from their conglomerate bondage. The 1930s were not a time of major capital formation in the air transport industry; the outstanding investment of the decade was C. R. Smith's purchase of a fleet of DC-3s, financed by RFC loans. Pan Am, and to a lesser degree United, were the only outfits that, late in the 1930s, went to the capital markets that in time came to dominate the finance and management of every airline but TWA. Infusing so much new capital at this point into the weakest of the Big Four, Hughes was in a class by himself. He was all the more powerful because these infusions were made expressly for new equipment.

If World War II obscured the importance of Hughes's intervention in the case of the first high-altitude airliner, it obscured even more his role in the development of the next big American transport plane. In 1939 Howard Hughes was thinking big. Even before TWA took delivery on its first Stratoliner, Hughes was talking to Frye about a bigger, stronger, and faster plane for their transcontinental service. Hughes had come on the scene too late to do more than underwrite the purchase of the Stratoliners. But the tremendous advances made by defense contractors like Wright and Pratt & Whitney, boosting engine horsepower for the Pentagon's new fighters and bombers, opened tremendous possibilities for a whole new generation of airliners. The two did not approach Donald Douglas, partly because Douglas seemed committed to unpressurized construction, but also partly for more personal reasons, later explained by Noah Dietrich. Back in the 1920s Hughes had bought a two-seat racing plane and sent it to Douglas's little plant at Santa Monica to have it rebuilt. He made such a stink about the changes — "No, that's not right. Tear it apart and do it differently," he would say — and then so outraged Douglas by accusing him of "trying to cheat me" on the final bill, that Douglas sent word: "I never want to do business with him again."

But Hughes and Frye made overtures both to Consolidated, then building the four-engine B-24 for the Air Force in San Diego, and to Lockheed, which was turning out a record peacetime order of twin-engine Hudsons for the British and French. To Bob Gross, Lockheed's president, Frye said, "Look, we want the fastest, high performance airplane in the world — and we want you fellows to build it, if you will." To this Gross replied, "The way to start is with the biggest engine in the world." This was the Wright 3350, bigger even than Pratt & Whitney's 2800: the engine that ultimately powered the B-29 bomber. But at that moment, the 3350 was still at the ground-test stage.

On June 22, 1939, Gross, chief engineer Hall Hibbard, and Kelly Johnson, Lockheed's talented young designer, went by car to Hughes's house on Muirfield Road in Los Angeles. They brought with them drawings and preliminary data for a four-engine airliner with a pressurized fuselage big enough to carry twenty sleeping passengers and a crew of six. According to Kelly Johnson's notes, either of two Wright engines might be used — the R2600, which was already in use on Pan Am's Boeing flying-boat plane, or the 3350, which was still a question mark. Hughes was fascinated. But Johnson spoke up. He told Hughes:

You are not asking enough for the state of the art. Basically, you want a plane with a 3000-mile range, and a big fuselage for 20 sleeping compartments — that's only a 6000-pound load. That would never pay. You would have to go to something bigger. If the plane could carry 20 sleeping passengers, then it could carry 60 sitting passengers.

Hibbard said later, "It was evident they were interested. We refined our drawings." Across town at the Biltmore Hotel sat Richter, Tomlinson, and Franklin of the TWA staff. Such was the secrecy Hughes imposed that they began speaking of their principals in code. Hughes was "God," Frye, "Jesus Christ." After hanging around, playing Ping-Pong, keeping out of sight, killing time for a couple of days, they were called in for the clincher. A contract was to be signed — not between TWA and Lockheed, but between Hughes and Lockheed. Hughes — that is, the Hughes Tool Company — would buy the planes for TWA. But who should draw up so secret a contract? Tomlinson said that his wife was a court stenographer. She was summoned from her room upstairs and duly typed out the contract that Hughes and Gross then signed.

The order Hughes placed was the biggest order for commercial airplanes up to that time, and in return he extracted certain promises. First, he was to have unchallenged right to the first forty planes built. Second, to give TWA the biggest possible competitive edge, the entire project must be kept secret. United and American must be kept from any knowledge that a new airliner was in the works. And indeed this was done. Gross put his project engineers to work in a remote hangar, harder to get into than Fort Knox.

Johnny Guy, who had been TWA's factory representative at Boeing, was called in by Vice President Richter. Closing the door, Richter said without any preliminaries, "Johnny, have you any intentions of quitting TWA?" When Guy said no, Richter said, "Before I tell you what your next assignment will be I want you to promise me that you will not leave the employment of TWA without my specific approval. Will you promise to do this?" Mystified, Guy said yes. Johnny Guy then became the fifth TWA employee to know that Lockheed had signed the contract with the Hughes Tool Company and also the fifth sworn to refrain from divulging either the name of the contractor or the ultimate operator of the aircraft. The plane, Richter said, had been designated the Lockheed O49, or Constellation.

Guy was to be the go-between, the coordinator of all messages between the three parties, passing all papers, drawings, and progress reports from Lockheed to Hughes while keeping Frye informed. Officially, he was to act as TWA's factory representative at Douglas, where more DC-3s had just been ordered. But he was to live in Burbank, near Lockheed, making sure that no one from Lockheed lived in any of the adjoining houses. The word "Lockheed" was to disappear from Guy's vocabulary, his wife and children were to know nothing of his Lockheed connection. He was to avoid any daytime meetings in his "home office" with anybody from Lockheed. He was to use plain envelopes for his correspondence, addressing letters to Hughes as "H," Frye as "F," and Hall Hibbard, his Lockheed contact, as "HH." Upon opening any mail from Lockheed, his first act should be to scissor out the Lockheed letterhead, the signature and any other references that might tie the correspondence to Lockheed.

For the next year Guy worked at Douglas, lived on Olive Street in Burbank, and passed countless papers between his bosses. Lockheed kept strictly to their part of the agreement. Guy said later:

I would receive drawings as they were released, check them against the Detail Specifications, then forward them to Howard with my comments and recommendations. In a week or ten days Howard woud call — always between 2 and 3 A.M. — and comment. His main interest was in the cockpit, instrumentation and cabin details. But once he called me and said he thought the Wright people were dragging their feet on the 3350 engines. He asked me to tell Mr. Frye to immediately call Guy Vaughan, the president of Wright, and tell them if Wright didn't get off their ass he would drop them and install Pratt-Whitney 2800s. That apparently stirred things up as he intended.

So it went, totally hugger-mugger, for a year and a half. Then, in the early fall of 1941, when construction of the first Constellation's fuselage, wings, and tail section was perhaps half complete, a War Department team arrived at Lockheed on a tour of aircraft factories. The purpose of the mission, on the eve of Pearl Harbor, was to determine the output capabilities of the manufacturers — and their existing commercial commitments.

With that the secret was lost. But as it happened, before signing with Hughes, Gross of Lockheed had struck a deal — without any such secrecy — to build a somewhat smaller four-engine landplane for Pan Am to be known as the Excalibur. Now the Office of Production Management in Washington told Gross that it could not possibly approve allocation of scarce materials for *two* Lockheed airliners. Thereupon Gross, explaining his predicament to Hughes, proposed that he talk Pan Am into dropping the Excalibur and joining the Constellation project. On the plea that the two lines were not competitive, Hughes gave his consent. Then, in a deal scarcely less secret than the original Hughes pact, Lockheed signed up Pan Am for forty Constellations — eighteen of them 73,000-pounders like

TWA's and twenty-two of a heavier (86,000 pounds) version for trans-ocean flying known as the Model 149.

Then war exploded, and all deals were off for the duration. Washington allowed quantity production of just one four-engine transport — Douglas's smaller, unpressurized DC-4 (military designation: C-54). The Constellation, which was scheduled for a first flight test in mid-1942, finally flew as the C-57 in December 1943 ("General Arnold stopped production on that plane seventeen times," growled Kelly Johnson). But Hughes and Frye were not to be denied. In June 1944 they flew the plane nonstop to Washington. With its powerful Wright engines, only introduced by the military that same year in the B-29 Superfortress bombers, the Constellation streaked across the continent in a record seven hours and thirty minutes. Thus Hughes served notice that he had the fastest airliner ever built and, come peacetime, his airline would be number one.

16

Death of a Senator

On the night of May 5, 1935, a DC-2, flying as the second section of TWA's transcontinental flight from Los Angeles to Newark, made its regular stop at Albuquerque. Senator Bronson Cutting of New Mexico climbed aboard. That brought the number of passengers to nine, including five movie-industry people from Los Angeles. During the Albuquerque stop a TWA radio man checked out a faulty transmitter on the plane. He reported that it would work on the daytime wavelength but not on the frequency used at night. But the U.S. Weather Bureau continued to forecast good weather along the way. At 9:15 Captain Harvey Bolton took off for Kansas City, about a half hour behind the first-section DC-2.

Within thirty minutes of takeoff, Bolton became aware that the weather was changing. Using the daytime frequency, he radioed and received permission to make an instrument approach into Kansas City. It was the last transmission heard from Bolton; apparently he could receive messages but could not make himself heard on his defective radio. When he passed over Wichita at 1:34 A.M., the TWA operator below could not get a response from him. But at 1:52 A.M. Bolton presumably heard the Department of Commerce's regular broadcast reporting that the weather ahead had got much worse. The ceiling at Kansas City had dropped to six hundred feet, less than the Department's seven-hundred-foot minimum for landing.

At Kansas City TWA dispatcher Ted Haueter directed the first-section DC-2 to land. The pilot first tried a visual landing. Losing contact with the ground, he veered back up into the overcast and went round in the laborious and time-consuming circuit necessary for an instrument approach — and finally got down at 2:24 A.M. Not until 2:50, twenty-six minutes after the first-section DC-2 was on the ground did dispatcher Haueter try to contact Bolton's plane. Notwithstanding the below-minimum ceiling, he told Bolton to try landing at Kansas City.

By this time it was evident that TWA's second plane was in trouble. His transmitter out, Captain Bolton was unable to ask for information or in-structions. He had fuel for one hour and twenty-six minutes of flying. Commerce Department regulations, however, required that he have a forty-five-minute supply of fuel in reserve at all times his plane was air-borne and carrying passengers. That meant that he had forty-four minutes to locate the Kansas City Airport, or any other, and land.

In the Kansas City office of TWA, Gordon Parkinson thought drastic measures were in order. He phoned across the field to the Phillips Oil re-finery and got them to flare off all their excess gas in hopes the pilot would see it. But nothing was heard or seen of the plane. At 2:55 Haueter called the emergency Kirksville Field 120 miles northeast of Kansas City; he was told visibility and ceiling were below minimums there too. At 3:00 A.M. Haueter sent new instructions: "Fly out northeast leg of Kansas City beam. Fly toward Burlington [the next field, 250 miles east, in Iowa]. Try land at first available field." At 3:12 A.M. at Haueter's request, the Com-merce Department station broadcast the same message.

In the DC-2 June Mesker, a TWA pilot's wife aboard as a nonrevenue passenger, suddenly noticed that she was no longer seeing stars out the window. Instead, clouds swirled round the plane. She called across the aisle to William Kaplan, one of the Hollywood passengers, to ask the time. It was 2:45 A.M., he told her, and the plane was late. He pointed to some lights in the distance, presumably Kansas City. Mrs. Mesker got a fleeting glimpse of them. Then she felt the plane circling and slowly losing alti-tude. She slipped her shoes on and watched for the seat belt light to flash on. It did not. Instead the plane leveled off suddenly and just kept going. Looking out, all she could see was "thick stuff."

Unable to find his way in to the Kansas City field, Bolton had decided to strike out for the Kirksville emergency field. By one way or another he contrived to bring his plane almost within maximum range of the Kirks-ville radio marker. There he decided to go down beneath the overcast and fly visually.

His fuel gauge already pointed to empty — but it was not a precision instrument, and he knew he had some gas. He also knew he was none too far from the emergency field. But if he ran out of gas while flying on the deck, he might also be able to pick out an open field. But the weather had closed in. There was not even the three-hundred-foot ceiling that Kirks-ville had reported earlier. In spots it was zero-zero. Everywhere over the rolling landscape lay a thin mist.

The seat belt sign June Mesker had been waiting for flashed on. Bolton soon picked up a concrete highway. Wheels up, he flew just above the treetops. Two miles on, he came to a fog-shrouded draw. Flying contact, he dipped down. Presumably he saw he was too low and started a turn to the left. The plane shot between a house and a barn, scraped the left wing on the ground, zoomed over the road, ripped into a sixty-foot embankment, and flipped over. It was 3:30 in the morning. Kirksville was sixteen miles to the north.

In the impact Senator Cutting was killed outright. Four others died, including the copilot. June Mesker acted the heroine, helping the other six injured passengers. Pilot Bolton, thrown clear, died, but not before telling rescuers he had "run out of gas."

It was the DC-2's first fatal accident, and TWA's first since the Rockne crash of 1931. But the death of Senator Cutting turned this into one of the most fateful crashes in airline history. Shock and grief swept the Senate. Bronson Cutting was one of its outstanding members. The son of a wealthy New York family, he had gone west after Groton and Harvard to overcome tuberculosis; he had stayed on in Santa Fe to make a political name, not as defender of privilege, but as one of the country's leading progressives. When his death was announced, Senators Borah and Norris wept openly. And along with the grief there was resentment — resentment aimed at President Roosevelt, who had forced an investigation of Cutting's election victory that was the cause for the Senator's trip home, and a more generalized resentment. His congressional colleagues, reacting to the loss of one of their own, demanded to know why.

The Department of Commerce conducted an investigation and shortly reported four lapses that contributed to the accident, only one of them chargeable to the government itself — failure on the part of the U.S. Weather Bureau to spot and warn of the weather change soon enough. The other errors were laid to the airline: letting the plane fly on from Albuquerque when the radio was not working properly; not rerouting the plane to an alternate field with better weather before it was too late; and — Captain Bolton's error — continuing the flight after finding he could not communicate effectively with the ground.

But the Commerce Department's finding absolved its own aviation bureau of all blame, and the Senate was not going to stand for that. Even before the report came out, the Senate voted its own probe into how "an honored member of this body" had died. Chairman of the Commerce Committee Royal Copeland, a conservative Democrat from New York and no friend of Roosevelt, led the inquiry.

It took nearly a year for the inquiry, which delved into subjects far from the Cutting crash and air safety, to find the whole conduct of government supervision of aviation wanting. Neither in the questions it asked nor in the conclusions it published did the Committee deal kindly with FDR's air policy improvisations. Of course the Department of Commerce was permitted to catalog still more airline "irregularities" for the record:

— failing to carry a supply of reserve fuel sufficient for forty-five minutes of flying.

— assigning Bolton, who ordinarily flew the Kansas City–Newark
route, to the Los Angeles–Kansas City leg, for which he lacked the
Department's sanction to fly.

— assigning Bolton to fly when he had not taken his last scheduled
quarterly medical exam.

— assigning a copilot to fly who did not have his scheduled air transport rating.

— undertaking an instrument flight without working, two-way radio
communications.

— trying to land Bolton's plane at Kansas City when the ceiling was
below minimum. (The Department proceeded to fine the pilot of the
first-section DC-2 $500 for landing at Kansas City that night against
the rules.)

But the Department had been lax in enforcing all these points, and as for
having a forty-five minute reserve supply of fuel at all times, TWA specifically denied that it had been notified of this requirement. To the senators, the fault for all these points lay as much with the government as with
the airline. They heaped heaviest blame for the crash on the government.
Senator Cutting might still be alive, they charged, had the government's
airway in Missouri been adequately manned and had the radio beam
worked properly that night.

Their report reverberated with wider questions: Ought not airway
standards be raised to keep up with the swift new airliners flying them?
Ought not the government to take over all air traffic control, together
with all the responsibility that belonged with that task? Should such re-
sponsibilities be entrusted to the administration's pip-squeak Air Commerce Bureau? And how, of all things, could such an agency be given
responsibility for investigating air accidents when that meant it was investigating itself? The bureau, moreover, the senators said, was shot
through with politics and seethed with petty intrigue. And the Senate
conclusions brought out that the Roosevelt administration (it's hard to
believe now, but those were its early, budget-balancing days) had slashed
funds so severely to this faction-ridden bureau that essential aids to air
navigation were often not available.

That was not all. During the long spell of Senator Copeland's inquiry
there had been a string of accidents which further sapped the confidence
of the flying public. "How safe is the air?" demanded *Fortune,* whose corporation executive subscribers were then the airlines' chief customers.
The question of whether he ought to take to the air had by then posed itself to every business traveler. His state of mind was not much improved
by the tendency of the industry, the government, the public press, and
even the pilots themselves, to blame the man at the controls for almost
every crash. In fact, pilot error entered into about 90 percent of the acci-

dents at this time; in one tabulation the Department of Commerce listed twenty-seven accidents and set down this factor as the primary cause of sixteen of them. This, of course, undermined the public confidence in flying. If the pilots were not up to the problems of modern air navigation, then on the old principle that a chain is as strong as its weakest link, the air must be unsafe.

On October 7, 1935, a United Airlines Boeing crashed in the middle of the night while flying through mountains near Denver; twelve were lost, including eight passengers. On April 7, 1936, a TWA DC-2 crashed near Uniontown, Pennsylvania, and all but two of the fourteen aboard, including ten passengers, were killed. On August 6, 1936, a Chicago & Southern Lockheed crashed trying to return after a night takeoff at St. Louis, with loss of all eight aboard.

On February 10, 1937, a United DC-3 crashed in San Francisco Bay with the loss of all eleven aboard. Since it was a straightforward night approach to the airfield, and a thorough check of the engines and propellers showed no sign of mechanical failure, the authorities were at a loss to account for what had happened. At the time, it was simply said that the pilot had made a tight turn while flying too close to the water, with the result that the wingtip touched the water and was torn off. In New York C. B. Allen, aviation editor of the *Herald Tribune,* heard shortly afterward of an American Airlines pilot flying a DC-3 who was unable to pull back on his control stick as he came in to land at Newark. Looking down, the pilot noticed that his radio earphones had fallen into the opening between the stick and floor. Allen suddenly remembered the unexplained San Francisco accident. Could the same thing have happened to the unfortunate United pilot? A microscopic examination of the control stick of the United plane disclosed that the same thing had happened. All airlines then fastened a canvas boot over the opening at the bottom of the control stick to prevent earphones or anything else from dropping in.

The San Francisco Bay accident was one of a string of five that winter that, coming after the earlier crashes, kept the public and airlines in a state of agitation. Most of these crashes involved radio — static that made it impossible for the pilot to identify the right beam to follow, or wandering or weak beams, or lack of an airway station transmitting a beam at a crucial point. And older pilots had been involved in most of these accidents. Younger pilots seemed, on the other hand, just as much at home with instruments as with contact flying. Clearly the industry was in a transition stage from the old to the new at this point. But the unhappy fact was that thirty-seven passengers had died in the winter's crashes, and many people were scared of flying.

And so the feeling spread in the industry and in Congress that fundamental aviation legislation was needed. The Air Mail Act of 1934 was a hasty piece of legislation, and Roosevelt accepted the provision for appointing a Federal Aviation Commission to report back with ideas for something better. The airline industry, which harbored deep resentment at the New Deal's abrupt intervention of 1934, saw its chance in Congress's changing mood to win back some losses and to obtain new legislation advantageous to its interests.

The opening shot in the campaign was the commission's report in 1935. Since the stench of the Post Office contract scandals still hung in the air, the commission members proposed setting up a new aviation agency altogether. But the commission would have left air safety in the hands of the same old underfunded Commerce Department, and anyhow those were the Depression days when Washington cooked up so many separate entities like WPA, NRA, and SEC that Al Smith growled the country was being fed "alphabet soup." So Congress and the President recoiled at first from the idea of yet one more independent government agency. Instead, Roosevelt proposed lumping the airlines with the rails, the bus lines and the other commercial carriers regulated by the Interstate Commerce Commission. That way, no new letters would be tossed into the broth. The recommendation was duly embodied in a bill before Congress.

As the pot stirred, the infant airline industry rushed to join the cooks. The insurance giants were accustomed to advising Washington through their national institute. The monsters of petroleum swayed Congress through their national council. The rails lobbied through their national association. At this time the airline industry, although still a "speck in the sky" in ICC Commissioner Walter Splawn's words, banded together to form the Air Transport Association and appointed Colonel Edgar Gorrell, a World War I airman who knew his way around Washington, as its president and spokesman. Gorrell brought in Pat Patterson of United, C. R. Smith of American, and Eddie Rickenbacker of Eastern to support a draft bill to "stabilize" commercial aviation by applying to it the same pattern of federal rate and route regulation enforced for the bus lines.

The airline industry may be the only industry in America that has ever demanded regulation. But it was hurting so badly after the Air Mail Act of 1934 that nothing less promised to keep it going. Despite the fact that passenger and mail traffic had tripled, all the domestic airlines lost money in the years thereafter. Half the capital invested in the industry had been "irretrievably lost," and "shaken faith on the part of the investing public in the financial stability of the airlines" was preventing them from taking advantage of the growth of passenger traffic. Supervision by Washington that would free them from the hazards of losing their routes to anybody's bid at the next Post Office auction — that was what they wanted above all, and soon.

As it happened, the mood was changing in Congress. Senators especially had been increasingly unhappy over the Commerce Department's handling of air safety since the Cutting crash. The tireless Colonel Gorrell insisted that civil aviation must have financial stability; when fifty died in eight airline crashes, he worked at stabilizing the accident rate too. He spent a lot of time persuading Senators Copeland, Truman, and McCarran that a large share of the blame for these accidents was traceable to the lack of funds for keeping airway radio and navigation aids up to snuff. Amid all the misgivings McKellar, the Senate's leading "spoilsman," vowed to filibuster against any bill depriving the Post Office Department of its power to hand out contracts. That ended the airlines' hope for relief in 1937.

But President Roosevelt was persuaded to appoint an interdepartmental committee on aviation legislation, and then things began, at last, to move. A new agency was the answer, after all. Although none of the half-dozen departments had been ready to concede any of their powers to one of the others, all quickly agreed to cede them to a new entity. Chairman Clarence Lea of the House Commerce Committee gave Gorrell and other industry spokesmen a chance to go over the latest draft to make sure it retained the desired provisions for "stability." Because all of Pan Am's mail contracts were due to expire that year, Henry Friendly was especially relieved that the bill covered international routes too.

There was a sharp fight in Congress. The chief issue — how independent of the President the new agency should be — was never finally resolved. Senator Harry Truman fought unsuccessfully to eliminate a provision requiring presidential approval for licenses granted by the agency to fly overseas — a power that Truman, as President, was to invoke more often than any other occupant of the White House. Others demanded that quasi-judicial functions such as the investigation of air accidents be sealed off from political influence, hence outside Commerce, Post Office, or any other executive department. Management of the airways and enforcement of safety rules through an Air Safety Board, it was decided, could be entrusted to a Civil Aeronautics Administrator who would be appointed and removed at the President's pleasure. At the top of this new mini-government five commissioners, nominated by the President but subject to confirmation by Congress, would serve six-year terms. In short, the new body would have a mixture of executive, legislative, and judicial functions.

The Civil Aeronautics Act of 1938 was framed to promote "the development of a commerical air transportation system to meet the needs of the foreign and domestic commerce of the United States, of the postal service and of national defense." Since the act went on the books at precisely the time air transport burst into spectacular growth, becoming one of the nation's half-dozen greatest industries, it would be pleasant to say that the law took its place alongside the Constitution and the Bill of Rights as one of the imperishable documents of the Republic.

In fact, less than two years went by before its careful balancing of powers was rudely upset. The instinct for trying to handle a dynamic new technology through a specially designated agency of government is probably a sound one, but allowing appointed administrators to "make law" is full of pitfalls. First of all, Congress did not take kindly to the agency it had created. Members kept up steady sniping. Some charged payroll padding, and one put out a report that the agency's staff was bigger than that of all the airlines being regulated. Even experts had a hard time keeping the respective functions of the agency's three parts clearly in mind. In 1940 the President, empowered by the Reorganization Act of 1939 to make sweeping changes in departments and agencies, took action. Declaring that "friction" within the three-man Air Safety Board and "differences" within the agency on budget compelled the move, Roosevelt abolished the board and transferred the Civil Aeronautics Administrator back into the Commerce Department. At the same time he gave the agency's five commissioners, whom he now redesignated the Civil Aeronautics Board (CAB), the added duty of investigating accidents. That ended the brief try at a semi-independent agency with separate but equal executive, legislative, and judicial branches.

One proviso of the 1938 act conferred immunity from antitrust laws upon the board's rulings. Another empowered the board to grant "certificates of convenience and necessity" — this was an old common-law phrase for public transport franchises — to the airlines that had been flying mail when the law was enacted. Under this authority the board proceeded immediately to confirm title to every existing mail contract route.

Thus the three big transcontinentals, established in such unseemly complicity with Postmaster General Brown, gained permanent possession of their lines. So indeed did all the eighteen contract carriers that survived the airmail cancellation crisis of 1934. All of them were confirmed in what they were pleased thenceforth to call their "grandfather rights."

Landmark legislation though it was, the Civil Aeronautics Act of 1938 in essence only ratified what had gone before. When World War II broke out shortly after and thrust commercial aviation forward by perhaps fifty years, the setup remained the same. The board almost invariably framed its new regulations for present conditions; in such a fast-moving business as air transport, this meant it merely ratified the past — "rear-view mirror vision," C. R. Smith called it. Thus, what was essentially de facto private regulation of the passenger system continued unabated.

In later years, as traffic mushroomed, the "reasonable competition" called for by the 1938 act became fact, as all the airlines began petitioning the board for certificates to fly somebody else's routes as well as their own. The pressures on the board to make awards favoring one airline or another were terrific. Although the arguments were economic, the pressures, of course, were political. Thus competition took a different turn, the self-regulatory nature of the industry declined, and the destiny and gradually the character of airlines came after all to be shaped by decisions in Washington.

Such were some of the longer-term consequences of the Civil Aeronautics Act of 1938. The immediate effect, however, which was the airlines' main concern at the time, was to drive the wolf away from their door. As soon as the new law took effect, the airlines broke out of the red. In 1939 they chalked up their first profitable year since the airmail crisis.

The new law recognized that mail pay was not just reimbursement for services but also a statutory device to gain national objectives beyond the wants of the Post Office Department. One objective, which became obvious in the war that followed, was to have a swift and dependable air transport service serving the needs of national defense. A second was to have such an air net serving the nation's commerce. Both goals presupposed that passengers had become a potentially bigger revenue source for the airline operators than mail. Hitherto the Post Office, by controlling mail contracts, had controlled the very existence of the airlines. The 1938 law, while requiring the airlines to carry mail as the Postmaster General directed, handed the new agency the job of setting mail pay. In determining the rates, the new agency instituted a carefully supervised subsidy system. This, with the permanent assignment of routes, gave the airlines their frantically sought "stability." At once air transport, awhirl with technological change, began attracting capital. Commercial aviation took off.

And what of the law's effect upon safety in air travel?

During the three-year senatorial inquiry into the state of air passenger safety, both the Commerce Department and the airlines made a big effort to pull up their socks. Behind the scenes, the air insurance industry — companies that insured airlines and their planes, not the ones that insured passengers, a wholly different business — were a powerful influence for tightening standards. One of the most valuable services of the Cutting inquiry was in drawing attention to many seemingly small factors (the fact that the field lights at the Kirksville emergency field had been dimmed to save money, for instance) that could conceivably contribute to such an accident.

Accordingly the airlines, the Commerce Department, and its field inspectors all moved thereafter toward strict observance and enforcement of the letter of the rules. Each airline had to submit for approval the operations manual it prepared for its air crews. Along with the detailed information about the airline's standard procedures, these had to spell out the detailed minimum altitudes and ceilings for each airfield along its routes, the approach and departure patterns to be followed, and the weather minimums along every sector of its routes. On the government side, many of its regulations had been issued piecemeal. Many could be found only through department or bureau correspondence. Their enforcement could be questioned because not all had been issued by the Secretary of Commerce, the only person authorized under the Air Commerce Act of 1926 to do so. In 1937 all regulations were brought together in the Civil Air Regulations, thenceforth the airman's bible.

That same year the Bureau of Air Commerce won its first new funds since the New Deal began — $7 million for modernizing and extending the radio beams that provided the highways of air travel. Thus the government was able to plant radio-range stations at such places as the Newfield Pass approach to Los Angeles where the airlines had lost two planes that might not have crashed if radio-navigational aids had been stronger. But the biggest part of the money went to convert all the stations so that the beam signals would not have to go off the air each time the periodic weather reports were flashed over the system. It was precisely the interruption of the beam for a crucial eight minutes for a weather report on the night of May 8, 1936, that may have cost Senator Cutting's plane its chance of locating the Kansas City Airport.

As early as 1922 the airlines had their first midair collision when an Air France Breguet smashed into a British De Havilland that was following the same railroad in the opposite direction along the route between London and Paris. By the mid-1930s there were 360 airliners flying in the United States, and traffic began to get thick around some of the bigger airports. These airports for the most part were owned and run by the cities. As early as 1930 Cleveland's Municipal Airport was using radio to direct planes to land and take off, and in the next five years twenty cities followed Cleveland's lead.

One of the first airport towers in the United States was at Chicago's Midway Airport. In 1932 it was manned by Chief Traffic Control Clerk John Becker. One foggy night that fall, three planes approached, all piloted by strong-minded characters who later held top jobs in commercial aviation. Homer Cole was flying in for Northwest from Milwaukee, Bert Lott was bringing his United Boeing in from the West, and Walter Braznell was headed up from Memphis in his American Airways Stinson. The custom was for the pilot to call in when he was five miles out. All three called and said they were five miles out. There was a pause, and then Becker said, "Lott first to land, Braznell second, Cole third." Braznell cut in and yelled, "I'm only two miles out now — I'm first." Cole piped, "I've passed the Maywood tank — I'm first." There was another pause, and this time Becker said, "Nuts to all of you. You're on your own. I'm going downstairs for a cup of coffee." And he did.

By the mid-1930s twenty big-city airports had radio in their control towers. Traffic control "clerks" manned their posts around the clock, assisted by a couple of radiomen and a fellow who flashed a light gun at planes waiting to take off — green for "go" and red for "hold" — and sometimes at those in the air.

By 1935 the airlines began to feel the need for someone to police and coordinate traffic on the approaches to such airports as Newark. By that time Newark was the world's busiest airfield, with a departure or arrival every ten minutes. The field had a control tower, but it was operated by the city. The various airlines were already exchanging information. It was American's Vice President Earl Ward who finally got action. He said, "Hey, you've got to have a boss — someone to say, 'You fly at 5000, you at 4000 and you at 3000.'"

On December 1, 1936, under Ward's leadership a half-dozen airlines formed an entity called Air Traffic Control, Inc. Their own men manned the first center, which was on the second floor of the Newark Airport tower. The idea was to coordinate the movement of airliners on the approaches to the field and to provide them with weather information. There was no direct communication with the planes. When the controller decided that American's incoming plane should hold at 4,000 feet, he telephoned American's radio operator, who would then radio the instructions to the pilot. In April the airlines started a second center in Chicago and a month later a third at Cleveland.

The only place where the incoming airliner could hold was right over the radio-range station. When he was told he was cleared to land, the pilot would fly on the beam to the cone of silence and then begin what the controllers called a "racetrack" pattern. That is to say, he would make a 180-degree turn that consumed exactly two minutes, then fly two minutes outbound, make another 180-degree turn, then fly back to the station. In another couple of minutes he would expect to arrive at the airport. Thus it would take ten minutes for each plane to land; often the time was longer than that. "Landing four planes an hour was pretty good," said old-time Controller Tirey Vickers later.

But the system caught on fast, and the airline people began to ask, "What about controlling nonairline traffic?" An authority was needed, a traffic cop. It was a good time to go to the government, which was anxious in the aftermath of the Cutting crash to show gains in mastering air safety. In July 1936 the Department of Commerce's Bureau of Lighthouses took over the three air traffic control centers and began opening others.

Now for the first time, the federal government began to get involved in airports, which had been the exclusive preserve of municipalities hitherto. And when the Civil Aeronautics Act of 1938 emerged from Congress, Washington was authorized at last to do something about airports. The newly created federal agency was directed to make a survey of the nation's airports and to bring in a recommendation on whether the government ought to start a program for a national system of airports. The Civil

Aeronautics Authority so recommended and, when war swept Europe and President Roosevelt called on the nation to build 50,000 planes a year, Congress voted to authorize a program that poured some $383 million into concrete runways and other improvements at no fewer than 535 airports.

There *was* a big gain in safety of air travel. On March 26, 1940, the Civil Aeronautics Administrator was able to announce that U.S. airlines had completed their first year without one fatal accident. They had come a long way since the day in 1919 when the passenger boarding the Lawson airliner to fly from New York to Washington sat right down and wrote his will before takeoff. But as every pilot and flight controller knew, they still had a long way to go before air became the safest way to travel.

17

Air Travel in the 1930s

In July 1932 the Democratic party met in convention at Chicago to nominate its candidate for President. Governor Franklin D. Roosevelt of New York was a strong favorite. On the afternoon the convention opened in Chicago, Guernsey Cross, a young man on the Governor's staff at the state capital, invited his friend and neighbor Max Pollet to take a drive in the country north of Albany. Pollet, even younger than Cross, worked for American Airways, which at that time operated one flight a day from Albany as far west as Cleveland. At that time no President, or candidate for President, of the United States had ever taken to the air. Once, on the spur of the moment, former President Theodore Roosevelt had gone for a brief spin in Arch Hoxsey's biplane racer at St. Louis.

Driving off on a gravel side road, Cross stopped the car. After swearing Pollet to secrecy, Cross said that the Roosevelt staff were pretty confident of the nomination and they wanted to get a jump on the campaign by having the Governor appear dramatically in Chicago before the delegates cooled off. Could Roosevelt be flown from Albany to Chicago on a moment's notice?

"We were willing," American's Albany manager Goodrich Murphy said later. "People were scared to fly — and getting the Governor on a plane might spread a little confidence." But the airline had to make a good many arrangements.

The planes on the route — the trimotor Stinsons and single-engine Pilgrims — were not safe enough. They were also too cramped, and the so-called "facilities" were all but useless. We had to engineer transfer of a Ford trimotor. We managed to have a Ford brought up from the Dallas-to-Los Angeles line. They flew it up at night about 36 hours before, and locked it in the only hangar. We got a couple of darn good pilots — Ray Wonsey, the chief pilot of the Eastern division; and Fred Clark, the copilot, had 3000 hours. We built a ramp to get him on the plane. The Ten Eyck Hotel fixed up a picnic lunch. As soon as FDR was nominated, he phoned the convention and said he'd be there tomorrow to address the delegates — and we polished up the plane.

Roosevelt's advisers Louis Howe and Henry Morgenthau wanted no part of the airplane scheme, and if the nomination had come a little sooner, Roosevelt might have caught the Twentieth Century Limited at Hyde Park. But the plane was waiting, and when the Governor went out to the little cow-pasture airport next morning at 8:00 there were seventy reporters on hand. Roosevelt was lifted in and put in the seat nearest the door. On the presumption that there would be a party of six, American was charging Roosevelt $300 for the trip. But no fewer than ten made the trip, including Mrs. Roosevelt and their sons Elliott and John.

The candidate was in high spirits, and pointed out Utica and Syracuse as the plane flew over. It was a rough ride along the Mohawk Valley, flying at low altitude against the wind, and as Murphy said, "that Ford was a balloon." John, age fifteen, was airsick almost all the way. But the Governor chewed gum and worked on his speech with Sam Rosenman. After a stop at Buffalo, Max Pollet, who had donned a uniform and become the cabin attendant, brought out the lunch. He said later:

The weather was improving. We still had storms to the south of us. Lunch was half a chicken, jelly-and-peanut-butter and cream-cheese-and-olive sandwiches, ice cream in small containers and fruit, and a large piece of chocolate cake. The cake was sticky, so afterwards I took the ice pail and filled it full of water, took some paper towels and gave all opportunity to wash their hands.

At Cleveland there was a crowd of five thousand. Again everybody but the Governor got off, and some climbed on the plane to talk with the Governor. Later, when NBC succeeded in hooking up with the plane's radio, Mrs. Roosevelt made her first nationwide broadcast from the sky. "Do you think flying should appeal particularly to women?" asked Pollet. "I think it should appeal to everyone," was her reply. Roosevelt sent a message ahead to Mayor Cermak of Chicago: "Sorry strong head wind makes us a little late."

The flight took seven hours and forty minutes, and the crowd at Midway Airport in Chicago had grown so large and rambunctious that FDR was jostled and his glasses were knocked off in the excitement of arrival.

Roosevelt said, "I regret I'm late. I have no control over the winds in the heavens, and can only be thankful for my Navy training. It could be symbolic that in doing so I broke traditions." "A political master stroke," said the *Syracuse Herald.* "Perhaps it was the Governor's physical condition — a topic that can hardly be ignored — that added to the feeling of uncertainty. The Governor's dash through the air acted as a convention tonic." Will Rogers wrote next day:

FDR made a good speech, and gave aviation the best boost it ever had. He took his family and flew out there. That will stop those big shots from thinking their lives are far too important to the country to take a chance on flying.

The flight unquestionably helped call attention to the fact that, five years after Lindbergh spanned the Atlantic, air transport was becoming available on a day-to-day basis in most American cities.

From the first, the way to get a ride on an airline was to use the telephone. If the rise of the railroads coincided with that of the telegraph, the airlines took off with the help of the telephone. Well before 1930 you could already book a flight reservation by phoning the airline's number. In bigger cities the message was taken at the airports because employees were on twenty-four-hour duty there. Around the middle 1930s, as the airlines began to get faster planes and more business, they began booking reservations at offices opened downtown. The charts for each flight would be sent to the airport when the downtown offices closed, and then brought back again in the morning. So far as the passenger knew, it was enough that he could call the number morning or night and get firm word that there was a seat for him on tomorrow's flight. This may seem commonplace enough, but it was all but unheard of then in the transport business. If you wanted a seat on the train to Chicago or St. Louis or Boston, you had to go and buy it. Airline accountants may have wished the old system prevailed for their flights too, but air travel simply happened too fast for such procedures, and the passenger gained an advantage he never lost. You decided to fly, and then you just picked up the phone. "As the telephone developed, we developed, as the switchboards got more proficient, we got more proficient," said Willis Lipscomb, the mastermind of sales for American and later Pan Am.

Just as airlines broke away from old practices to let passengers book seats by phone, so they cut loose from old-poke ways of paying for them. For a time in the early 1930s, travelers on American Airlines could save time and trouble by availing themselves of the airline's "scrip" plan. Hard up as it was for riders, the airline began sending salesmen to call on their more frequent passengers in their offices with little books of scrip. In return for his check for $350, the customer received a booklet containing $425 worth of slips. He could present his little book to the airline ticket clerk each time he flew, and the clerk would fill out another of the slips until the scrip was finally used up.

Thus the passenger not only escaped the hassle that went with getting a railroad ticket but he got 15 percent off the regular air fare. But something else happened as well. The scrip books hooked their holders into what in the early 1930s was a very select company of travelers. In clubhouses and board rooms along American's route, a scrip book became a badge of distinction. One Dallas insurance company president waxed so proud of his little book that he pulled it out to show his friends. It was his way of saying, "I travel so much that I bought one of these things."

But the owner of the scrip book still had to pay for his flight in advance.

The feisty young airlines, growing up in the age of consumer credit, soon found a better way. In 1936 the industry turned a big corner. Cutting loose from the apron strings of the plane manufacturers, they formed the Air Transport Association. One of its first acts was to create the Air Travel plan, the main feature of which was the Air Travel Card. The credit card, which soon became a way of life in America, was pioneered by the airlines. With the Air Travel Card, the traveler could fly now, pay later. The card was good on any scheduled airline. But the Air Travel Card of the 1930s got you something that no credit card now offers — a 15 percent discount on all tickets bought with it. Like nothing before, these small cards built up the flying habit. Up to this time an executive liked to impress his luncheon companions by saying, "I flew in from New York this morning." Now he simply flashed his Air Travel Card as he pulled out his wallet to pay the tab.

But the greatest prestige flying conferred in these years went with membership in the Admiral's Club. It was C. R. Smith's idea. American's new DC-3s were dubbed flagships; the copilot fastened a multistarred pennant outside the cockpit window before every takeoff. Why not honor the airline's friends, the politicians, bankers, members of local chambers of commerce aviation committees, and the most stalwart businessmen-travelers by making them "Admirals" of the flagship fleet and giving them certificates of recognition?

When told of the plan, American's man in Boston, Bill Bump, was none too sure of its reception. But he was a good deckhand and willing to give it a try. He decided to start with Charles Francis Adams, former Secretary of the Navy and undoubtedly Boston's leading citizen. Arriving at Adams's office, he asked him to stand up, then presented him with the certificate that made him an Admiral of American's fleet. "He loved it," said Bump, still mildly amazed, years later.

It was the same all along the line. When La Guardia Field opened, American's Vice President Red Mosier readied a small and exclusive lounge where the honorary Admirals could await the departure of their flagships in commanding comfort. Inside were easy chairs, a bar, magazines, and, soon, a hostess. And it was all free. Overnight, as businessmen scrambled to be admitted (there are 100,000 framed Admirals' certificates scattered around the land), rival lines tried to catch up by opening President's clubs and Red Carpet lounges to curry their best customers' favor.

That was after the DCs arrived, and as many as a hundred airliners a day were taking off from an airfield like Newark. In the early 1930s things did not happen so fast when you went to get your ticket. All tickets were written up at the airport terminals, and a ticket from Tucson to Dallas to Nashville to New York was a four-coupon affair four fingers wide and two feet long. As many as twelve separate coupons might be required, each one specially stamped, with names of cities, numbers of flights, and fares written out, railroad-fashion. Passengers were urged to get to the airport at least fifteen minutes beforehand — obviously the airlines didn't expect many in those days.

But when the sleek new airliners began entering service, the airlines knew that to keep up with the planes they would have to cut the last-minute paperwork. American Airlines resorted to carbon paper, started writing tickets partially in advance with every telephone reservation, and typed out names in lists so that the agents did not have to write out passenger manifests on the spot. With the founding of their Air Transport Association in 1936, the companies jumped to the uniform airline ticket that, with minor changes, has been in use ever since. With this the air traveler could journey almost anywhere on a ticket that would fit in his breast pocket. C. R. Smith fumed and demanded something no bigger than a cross-town bus ticket. The bookkeepers dug in their heels, and he never got it.

Americans had not yet dispersed in millions to suburbs when the airlines were getting started, and certain sharp-witted fellows thought there might be sizable business in hauling passengers from the cities to their outlying airports. John D. Hertz, the Yellow Cab magnate who took control of TWA for a time in the early 1930s and tried to run the airline like a taxi business, was one of the first to offer transportation to the terminal. Another who thought to combine his automotive interests with air travel was the Wall Street plunger E. L. Cord. After he used his Auburn Motor profits to seize control of American Airways in 1932, passengers rode to his planes from downtown hotels in Auburn cars.

General Motors was so deeply into air transport that in 1931 Eastern, TWA, and Western Air Express were all headquartered in its Manhattan offices. But it preferred to concentrate on making cars. It was content to sell its Cadillacs to local enterprisers who saw a need for transporting people to and from airports. In Phoenix this meant catching McCarran's cab. McCarran was American's agent there. After you bought your ticket at his hotel stand, he took you to the airport in his Cadillac. In Tucson you rode to the airport in one of several dusty Cadillacs operated by Tanner Motor Services.

The father of the business, however, was John F. Carey. Carey ran a haberdashery in Manhattan's Hotel Commodore, where there were lots of calls for transportation to Newark. As a sideline Carey began operating limousines from the hotel door through the Holland tube and across the new Pulaski Skyway to the airport. To make sure his Cadillacs did not return empty, he gave shirts advertising his service to the men at the Newark ticket counters. Specialty manufacturers began building twelve-passenger fuselages for airport limousines with Chrysler, Buick, and even Pontiac chassis and engines, but a good many passengers continued to call them Carey Cadillacs. Still later, many called the diesel-powered buses (forty-eight seated, twenty standing) that increasingly took over as the airliners disgorged thousands, airport limousines.

At no time in the mid-1930s did the airlines cover more than two-thirds of their costs. To this day perhaps half of the population has yet to take to the air; in the 1930s air travel was most extraordinary. What an inspiration it was, therefore, just when improved engines at last gave promise that regular air travel was possible, to place aboard the pitching, rolling airliner a pretty girl to buck up the quaking mortals and persuade them that the miracle of flight was routine. We know the genius who did so. His name was Steve Stimpson, a shrewd, back-slapping Rotarian, manager of Boeing Air Transport's San Francisco office. Stimpson found the symbol of America's organized venture into the clouds — found it in the girl next door.

Early in 1930 Stimpson was barraging his bosses with telegrams. The burden of his messages was that the airline simply must put another member on its flight crew if its passengers were to put up with the racket and buffets of air travel. One day he shot off a wire saying Boeing might try hiring males — Filipino youths who were small, but wiry and willing. Then Ellen Church stopped in at his office. Ellen Church was a nurse at French Hospital in San Francisco, poised and petite. She had seen Ruth Law, the first woman stunt flier, perform at an Iowa fair. She had taken a few flying lessons herself. And as she unfolded her idea, which was that Boeing should hire young nurses as flight attendants, Stimpson forgot about putting males on the job. Of course, it was women, young women as fetching as Ellen Church, who should be looking after passengers on the airline's trimotors.

People were apprehensive about getting in airplanes, and Stimpson figured that being apprehensive about health was a kind of parallel concern. The company wouldn't have nurses on its planes for treatment as such, but half the passengers got airsick, and having someone at hand used to caring for people would help. Stimpson fired off a memo to his boss that wrote the job description for a new profession:

It strikes me that there would be a great psychological punch to having young women stewardesses or couriers or whatever you want to call them. I am certain that there are some mighty good ones available. I have in mind a couple of graduate nurses that would make exceptional stewardesses. Of course it would be distinctly understood that there would be no reference made to their hospital training or nursing experience, but it would be a mighty fine thing to have this available, sub rosa, if necessary for airsickness.

Imagine the psychology of having young women as regular members of the crew. Imagine the national publicity we could get from it, and the tremendous effect it would have on the traveling public. Also, imagine the value they would

The first eight stewardesses pose in 1930: from upper left, Ellen Church, Alva Johnson; from lower left, Margaret Arnott, Inez Keller, Cornelia Peterman, Harriet Fry, Jessie Carter, and Ellis Crawford.

be to us not only in the neater and nicer method of serving food but looking out for the passengers' welfare.

There were two W. A. Pattersons working for the airline in the 1930s. One received the memo as Stimpson's immediate boss and sent a prompt, one-word reply: "No." The memo also went to the other W. A. Patterson, then assistant to the president and shortly to become United's president for thirty years. Pat Patterson bought the idea.

He authorized Stimpson to hire Ellen Church as chief stewardess, and Ellen Church hired seven more nurses from San Francisco and Chicago hospitals, Ellis Crawford, Harriet Fry, Cornelia Peterman, Jessie Carter, Inez Keller, Margaret Arnott, and Alva Johnson. Ellen Church wrote the first specifications: they must be nurses, not over twenty-five years old,

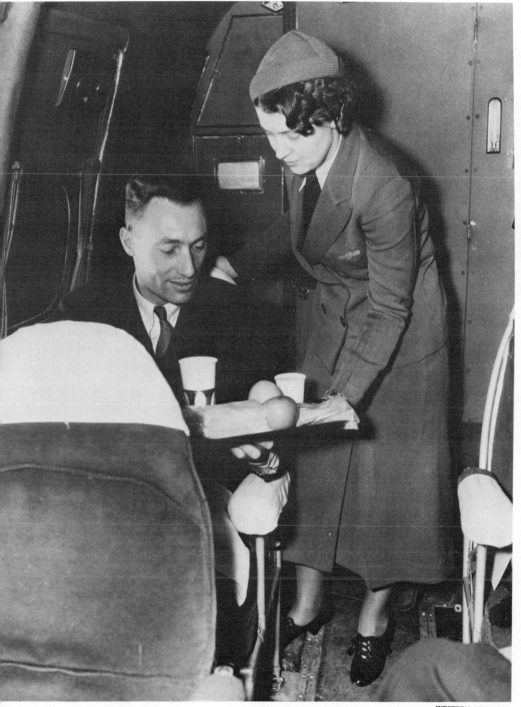

A stewardess serves a box lunch to a Western Air passenger in 1935.

not over 115 pounds, not over five feet four inches. She designed their first uniforms: green wool twill double-breasted jacket and skirt, with a collar, below-knee coat, sensible shoes, hat, and gloves.

While four of the pioneer stewardesses began flying between Chicago and Cheyenne, Ellen Church and the three others started out on the 950-mile run across the mountains between Cheyenne and San Francisco. Their Boeing trimotors made five scheduled stops. Inez Keller, who later became Mrs. Richard Fuite, recalled, "It was supposed to take 18 hours but it was usually more like 24. If the weather got bad, we would land in a field and wait for the storm to clear up." The girls often had to carry all the luggage on board, and if the seats were not fastened down tight, they had to screw them down themselves. "Then we had to dust the whole plane," said Mrs. Fuite. At times they had to join bucket brigades to help fuel an airplane, and on occasion they helped the pilots push planes into hangars.

Aboard the planes, the stewardesses took their lumps along with the passengers. If all twelve seats were full, they sat on a mailbag or a suitcase at the rear of the plane. An important responsibility was to make sure that passengers did not open the exit door by mistake when they were going to the washroom. During flight, these first stewardesses wore white nurse's uniforms. They earned $125 a month for a hundred hours of work — in 1977 flight attendants worked seventy hours for an average of $1,200 a month. Regardless of the time of day, they served the same meal — coffee, fruit cocktail, rolls, and fried chicken lifted out of steam chests that were supposed to keep the food warm. Once Mrs. Fuite's plane ran out of gas and landed in a wheatfield near Cherokee, Wyoming. She said later, "People came in wagons and on horseback to see the plane. They'd never seen one before. They wanted to touch it and touch me. One of them called me 'the white angel from the sky.' " Another time, when the plane was full, the pilot could not gain enough height to clear the mountains outside Salt Lake City. "He finally flew back to the airport and dropped me off," Mrs. Fuite said. "Then the plane made it over the mountains."

Although the passengers took to the new crew members, the pilots did not. Mrs. Fuite said, "The pilots did not like us at all. They were rugged, temperamental characters who wore guns to protect the mail. They wouldn't even speak to us during the first couple of trips." Pilots' wives liked the change even less and started a letter-writing campaign to Boeing charging that the stewardesses were trying to steal their husbands, and should be removed. "One pilot's wife in Salt Lake City was so jealous she always met her husband at the plane," said Mrs. Fuite.

The planes flew at an altitude of two thousand feet. Apart from the forced landings, none of the first stewardesses experienced an accident. Mrs. Fuite found the lack of pressurization hardest to bear. "One day I lost the hearing in my left ear when the plane hit an airpocket near Reno and dropped 500 feet. I quit a few days later because I didn't want to go deaf in my other ear too." All the first stewardesses left to get married. Ellen Church, who married a doctor, died in a horseback riding accident in 1965.

Not for three years did the other airlines follow United's lead. But, Hal Blackburn of TWA said later, there was really no other choice.

United was getting the business. We gave out cards to our passengers to check their preferences. Seventy percent answered they preferred the existing setup of having copilots look after passengers' needs. But they didn't really mean it. They said one thing, but they did another. They took to flying with United. They really preferred having stewardesses.

American accepted the idea in 1933. TWA bowed in 1935. President Jack Frye insisted on calling TWA's girls "hostesses." "They're taking care of our guests," he said. Eddie Rickenbacker, after hiring a few nurses and then cutting out all cabin attendants briefly when Eastern lost its mail contracts in the cancellation crisis of 1934, became the industry holdout. Alone among domestic airlines, Eastern planes carried only male stewards until 1944, when so many males had gone to war that Rickenbacker was obliged to hire women again.

Such planes as United's Boeing 247 were low-ceilinged; on all the early airliners space and weight were at a premium. Some of the first cabin attendants bordered on the midget in size. Yet they had to be husky enough to pull out and stow the sleeper berths in minimum time. TWA's Philadelphia manager, sending on a group of recruits to headquarters in 1935, hailed "the beauty and uniformity of their appearance." In 1937, taking over as American's "bull, stewardess," Newt Wilson ordered the first break from mannish-style uniforms. "For confidence, women have to know they are well-dressed," he announced. The new touch was to make the stewardess's get-up blend with the colors of the passenger cabin, which usually meant outfits of blue or gray.

Too often travel on trimotors verged on torment. At the start, copilots on American's early Stinsons passed out little packets of "comfort." These contained chewing gum, a bit of cotton, and an ampule of ammonia. The first helped equalize pressure on your ears during landings and takeoffs. The second helped you endure the noise. The third was to be sniffed when you felt airsick. The trimotors flew low, underinstrumented and lurching, as Irvin Cobb said, "from airpocket to airpocket." Tommy Tomlinson of TWA recalled, "Airsickness was so general on our southwestern flights that we used to say we had pictures of the Grand Canyon on the bottom of our erp cups so everyone could see the Grand Canyon." In the east the airlines thought it was too bumpy over the Alleghenies to tote passengers at all.

But even when Steve Stimpson was driven to the hospital in his search for succor for the suffering air passenger, some relief was on the way. Engineers and other experts were called in to identify and attack the worst enemies of human comfort aloft. That was not so easy as it might seem. It took years to find out that the propellers caused more racket than the engines; although in all conscience, the engines were bad enough when one of them was mounted, as in the trimotors, on the same part of the plane that carried the passengers. In 1932 the Curtiss company, planning a new biplane, got hold of an acoustician named Dr. Stephen Zand. Zand took sound readings and installed fiber linings in the walls and baffles at the front of the Curtiss Condor, the first "soundproof" plane. When he was done, the cabin filtered in only 1/50,000 of the former sound energy. That was well below the 85 decibels of noise that Dr. Zand said people could handle and sensationally less than the ear-splitting 117 decibels registered aboard the Ford.

By the time the Douglas airliners came along, the experts professed to know the acceptable limits for noise; they had also established a few other rudimentary minimums for airborne well-being. The Sperry Corporation, which had considerably allayed malaise at sea by installing stabilizers on ocean liners a few years before, said that five degrees was all the roll the human stomach could tolerate, whatever the element. In flying machines these limits were proclaimed: acceleration — no more than 0.1 G; vibration — 0.008 inches at the outside; ventilation — at least 30 cubic feet of air per minute per passenger; and atmospheric pressure — about 2½ pounds per square inch inside the cabin at 5,000 feet. Smells, traceable usually to oil vapors or to air heated by the exhaust manifold, proved impossible to reduce to figures, but were very important. Any variance in any of these factors robbed the traveler of his capacity to relax and caused his muscles to tense. The source of most acute discomfort for air travelers was the type of acceleration the engineers identified as "pitching," which passengers complained of as "turbulence" or "bumpy air." In human terms, this sensation excited the fear of falling. Fear was the greatest cause of airsickness, P. R. Bassett declared. Airline medics — Eastern took on the first one in 1935 — unanimously agreed. More than anything else, it was fear of falling that turned air travelers green.

The new Douglas airliners had cabins scientifically noise-shielded by Dr. Zand. Their engines were out on the wings where they were less likely to shake up the passengers. The seats were deep-cushioned and even the arm rests were padded to minimize the bad vibrations that might be transmitted to the traveler. At each seat were individual vents delivering a steady supply of fresh air. The pilots not only walked back frequently to ask you if everything was all right, they were also under orders, by now enforced by federal regulation, to fly sedately. Gone forever was the day when a Colonial Air Transport pilot looped a Ford trimotor with a load of passengers over Boston. And still the silvery Douglas airliners delivered something less than a smooth ride. To be sure, they flew higher than the trimotors, but they still had to fly through, rather than over, the weather.

Today, when a jet carries you as if you were lolling on your home carpet, it seems almost inconceivable that those cardboard receptacles that flight crews called "erp cups" once saw use on every airliner. To the airline and to the passenger, they were a great source of embarrassment. A Pennsylvania Central traffic manager ordered them placed on the floor out of sight because they were "disturbing" to the traveler. Stewardess Margie Giblin read him the riot act and got the cups put back in the pocket on the back of the seat ahead where the passenger in distress would be sure to find them. All too often passengers tried, in their embarrassment, to get to the lavatory instead. Then Florence Nightingale had to know when to intervene. Once TWA hostess Gay Smith had Carole Lombard and Clark Gable for passengers in her DC-3 cabin over Arizona. She saw the man seated just ahead of Gable go green. The man tried to rise, but Gay Smith knew he would never make it. Pushing him back down, she thrust out the erp cup and with perfect timing caught the shower in midair. From the seat just behind, she said later, came "the booming voice of Rhett Butler, 'Thank — you — very — much — Miss — Smith.' " Gay Smith turned and replied, "You — are — welcome — Mr. — Gable."

The heating system on the Douglas airliners was an improvement on that of the trimotors. But it was fitful. On the other hand, the fancy new all-metal airliners could get awfully hot while standing under an August sun before an airport terminal. Passengers complained loudly that they could not open the windows and cool off as they did on the old trimotors. In winter they were just as uncomfortable waiting for the plane to start at an icebound Buffalo or Detroit airport. Finally American decided to take action. At Newark Airport one July day, a resourceful maintenance man named Benny Fuller rigged up a truck with a refrigeration unit and ran a little hose from it into the DC-3 to help cool things off. Before long, units were regularly brought alongside that gave waiting passengers air-conditioning on the ground in both summer and winter.

While copilots continued to hand around sandwiches on other lines, United's stewardesses in their white nurse's uniforms served meals on trays from the first. But what the passenger got was cold chicken. On the first Douglas transports, meals were also brought to passengers on trays, but there was no hot food, only cold sandwiches and many a cold pickle. With the arrival of the DC-3 and DC sleeper in 1936, American was determined to change all that. For the first time on such flights, passengers could have hot food.

The trays came on board, each with its real plates and cups and real knives and forks in place beneath a wax-paper cover. Then, in her little "kitchen" near the front of the cabin, the stewardess ladled the hot food onto the biggest plate from one of four two-foot high Jumbo Jars that had been placed aboard just before departure. These big black thermos bottles made the new service possible. They were supposed to keep food hot for one to one and a half hours without need for further heating. Four of them held enough hot food for all twenty-one passengers and three crew — baked chicken, meat stew, hamburgers, steaks, or even lobster, although that was abandoned because not enough passengers asked for it. Coffee was supplied from a couple of two-quart containers specially designed by Thermos for American's tiny galley.

On the DC-3s the stewardess placed the tray on your lap, usually atop the little pillow that was supplied at each seat to abet your nap. On the DC cross-country sleepers, arrangements were much fancier. The stewardess served your meal on a table, with real Syracuse china and real Reed & Barton silverware. When the berths were stowed away, this plane's passengers sat facing each other knee-to-knee; there was a table fastened to the wall. When it was time for the meal (on the westbound that was 7:00 P.M., following the Washington stop), the stewardess swung the table out to lap level between the facing passengers and dropped a single leg down to steady it on the aisle side. The food, however, came out of the same big black Jumbo Jars.

Well before the next stop at Nashville, the stewardess had pulled the upper berths down from overhead, Pullman fashion. Then, removing one of the seats and swinging the upper part of two seats down so that they made a flat surface, she arranged the lower berth and proceeded to make up the two beds from sheets and pillowcases stowed overhead. By the time she had performed these feats, the passengers had used the *two* lavatories and dressing rooms to prepare themselves for the night. Some passengers insisted that there was no sweeter lullaby than the drone of a pair of Wright Cyclones close by. Others wished for a nightcap — but no liquor was served on planes in those days.

Another stewardess got on at Dallas to assist in the morning excitements. These began when at 6:00 A.M., if all went well, the plane landed at Tucson.

In those days American's hot meals came from wherever the airline could find them. Usually that meant the airport restaurant. At Tucson in the 1930s the airport restaurant was a four-stool lunch counter run by an enormous woman named Mary Wray. When American's sleek new sleepers began flying into Tucson, Mary Wray was nominated to fix the luxurious breakfasts bestowed upon passengers after the Tucson stop. Each morning at four Mary Wray arrived on the field to cook the breakfast. By one of the triumphant advances of airline communication, Dallas had already radioed the number of mouths she would have to feed. Soon as the airliner touched down, Mary and her son lugged aboard boxes as big as

TV sets. In these were the trays with silver and china all in place. Finally they trundled out the four Jumbo Jugs and the thermos containers.

There also had to be hot water for shaving on board when the passengers woke up. Just above the toilet, on top of the plane's fuselage, was an opening through which water was poured into a special tank. It was the station manager's job to see the tank properly filled, and this he did by propping a ladder against the side of the plane, scrambling up with his bucket, opening the panel, and emptying the contents while teetering atop the ladder. One icy January morning in 1939, when it was still dark, Manager Bob Scruggs climbed up, slipped, and poured all five gallons of scalding water down his front. Heedless of onlookers he stripped off his pants; eighteen inches of skin peeled off too. Before takeoff the passengers got their shave water but Bob Scruggs went into the hospital with blood poisoning.

Aboard the plane, the stewardess turned first to rousing her passengers. On Eastern's leisurely Condor flights to Florida, she handed the passenger a glass of orange juice right away. The transcontinentals left no time for such amenities. After Tucson, the stewardess had to take down the bunks, set the seats in place, arrange the tables, and serve breakfast — all in time for the passengers to be rested, revived, shaved, and fed, ready to sally forth fresh and eager into Los Angeles seventy-five minutes later. You could depend on it: the breakfast was scrambled eggs. But there was a choice of bacon or ham, and coffee or tea. Rolls were also served — toast was beyond the state of the art.

Both the Pennsylvania and Santa Fe railroads were numbered among TWA's many ancestors, and for a time TWA's passengers got meals on the ground and in the air prepared by the Fred Harvey Restaurants established in Santa Fe depots. But in the face of such competition, TWA hired Dave Chasen's posh Hollywood restaurant to cater its sleeper flights out of Los Angeles. If nobody was equal to toast, TWA could brag about its sourdough bread with sweet butter in little rolls, a delicacy practically unknown farther east at the time.

United made its culinary contribution too. Having invested heavily in the ten-passenger Boeings only to see them outclassed by the faster and bigger Douglases, United found no choice but to join the parade to Douglas. Pat Patterson's strategy, coming in late, was to get Douglas to give his airline DC-3s that were different. So the DC-3s that Douglas built for United had only fourteen seats in them — swivel chairs like those in railroad parlor cars — and the fare to ride in them was a mere two dollars extra. There was a manifest air of luxury about such travel, and United went to some trouble to give passengers something special in the way of food too.

The airline hired Don Magarrell, a hotel man who set up the kitchens on the United States Line's S.S. *Leviathan.* Meals could be served on planes in the room-service style of a good hotel, Magarrell said. To ensure quality, he decided to prepare the meals in the airline's own kitchens on the ground. America's first flight kitchen lighted its burners in Oakland in August 1937; a Swiss chef plucked from San Francisco's Clift Hotel minded the stove. Though Magarrell wanted something more than a fried chicken wing, there was no chance of matching the seven-course spectaculars his Swiss whipped up at the Clift. The DC-3, moreover, was no *Leviathan*, nor even a Boeing flying boat with stewards moving in and out among dining tables with successive courses. The DC-3 meal had to fit the dimensions of a passenger's lap; it had to be served quickly and in one sitting.

Thus was born the three-course meal that became standard in the air, and at just about the same time, on the ground. The dimensions of the tray determined the size, and within these severe limits Magarrell designed the meal that air travelers have been eating ever since. In the center of his twenty-four-by-thirty-inch mock-up was a twelve-inch depression — that was for the main dish. Laid on at the right were coffee, cream, and sugar. Salad in a paper cup took the upper center, dessert the upper left corner, and appetizer the upper right, and there were little holes for salt and pepper shakers. What then emerged from the atelier of Industrial Designers, Incorporated, was a flat, rigid artifact of pressed pulp, light and disposable, but strong enough to hold things in place in choppy air. A cunningly shaped cover, in matching white with blue and buff stripes, fitted over it.

United's service introduced other refinements — dry ice to keep salads cold, chemically warmed grills to make sure the entrée was hot enough. Because DC-3s flew higher, United's kitchens cut out all fried foods; in the thinner air eight thousand feet above sea level, internal grease was forced to the surface and made food so prepared appear soggy and unappetizing. Whipping cream was used sparingly after stewardesses found it would not stand up at changing altitudes. Coffee that was perfect at sea level was found to lose most of its aroma and flavor when served a mile and a half above ground. The airline experimented with little vents on its coffee urns, but who really expected to get hotel-type room service in the sky when the food came without charge and — by firm airline rule — without tipping the waitress? As long as the DC-3 flew, meals on United's club-car Skyliners, as on all trunk-line planes, were hauled aboard in those big black Jumbo Jars.

Who flew? Most of the people who traveled on planes in the early 1930s "had damn good reason to travel," said C. R. Smith. "Their son fell off a horse, or they had to go to Mayo's — that kind of thing. There wasn't much discretionary about that kind of travel." Movie people, who had business reasons to cross the country frequently, patronized the transcontinentals. Supposedly most of them flew TWA because Howard Hughes was in the movies and TWA was his airline. But at American C. R. Smith was happy to have them too. "There were damn few of *them* afraid to fly," he said later. They were flamboyant, and they were influential. Smith made a point of visiting Louis Mayer, head of MGM, and Harry Kohn of Columbia Pictures, an early New York–Hollywood commuter.

The well-to-do flew. Since flying cost more, air travel was elite travel all through the 1930s. And of these, only the brave flew. A few might take a trip to "see what it was like." Others flew for the exaltation earthlings were still discovering in the sky. At southwestern stops like Tucson and Phoenix, students returning home from eastern colleges at Christmas sometimes arrived by plane. People whose doctors recommended desert air for their lung ailments also took the long flights.

But the dominating motive for the 474,000 passenger flights taken in 1932 was speed. It could not have been anything else, *Fortune* said, because planes were not as safe as trains, and far less comfortable. One in every 2,200 who traveled that year was involved in a flying accident. That was enough to give anybody pause. Of course only twenty-five of these actually got killed — one in 20,000, a somewhat more reassuring figure. Still, in 1932 a $5,000 insurance policy for a plane trip cost $2, for a train journey, twenty-five cents. Wives were still a powerful influence — they swayed men to stay off airplanes after every crash.

Manufacturers' representatives were the backbone of air travel in the 1930s. These were men who had to travel to sell, and the airlines sought their patronage. By 1933, when nearly half of all mileage was flown between dusk and dawn, they claimed that a traveler could save an entire business day by riding a late flight. For a time United's midnight plane from New York to Chicago acquired the name of the Drummer's Special. Those who missed the Twentieth Century Limited's late afternoon departure simply continued their celebration on company time, reported at Newark at midnight, and slept heavily until set down at 7:00 A.M. at Chicago's Midway Airport. The record was achieved by a Parker fountain-pen salesman who missed not only seven Centuries but seven midnight planes in a row. When United's men then rounded him up for the Drummer's Special, he displayed his gratitude by sending gold-banded fountain pens to each and every United clerk in New York. Catching the midnight special did in fact save a full day that would otherwise have been spent on a Pullman.

Willis Lipscomb, American's Dallas manager, said later that, beyond the salesmen, some people in banking traveled by plane. Among the professionals he noted a fair number of lawyers. Oil company salesmen and executives flew almost daily. As Southern Air Transport and later as

American, his airline made so many small-town stops in western Texas in those years that these men became practically dependent on its planes. Individuals in the oil business, wildcatters were also prone to fly. But not politicians, Lipscomb said. And women? "They were practically non-existent — too timid."

All the same, probably the most famous of all airline passengers of the 1930s was Eleanor Roosevelt. Her column "My Day" read like a flight log as she hopped about the country visiting miners' wives, addressing school children, taking tea with mayors. When a night-club comic cracked that she had alighted in every state in the union, the Air Transport Association announced that she had touched down in all but South Dakota — and "My Day" shortly recorded her repair of that omission.

Half the pilots of the airlines had the First Lady for a passenger. Once she was flying to Fargo when Captain Lee Smith of Northwest had to hold his Lockheed Electra on the ground for hours at the St. Cloud, Minnesota, airport. Mrs. Roosevelt sat imperturbably knitting and chatting in the waiting room till the weather cleared. On a trip to Arizona in 1932, Captain Mo Bowen had to make two forced landings, one at Waynoka, Oklahoma, and the other at Tucumcari, New Mexico. "How nice not to have a crowd around," was Mrs. Roosevelt's comment as she ate a snack brought out by the TWA station manager's wife at Tucumcari.

In November 1934 it was necessary to schedule a special night flight to enable her to join the President's TVA site inspection at Harrodsburg, Kentucky. Shortly after the Ford trimotor took off from Newark, she entered the cockpit to say, "Franklin told me not to fly on any unlighted airways." Her pilot was Ernie Cutrell, old-time American captain. "We'll get right down to it," he replied. As Cutrell told it later:

The airway was lighted from Philadelphia to Pittsburgh, and we had to gas up at Pittsburgh. There were beacons also as far as Cincinnati. We got into Cincinnati at 3 o'clock in the morning. We found only a night watchman in a shack. He built a fire. It was pretty cold at that field by the river.

I told her I could go on either to Lexington or Danville, and she wanted Danville. So I called the mayor of Danville at three in the morning. "I have the wife of the President of the U.S. here," I said. "Do you have any lights at your field?" He said no. I said, "We'll be down at dawn. She wants a car." The wife of the president of the U.S. waited in the shack till dawn. When we got to Danville, the mayor was waiting with three Cadillacs. She got to Harrodsburg, all right.

Not many other women made the chancy flight in a Ford over the perilous Pennsylvania mountains, but one was the heiress Barbara Hutton. Many years later Captain Weldon Rhoades remembered the flight that took her from Newark to Chicago. What was unforgettable was not the rough weather but the tremendous pile of excess baggage — "high as an Allegheny" — he put aboard for her.

*   *   *

With the growing number of passengers, it was inevitable that there would be some who created problems. By strict regulation airlines were forbidden to permit anybody intoxicated aboard, and if this should happen it was the captain's express responsibility to get the person off the plane as soon as possible. In those days liquor was not served on any domestic airliner, but of course passengers could not be prevented from bringing a bottle aboard. One evening in 1938 Eleanor Holm, Olympic swimmer and belle of Billy Rose's World's Fair Aquacade, flew into Tucson on a DC-3 and got off the plane without any clothes on. An American Airlines staffer ended up tying her to a palm tree, after which the plane went on without her.

Getting drunks off the plane was probably the chief disciplinary activity of flight crews. Some stewardesses excelled at it, but usually it was up to the captain. In March 1940 TWA Captain Harry Campbell reported that he had been compelled to take measures with a Mr. McPherson. When McPherson boarded the plane at Wichita, he had walked quietly and perfectly upright. However, when Captain Campbell was back visiting in the cabin he observed McPherson rise and relieve himself against the wall of the plane. Captain Campbell reported he then held a blanket around his passenger until he was finished.

By the end of the 1930s, air travel was no longer the world of showoffs, freaks, and tremulous first-timers. In 1940 2,225,000 passengers flew on the domestic airliners. Between January 1939 and August 1940 there was not a single passenger fatality in 150,000,000 miles of scheduled flying. For the third successive year, trip insurance on an airliner cost the same as on a railroad train.

The typical traveler who now took his seat in the DC-3 was Henry Gunderson, a young flour-mill representative from Minneapolis. He traveled to sell. He wore a dark suit. In his firm he was marked as a comer. He had landed his job in the Depression, and he was on the make. He was approaching thirty, not yet married. He was used to getting up early. He took his chances because he wanted to get up and go.

He knew the new Lockheed Electras, and he had an opinion of Stinsons and Pitcairns. He flew a couple of times a month — to Kansas City, to Buffalo, to Cincinnati, and of course to Chicago. He knew the times of departure and arrival. He knew the voice at the telephone, the young man at the airport counter.

He could be in Kansas City tomorrow. He calculated his chances. He noted the stories of crashes in the newspapers, and was not put off. After a real crash, in fact, he flew with more confidence — because he figured there wouldn't likely be another for a while.

He made a good salary, more than a hundred dollars a week. He traveled on an expense account and stayed in good hotels. Already he had an Air Travel Card, and used it on Northwest, Mid-Continent, American, and United alike. The card enabled him to fly on credit and at a discount. The card was for regular travelers — and young Henry Gunderson was a regular. He was the reason why the domestic airlines made their first profits in years as the decade ended.

18

Heyday of Pan Am

On the morning of August 12, 1933, a savage revolution erupted in Havana. Mobs roamed the streets. Noncom-led soldiers poured into the palace. General Gerardo Machado, the Wall Street–backed dictator, was out. A group of young men calling themselves the ABC revolutionaries appeared to be in.

At the harbor Pan Am's twenty-four-year-old acting manager, Sanford Kauffman, learned to his surprise that the ABC was led by his tennis partner. Kauffman called Miami: "We have a revolution."

A flying boat was due with the mail. Out at the airfield, where Pan Am's CUBANA subsidiary operated, Captain Bill McCullough had a couple of little twin-engine Curtiss Robins. At headquarters Kauffman told his people to take reserved passengers only, and not to sell any tickets. Then the phone call from the embassy came: "Get McCullough to take Machado out. He's already on his way to the airport." Kauffman replied, "Why don't you take him out on the destroyer in the harbor?" The voice said, "The United States can't be taking sides with Machado." "Who decided this?" Kauffman shouted. The embassy voice said, "Everything has been arranged between the Secretary of State and Trippe," and read Kauffman a telegram. Kauffman rang McCullough and told him. Machado and a group of senators crying like babies flew off to Nassau in McCullough's Robin.

Word of Machado's flight on Pan Am spread fast. A mob started for the waterfront office. Kauffman had already ordered Captain Leo Terletsky, an ex-Marine, to berth his big plane at the buoy on the far side of the harbor. Kauffman would bring passengers and mail out by launch, so that the plane could take off directly. Kauffman said, "If I wave the checkered flag, I have passengers for you. If you see me waving the red flag, that means no passengers — just take off without 'em."

Kauffman received a startling report from crewmen on the plane: "We have a customs man here and he has Orestes Ferrara with him." Ferrara was Machado's secretary of state — a notorious killer. He had arrived by taxi with his wife, hiding under her skirts. The customs man, obviously bribed, had unlocked the gate and taken them aboard when the plane was still at the dock. Kauffman went out to the plane. "Have you got a ticket?" he demanded. Ferrara said, "No, but I'm in danger of my life. I will not get off. Under international law I'm under protection of the American flag." The man reached under his shirt and pulled out a gun. Kauffman turned to Terletsky and said, "Let's go."

Kauffman wanted the plane to take off on schedule lest Pan Am be charged with collusion. When he returned to the dock, a dozen passengers had arrived — American schoolteachers, tourists. There was just time to get them on the launch when yelling revolutionaries appeared. Kauffman rushed out, waved the red flag. The angry crowd cornered Kauffman in his office, but luckily his old tennis partner suddenly appeared and freed him.

Across the harbor the plane began taxiing for takeoff. But the crew could get only two engines started on one side. Helplessly Kauffman watched as the plane made a big circle that brought it, taxiing fast, close to the dock. The mob was yelling. Shooting started. Bullets pierced the flying boat. One passed right behind Terletsky's head. Another went through the toilet behind which a steward crouched. The crew got the other two engines started. The plane roared across the bay and into the air.

Though the revolutionaries telephoned friends in Miami to knock off the fugitive when he landed, the move failed. Ferrara got away. To each S-40 crew member he gave a gold watch, but another captain flew the mail to Havana next day. It would have been worth their lives for either McCullough or Terletsky to do so. The plane brought a load of returning exiles — Kauffman watched them borne off on the shoulders of celebrating revolutionaries. Pan Am, its cool Havana manager, and its even cooler chief in New York, were still in business. In fact, the next strongman to emerge in Cuba was a young Air Force clerical sergeant who had done typing for Kauffman at the landing field — Fulgencio Batista.

Only heaven knows how many revolutions Pan Am survived as it spread its wings through Latin America. When the State Department arranged for latino chieftains to visit the White House, Pan Am flew them northward, and when the rulers fell, it was usually Pan Am that carried them to foreign refuge. Pan Am lifted oilmen to Venezuela, mining engineers to Brazil, hide traders to Argentina. And when hurricanes flattened Caribbean ports, the predictable first arrival a morning or two later was Pan Am's plane. The 1930s were Pan Am's great years, the years when the airline overcame everything that man and nature put in its path and went on to fly the oceans. It was the heyday of the flying boat.

At the time, Pan Am scheduled one trip a week down the South American coast to Argentina, and it took at least a week to complete the journey. "Very wisely," an English passenger noted, the airline rested content with "beating the speed of ground and sea transport by an attractive margin." The flying boats did not fly when weather threatened, and they never flew at night. Of course clouds and fog occasionally crossed them up, causing crises on the coastal service. Once the flight south from Rio ran into such thick stuff off Santos that Captain Bob Fatt could not locate the mouth of the harbor. Finally he decided to land by the shore outside. It was a hard landing, the plane shipped a lot of water, and Fatt drove it up the beach. He told First Officer Joe Fretwell to go back and tell the passengers where they were. "But what shall I say?" asked Fretwell. The captain said, "Tell 'em any goddam thing you can think of." So the first officer, sloshing in water up to his knees, opened the cabin door and said, "Ladies and gentlemen, we have arrived at Santos, one of our regular stops."

Of course passengers were not the chief source of revenue on these intercontinental flights. At no time between 1930 and 1939, when Pan Am's passenger traffic was rising from 44,000 to 246,000 per year, did mail pay account for less than half the airline's income. In the early 1930s Pan Am's planes sometimes carried as many as 4,000 passengers per week. These were mostly tourists flying to Havana and Nassau — the airline was already operating six flights daily to Havana in the winter of 1931. Four times a week twin-engine S-38 amphibians flew all the way from Miami across Cuba, Haiti, and the Dominican Republic to Puerto Rico. The trip required the entire day, and at the destination there was only one hotel. Alfred Sloan of General Motors, Walter Chrysler of Chrysler, and publisher Nelson Doubleday, all wearing natty plus fours, were among nine intrepid air voyagers from faraway New York who arrived safely in the sunshine of San Juan one afternoon in April 1930. The following year Pan Am pushed the sunshine frontier farther south. Fueling for the big leap at Cienfuegos, a hundred miles west of Havana, Pan Am's Sikorsky achieved the first dawn-to-dusk crossing from Cuba to Panama. This opened up further vistas for the venturesome. Flying over the emerald and azure waters at five hundred feet and a speed somewhat less than a hundred miles an hour, Miss Baracara Lloyd of Miami and five other first-flight passengers got a good look at a giant sea turtle asleep on the sea bottom south of Jamaica.

With an old Navy hand like Juan Trippe at the helm, Pan Am's enterprise was nautical from the first, and why not? Its planes flew out of sight of land. So while budding domestic airlines learned how to fly over farm and field, Pan Am was coming to grips with another element. The airline grew up as an operation apart. It was a seagoing enterprise. As such it uniformed its flying officers in gold-braided navy blue. Ship's bells tolled the time on Pan Am's flights. In the tail of its amphibs were drums of oil to quiet the seas in a rough landing.

And yet at no time in the great years of the flying boats did Pan Am cease to fly landplanes. Between the Texas border and Panama the airline operated Fords, Fairchilds and, before long, Douglas transports as speedy as any flown on domestic routes. Battling the clouds that ringed Mexico City, Pan Am's regional chief pilot Ed Snyder organized instrument flying sooner than many at home. Farther south, Panagra blazed overland trails between South America's west coast and Buenos Aires.

One of the most spectacular routes led through the twelve-thousand-foot pass on the airline's Andean route between Santiago and Buenos Aires. Smack on the border between Chile and Argentina stood a celebrated landmark statue, and Panagra pilots claimed that the tailskids of their planes combed the hair of the Christ of the Andes almost every time they flew over. The airline lost several planes in this pass. Winds that had swept all the way from Australia smashed like ocean waves against the Andean wall. Penetrating this narrow pass, the winds sometimes played strange tricks. On the Argentine side they plunged in terrific downdrafts. But on one sparkling day in 1932, Panagra manager Harold Harris piloted his single-engine Fairchild six-seater into the pass, and met just the opposite. Flying straight and level toward the east, he watched his rate-of-climb indicator. The instrument showed his plane rising 1,200 feet a minute. A glance at the cliff alongside confirmed this. There was room enough, so Harris gunned his engine, and the plane shot forward like a surfer riding the crest of an ocean wave, so high that it was hurled right out of the pass and escaped the usual downdrafts altogether.

Unable to fly over the towering mountains, Panagra pilots had no alternative to braving the pass. A man and his wife were hired to live at the summit where they could watch the snowstorms forming on both sides. Telegraphers both, they tapped into a government line and flashed word each winter day to the airports below. "Each spring when they came down she would deliver another baby," Harris said later.

Next, Panagra's landplanes established a shortcut from Lima across Bolivia to Buenos Aires. The concession, won in 1934, involved setting up landing fields across the Bolivian *altiplano*. As the contract deadline neared, one field remained to be built — in the Aymara Indian country at Uyuni (altitude: 13,000 feet). In spite of the Chaco War then raging between Bolivia and Paraguay, Panagra's Dutch engineer laid out the field in time and reported this by radio. Later he explained how this was accomplished:

I got it done for two bottles of whisky. I gave the police chief a bottle when I got there. We climbed into his police car, and he lassoed every Indian he saw. He piled them in a truck and carted them to a fenced-in compound. Then he told the Indians to go out and pick up rocks.

There wasn't any question of a runway. This was a plain, really an old lake bed, nothing but solid clay mixed with salt. No trees, no grass. It was a matter of removing those rocks. They picked up the rocks and threw 'em in the truck. The Indian women came and put food through the barbed wire fence to the men inside. When the rocks were all picked up, we had our airport. The police chief turned the Indians loose, and I gave him the other bottle of whisky.

Audacity was a quality not lacking in Juan Trippe. Having fought to gain possession of an eighty-mile airmail route between Florida and Cuba, he had transformed it in less than three years into the world's longest airline. Even before the Latin American network was completed, he proposed to branch out across the world's oceans. In 1930 Trippe was already holding talks in New York with Colonel George Woods Humphrey of British Imperial Airways with a view to joint development of a transatlantic route. But isolationism was the U.S. policy between the wars. In those years the U. S. government simply refused to sign any agreement that would give other countries the right to land planes on American soil. Since the British demanded two-way rights, Trippe's Atlantic ambitions hung fire. He turned to the Pacific.

Here, too, Trippe had to shape his bold plans within political limits. When you thought to girdle the world through the air, you discarded ordinary Mercator maps. In his office Trippe kept a globe. On this he was pleased to demonstrate with a piece of string — Lindbergh had shown him how to do it — that the shortest route to China ran through Alaska. This was the great-circle track — a notion not grasped by earthlings in those days — that traced the shortest distance between two points irrespective of sea or land barriers. Indubitably, the great-circle route over Alaska was, by a thousand miles, the shortest line to the heart of Asia. Accordingly Trippe began his campaign by taking over two small Alaskan airlines. But the Soviet Union and Japan also sat astride the great-circle track, and when Trippe sought rights, trouble blocked his way.

He turned to an alternative farther south, longer, and more hazardous, challenging even to Trippe's audacity. This was a route across the very middle of the Pacific, leaping over its most open spaces, bridging the biggest of all oceans at its widest point. It had the merit of avoiding the ice and storms of the foggy north. But its principal advantage to Pan Am was that no foreign permission was needed. Instead, as the government's chosen instrument of international air transport, Pan Am could count on Washington's help all the way across.

As early as 1931 Pan Am sent letters to U.S. plane manufacturers inviting them to build a four-engine flying boat capable of carrying mail and passengers on transocean flights. In 1932 Trippe entered into one contract with Sikorsky to build a seventeen-ton flying boat, and a second with Glenn Martin to construct an even bigger one — a twenty-six-ton giant capable of carrying a payload across the twenty-five-hundred-mile distance between San Francisco and Honolulu. In 1933 Trippe paid a visit to

President Roosevelt, who pored, fascinated, over the maps and charts that Trippe spread before him.

At the time Trippe called on Roosevelt, only a handful of planes had ever flown from the mainland to Honolulu. In the Dole race held a few years before, half the contestants fell in the ocean. To the American public Hawaii was then a remote and little known outpost. But on the principal island of Oahu there were at least a fair-sized city in Honolulu (population: 50,000), a newly opened Navy anchorage at Pearl Harbor suitable also for flying boats, and such amenities as hotels and tourist services. Compared with the way stations farther west, Hawaii was a center of civilization.

Something more than 1,380 miles northwest of Honolulu lay the first of these stepping stones, a sandspit called Midway. At the time Midway was nothing, although the geopolitical importance of the kind of airline-building Trippe was embarked upon became clear a few years later when the decisive battle of the Pacific war was fought for its control. On one of Midway's sandbars was a cable station. Another was the nesting ground for "gooney," (guano) birds. But for Pan Am's purpose Midway was American, one day's flight northwest of Honolulu, and an atoll with a lagoon in which flying boats could nestle. The next landing place in Trippe's skein was even more insignificant. Wake Island, 1,280 miles southwest of Midway, was utterly uninhabited, unshaded from the blazing sun by a tree. Nobody had heard of it; hardly anybody had been there. But the United States had claimed the place ever since Lieutenant Charles Wilkes's *Vincennes* called there in 1839. As nearly as Pan Am could tell — its planners worked from a recent *National Geographic* article — Wake was another atoll with a sand-ringed lagoon, another day's journey for a flying boat beating its way across the Pacific.

From Wake the track led west across another 1,300 miles of open ocean to Guam. A comparatively well-developed island in the Mariana chain, Guam had been administered by the Navy as a U.S. possession since the 1898 war with Spain. This was the last of Trippe's stepping stones, but woe would betide any Pan Am pilot who missed it. Beyond lay nothing for 1,300 miles but the deep blue sea. Then, however, the flying boat reached the Philippines, a populous archipelago of a thousand islands and the principal rampart of the U.S. empire since 1898. At Manila the American flag flew closest to Asia. There the Pan Am route officially ended, although as Trippe told Roosevelt, it was only another 1,150 miles from Manila to Shanghai on the China coast.

Such was Trippe's vision, and no one knew better the potentialities of advancing air technology that would make it practically possible. The link he proposed would tie America's overseas possessions, facilitate the nation's Far Eastern commerce, and foster the never-ending U.S. drive toward westward expansion. With the White House's blessing, the Interior and Navy departments cleared the way for Pan Am's pioneering. The trans-Pacific route took shape as an all-American venture.

While Trippe laid his vaulting plans, others were building the organization to accomplish them. As much as any individual, Andre Priester had a hand in this endeavor. Up to 1928 no U.S. airline had so much as an operating manual. One day an aide handed him Lufthansa's. Priester said, "This is what we want — but ours will be different." Pilots as well as airline engineers will tell you that Priester was the man who set the high standards, who built the high performance, of Pan Am.

Priester saw the sea as Pan Am's element, and he strove to establish for its air operations the traditions of reliability, discipline, and mechanical perfection of the ocean liners. He was never happy with a twin-engine plane. When the Douglas DC-2s and DC-3s arrived and took over almost all domestic airline flying, Pan Am bought them only because four-engine landplanes were not available. Priester started out with the short-range Sikorsky amphibian, but he took the landing gear off (except to beach the plane when it flew to Miami for overhaul). By 1931, when four-engine planes were all but unknown in domestic air transport, Pan Am got its first Clipper. It was the Sikorsky S-40, delivered as an amphibian but soon turned into a flying boat. With this plane, which could fly nonstop across the Caribbean, Pan Am employed its direction-finding radio for guidance as much as two hundred miles from shore and thus acquired indispensable experience for ocean flying.

On Pan Am, as on all early airlines, the men who rose to captain airplanes had learned their skills barnstorming through the school of hard knocks. A few of these men commanded Clippers to the end of their flying days. One, Captain A. E. LaPorte, flew the first passenger flight across the Atlantic in 1939. But to fly the ocean, Priester reasoned, it would be necessary to create wholly new standards for pilots. Furthermore, on the big ocean boats due for delivery, as many as a dozen air crewmen would be needed, at least four of them pilots, to spell each other over the long hours the planes would remain in the air. On such exacting missions, no crew member could possibly know too much.

Priester therefore instituted a broad and rigorous system of training for all. A hierarchy of flying officers came into existence, much like that which prevailed on ships at sea. Four classes of pilots were set up — apprentice pilots, who had everything to learn; junior pilots, who were still learning the ropes; senior pilots, who could be entrusted with command of coastal flights; and finally, masters of ocean flying. These last stood at the summit of the profession. They were men who could take the kind of responsibility for an airliner and its passengers that the skipper of a Blue Riband ocean liner assumed when he took his ship to sea.

The impact of the new setup was immediate. March 1933 was the depth of the Depression. An excited ex-Navy flier named Bill De Lima jumped into a Manhattan phone booth and rang up his Pensacola class-mate Hack Gulbransen. "Call Pan Am," he said. "They're hiring anybody. They just hired me." College graduates, the two were sent off to Miami to join others as the first apprentice pilots. "When we entered training we were hotshot naval aviators," Bill Masland said later. "They handed us a pair of overalls and sent us into the hangar at Miami to stay there until we had earned our engine and aircraft mechanic's licenses, even if it took two years — and in some cases it did." One day Gulbransen was chipping cor-rosion off a flying-boat hull when a mechanic said, "You young ones are stealing our jobs." Gulbransen replied, "But we're going to become pilots." The mechanic looked at him hard and said, "That's what I mean."

Before Pan Am attempted regular service, Priester took the first Si-korsky Clipper, stripped out the luxurious furnishings, installed extra gas tanks, and sent the big plane on a series of survey flights. For a starter chief pilot Musick flew it from Miami to the Virgin Islands and back non-stop — twenty-five hundred miles. This was equivalent to the distance from Honolulu to San Francisco and longer than the shortest North At-lantic segment, Newfoundland–Ireland.

Now began ocean flying. Consider the risks. The big jump to Honolulu was longer than any passage across the North or South Atlantic. Only a handful of planes had ever flown it before. Grover Loening, one of the leading figures in U.S. aviation, resigned from Pan Am's board convinced that the venture would fail. John Leslie, sent by Priester to organize the effort, said later, "The only way to do it was to do it. We pushed the state of the art as far as the space engineers did when they sent men to the moon. We were at our extreme limit, in fact a little beyond."

With Leslie calculating the loads, Musick flew first to Honolulu and back, right on the nose. Next time the Clipper flew to Hawaii and on to Midway and returned. Again they went over the results — carefully, de-liberately. After that Musick took his flying boat to Hawaii, Midway, Wake, and Guam and returned. Finally he flew a survey all the way to Manila and back. "That's how we found out what we could do," Leslie said later. "We kept track for each trip and for each leg — on radio com-munications, on direction-finding, on weather, on crew coordination, on fuel consumption." When all was complete, Priester flew out on an in-spection trip. Later Midway boss John Boyle got a letter: "The base ap-peared to be in operating condition . . . but we do not understand: the bell on the launch was not polished."

For the first scheduled mail flight Trippe brought up the first of Pan Am's great new flying boats. It was the China Clipper, a $417,000 beauty built by Glenn Martin at Baltimore. The plane's huge wings spanned 130 feet, the equal of a Boeing 727 jet of two decades later. Its four 830-horse-power Hornet engines delivered power enough for the plane to cruise at 130 miles per hour with a load of mail and sixteen to thirty-two passen-

Captain Eddie Musick

Pan Am's China Clipper (a Martin M-130) passes over San Francisco on the start of the first airmail flight to Manila on November 22, 1935.

gers for distances up to twenty-five hundred miles. Aboard this big plane in San Francisco Bay on November 22, 1935, were Captain Musick and his ten crew members. Standing at the Pan Am dock at Alameda were Trippe, Postmaster General Farley, and a large crowd.

"The greatest pilot of his era," was what they said about Musick, Pan Am's first master of ocean flying. He was a quiet man, anything but flamboyant. He was handsome, unpretentious, with a dry sense of humor. Though he was scanty with praise, his crew loved him. He had a way of getting into the pilot's seat — wriggling and squirming until he was seated to his satisfaction. Then the crease of his trousers had to be adjusted to his satisfaction. He reached out and pulled at the switches until he was satisfied that they were all exactly right.

Then, ready for takeoff, his demeanor changed. Even his face altered and took on a thrust-forward look. His eyes narrowed — there were plenty of crow's feet in the countenance of a man who had, at thirty-four, already flown 15,000 hours.

As Musick moved the big plane into the bay, a band broke into the "Star Spangled Banner." Schoolchildren given the day off were among a quarter million who watched as the China Clipper roared between the piers of the uncompleted Bay Bridge and thundered out the Golden Gate. The training and the long preparations triumphed. The Clipper landed in Hawaii twenty hours and thirty-three minutes later, three minutes behind the scheduled time.

For a year the three Martin flying boats carried only mail. Then Pan Am flew its first passengers across the ocean. Trippe himself led the first party of senators, bankers, and publishers. At Honolulu they were welcomed as harbingers of a new era of travel and put up at the Royal Hawaiian Hotel. The next night they found themselves in Pan Am's new hotel, a low, screened structure in two wings beside the lagoon at Midway. The next morning all trooped down to the dock and rode the launch out to the big flying boat at its mooring. Just as the sun rose, the Clipper's engines roared; the big boat strained and rose on the "step"; skimming past the coral reefs, it lifted into the blue and turned for Wake. The Clipper flew all day, but when it alighted on the lagoon at Wake, it was already tomorrow, because the flight had crossed the international date line. At Wake, the party found another neat, roomy, prefab hotel on the beach. Welcomed by Mr. and Mrs. J. L. Pinkham (travelers were always met by a woman at these lonely stops), they had drinks and an excellent steak dinner prepared by Pan Am's resident cook. The early morning takeoffs were a necessary precaution in case the flight ran into head winds, but when all went well the Clipper arrived well before dark, and the passengers had a chance to relax and relish the amenities that Pan Am had readied for them. At Guam winds were light and Captain Jack Tilton said later that the flying boat cleared a coral reef at the end of the takeoff run by a scant two feet. But by five in the afternoon the Clipper was circling Manila and making its descent at the Cavite naval base.

For a brief time the Philippines were the end of the line. Trippe blamed "disturbed political conditions" in the Far East for holding up flights to Asia. Passenger traffic beyond Honolulu, he told stockholders, was "small." Rich Filipinos like the Elizaldes and Sorianos were early trans-Pacific passengers. When it proved impossible to obtain landing rights in China, Trippe won permission to use Portuguese Macao as Pan Am's first mainland terminal. That helped him gain access to British Hong Kong, where passengers could board the Sikorsky amphibians of Pan Am's Chinese subsidiary CNAC, and fly to Canton and Shanghai. For that brief hour before the Japanese militarists occupied the China coast and prepared their blows at Hong Kong and Pearl Harbor, Pan Am carried the U.S. flag by air to the farthest shore of the Pacific. At that moment of high achievement, Priester told young Gulbransen, "Seaplanes are the only way to cross the ocean safely. There will never be a landplane operation."

To keep the fortnightly trans-Pacific schedules (every missed trip cost Pan Am $10,000 in mail pay), Pan Am's air and ground crews performed prodigies. Once Captain Steve Bancroft plugged a hole in his flying boat's hull at Guam with a thousand pounds of concrete and flew on. Another time Tilton, the skipper who had also been U.S. vice consul at Liverpool, scraped bottom at low tide at Midway. A half-dozen native stewards were posted on one wing to tip the big plane over just far enough so that mechanics, lying on their backs in the water, could affix patches on the hull in the night. The five passengers, held up for two days, rounded up the year's most prized travel mementos — glass balls torn from Japanese fishing nets seven thousand miles away and cast up on Midway.

That year the Martins carried 106 passengers, a very special group of travelers indeed, quite literally the chosen few. Captain Tilton, preparing to fly to Hawaii, would draw up a number of flight plans, then go to the meteorological office and pick out the plan whose route and altitude would require least fuel. Finally, after ordering the fuel, he would tell the traffic manager how many passengers he could take. Then the agent would go into the lounge where the passengers sat waiting and point: "You, and you, and you can go — and the rest of you can go on the flight next week." On one trip from Honolulu in November 1936 Tilton had room for only one. The young woman chosen was understandably nervous. After the takeoff she did not go to bed until far into the night. While she slept, head winds got stronger and stronger. Fourth Officer Hack Gulbransen, understudying the navigator, took star sights every fifteen minutes. The engineer reported a heavy drain on fuel. At last, fifteen and a half hours out of Honolulu, Captain Tilton decided to turn back. With the wind behind them, it took only eight hours to return. The purser did not rouse the passenger until the plane was passing Diamond Head. Gulbransen said later, "When we landed at Pearl Harbor, she went out like a scalded cat. She went to the mainland by boat. We went out the next night — empty."

The pioneering was costly. In 1938 Pan Am's Pacific losses devoured all but $75,000 of the $1.2 million profit amassed by the Latin American division. In July one of the three Martin Clippers was lost without trace between the Philippines and Guam. And an even heavier loss was suffered when Captain Musick's Sikorsky S-40 flying boat exploded off Samoa.

Taking off from Pago Pago harbor during one of his survey flights on the southwest Pacific route, Musick reported an oil leak and said that he was turning back. It was the last ever heard from Eddie Musick. Witnesses told of seeing a flash and hearing an explosion out to sea. Later, Samoans went out and recovered pieces of the wreckage. Almost certainly, Musick had been dumping gasoline to lighten the ship for a landing. In those days gasoline dump valves were flush with the bottom of the wing. If conditions were right, as Pan Am found out by reenacting the maneuver later, gasoline (either as liquid or vapor) would wet the wing and ignite when the flaps were lowered for the landing approach. From that day on, all planes were refitted with dump valves that protruded far enough to keep gasoline from streaming back along the wing.

Pan Am never had another pilot as famous as Musick. In his day his renown equaled that of the astronauts of the 1960s. But all of Pan Am's early masters of ocean flying were redoubtable figures. At Treasure Island, Pan Am's base throughout San Francisco's 1939 World's Fair, the Clipper's 3:00 P.M. departure each Thursday was a top spectacle, and people crowded to get the captain's autograph. Barbara Hutton, the Woolworth heiress who flew often to Honolulu, would call the Pan Am operations manager and say, "I want to go today." She would always ask, "Can I have Captain Bancroft?" Member of a prominent San Francisco family and nephew of actor George Bancroft, Captain Steve was a handsome and dashing fellow who had been an all-Pacific Coast tackle on the great Berkeley teams of the early 1920s and a man-about-town from Rio to Honolulu. When Ernest Hemingway flew to the Islands, he asked for Bancroft or Bancroft's pal John Hamilton. Hamilton, it was said, looked and drank — off duty, of course — like Hemingway. Not all of the masters of ocean flying were supermen, however. After the Caribbean, Captain Ralph Dahlstrom found Pacific flying unsettling. Such long trips frayed his nerves. Younger pilots said that when he strolled back to the tail and the plane took a little dip, he would rush to the front and shout, "Keep it level, I told you." Dahlstrom transferred back to South American coastal runs.

By the mid-1930s all the energies of the Pan Am planners had turned toward building an Atlantic air bridge to Europe. But the British still withheld landing rights at Bermuda. The French gained an exclusive franchise from the Portuguese to land at the Azores. And on the great-circle route farther north, the indispensable jumping-off point was Newfoundland, then a British crown colony. No U.S. plane could hope to fly there without London's permission.

A deal was in the making. By an exchange of letters between Secretary of State Cordell Hull and Sir Ronald Lindsay, British ambassador in Washington, an agreement for reciprocal landing rights was achieved without having to seek Senate ratification.

Flying boats had been the first commercial aircraft to span the Pacific, and now they would lead the way across the Atlantic. During this same exchange the two sides struck a gentlemen's agreement: the British and American companies would develop transatlantic service together. When Pan Am, under this arrangement, promised to pool passengers and cargo with Imperial Airways, the British government promptly authorized Pan Am to fly the Atlantic. All was on a reciprocal basis, and of course each airline would be its country's only participant. The deal was not exactly like the pact made nine years earlier when the chief of Standard Oil met in Scotland with the European oil barons to carve up the world petroleum market. After all, the Atlantic air transport business had not even come into existence as yet. Still, about this cozy compact there was an air of the big shots' sharing out the plums between what were already the world's two biggest airlines, and it made trouble later, especially for Pan Am.

So the survey flights began, with great attention to making sure that the British and American planes should carry out their crossings at exactly the same time. It was even arranged that they should pass in mid-Atlantic. For experimental purposes the British had taken one of their famous Empire flying boats (admirable planes, but with a normal range of only five hundred miles) and crammed it with 2,320 gallons of gas. Pan Am again sent out a Sikorsky S-42 Clipper stripped bare of all fancy fittings and with the cabin full of extra gas tanks.

Confident after a few months that its pilots could manage the long jump, Pan Am pressed for launching transatlantic service. The British held back. They were not ready. Their Empire flying boats, suitable enough for coastal flights to India and Africa, simply lacked the range for flying mail and passengers across the ocean. But until their own airline could start full operations to the United States, the British would not allow the Americans to serve the British Isles. To let the Americans fly alone would have been damaging to Britain's prestige.

For the opening of Atlantic passenger service, Pan Am had ordered a huge flying boat from Boeing. Boeing's Clipper was the creation of Wellwood C. Beall. It has been said that the five aeronautical engineering schools set up on university campuses by the Guggenheim Fund for Aeronautics in the 1920s won American air supremacy in World War II. Beall from New York University, Kelly Johnson from Michigan, John Stack from MIT, made tremendous contributions. As Boeing's engineering chief, Beall produced the B-17 Flying Fortress, the B-29 Superfortress, the postwar B-47 and B-52. But he got his start building superplanes when he designed Pan Am's biggest Clipper.

In 1936, returning from selling Boeing fighters in China, Beall heard that the company had turned down the chance to build Pan Am's big plane. He was twenty-nine, a naval reserve pilot, and he had thoroughly examined Pan Am's first ocean-going Sikorsky Clipper with Musick at Shanghai. He said, "Look, we've got this four-engine XB-15 bomber design for the Army. Take that wing and put a hull under it." At home on his dining room table Beall drew the design, "and I hooked that wing on it." Boeing's top man Claire Egtvedt liked it. Pan Am liked it, though the price of $485,000 "really blew 'em." However,

our plane was bigger, more comfortable. Strength, sturdiness, stamina were my criteria for the structure. Seaworthiness was the big requirement. This was the first airplane for which I was responsible. I signed every drawing.

They wanted it luxurious, spacious. I conceived of the door as the door to a salon. I put in a spiral staircase. Hell, you can't do that, they said. Had to put in railings on the side.

If the Martin 130 had been a giant, the Clipper that Boeing built for Pan Am was bigger by a whole order of magnitude. It was the biggest commercial plane to fly until the advent of jumbo jets. Its four 1,500-horsepower Wright Cyclone engines were the most powerful fitted on any prewar commercial plane. The wing, spanning 152 feet, was built like a bridge; there was a walkway through it with room for mechanics to go out and work on any one of the four engines. The main body of the plane was double-decked, the flight deck on the upper level surpassing the cabin of a DC-3 in size. Interior spaces of such proportions were never seen in an airliner until Boeing's 747 thirty years later. On the flight deck were positions for eight flight officers including a navigator's table six feet long. Junior pilots sometimes spread a blanket and sacked out beneath the table between watches.

Passenger quarters resembled those on a Cunard ocean liner. In Beall's salon were tables at which passengers were served their meals, the captain presiding at one and the first officer at a second. Other crew members joined them, depending on the watch schedules. Leading aft was a companionway sometimes called the promenade deck. Seven commodious compartments opened to the sides, each with straight walls, flat ceilings, soundproofing, and fresh-air ducts. Each was fitted out like a drawing room, furnished with big lounge chairs and four sleeping berths. At the rear was a "honeymoon suite."

Such a huge plane took time to build. When Beall and test pilot Eddie

Allen first took it out, they could not get it more than fifty feet into the air. Beall then added two more huge fins to the tail surfaces, and by early 1939 it was ready for delivery to Pan Am.

One feature Beall insisted on introducing in the Boeing Clipper was flush toilets. Up to that time airliner toilets simply discharged through a hole in the bottom of the plane. Determined to end passengers' exposure to the elements, Beall personally designed the first aerial water closet. A sort of drum mechanism held a bucket inside. When the lid was raised, washwater that collected in a catch tank flowed into the bucket. When the passenger then put the lid down, the drum turned upside down and dumped downward into the ocean. Having tried the device out at the plant for a year, Beall anticipated no problem. But when they took the big boat on its maiden spin to Honolulu, Andy Priester woke Beall in the middle of the night. Both toilets were stuck. Beall and his mechanic got to work. But again on the return flight both toilets broke down. Finally the mighty Boeing engineering department came up with the answer: Boeing's toilet paper was three thousandths of an inch thick, but Pan Am's was ten thousandths of an inch thick and ruffled besides. Pan Am's paper was just thick enough to catch on the rim and stop the bucket from turning. Beall said later, "We changed the paper, and the toilet worked."

The delay the British imposed on the Atlantic start cost Pan Am dearly. The gentlemen's agreement Trippe made with Imperial Airways weakened his long campaign to keep all U.S. overseas commercial flying for his airline. For a decade his drive to fend off competition had succeeded on all fronts. An isolationist Congress and country, hostile to entangling pacts with foreign governments, had been quite content to let Pan Am chase after landing rights on its own. If any U.S. business interests thought to join the pursuit, Trippe shrewdly headed them off. At the beginning, when the three big combines controlled domestic airlines, he invited them in as Pan Am shareholders and thus dampened their zeal for branching out overseas. Later, when Pan Am expanded further, he kept shipping interests from starting rival airlines by making the Grace and Matson lines his partners in Latin America and the Pacific. But in the Atlantic the deals and delays created an opening for what Trippe had hitherto managed to prevent. Competition reared its head.

American Export Lines was an up-and-coming shipping company that in a few years had built up a profitable business operating a handful of ships to the Mediterranean. In 1936 the company was on the point of expansion. But the next step was a big one. Did the firm really want to make the investment that building a superliner like the *Queen Mary* or the Italian Line's *Rex* would entail? The U.S. Maritime Commission had just made an arresting study. Of the twenty thousand passengers crossing the Atlantic yearly, the agency calculated that at least one-fifth might go by air if there were such a service. The canny people in charge of Amex, led by President John M. Slater, thought it might make more sense to forgo the grand ship and develop a transatlantic air service instead.

So American Export Airlines came into existence, acquired a long-range twin-engine Consolidated Catalina flying boat, and began survey flights across the Atlantic. Trippe tried to buy out the new company, but to no avail. By May 1939, having ordered big new four-engine Sikorskys of transocean range, American Export applied to the Civil Aeronautics Board for a certificate to fly the Atlantic. Slater's lawyers pointed out that the Pan Am–Imperial pact for dividing Atlantic air traffic between them flew in the face of U.S. antitrust laws. That caused a stir at the Justice Department. Then Slater roused the House Merchant Marine Committee to wrath by reading from Pan Am's contract with Portugal, barring anybody else from landing on the Azores for fifteen years. Now it was Congress that began asking questions. Under such pressures, Trippe made it clear that Pan Am would not agree to the pooling of passengers with Imperial, and Pan Am surrendered its exclusive rights in the Azores. The Aeronautics Board granted Amex a temporary certificate to fly to Lisbon.

Just in time for defense of its heavyweight championship, the first of Pan Am's Boeings arrived on the East Coast. Not only American Export but also Air France was getting ready to fly Atlantic seaplane tests. The Germans had already jumped ahead by sending over a four-engine Focke-Wulf landplane that flew nonstop to New York and back. The gentlemen's agreement faded away, although Imperial Airways, stripping two Empire flying boats and refueling them in the air, gamely flew fifteen transocean mail flights in the summer of 1939.

Still, Pan Am was not to be hurried. Priester went ahead with his customary thoroughness. Since it was a little early in the year to fly through North Atlantic weather, the Boeing Clipper's first proving flight, on March 27, 1939, began at Baltimore. It was to cross by way of the Azores, bypassing Bermuda altogether.

Once again, planning and training made the performance look easy. Bystanders asked how such an enormous machine — the *New York Times* called it "the largest plane in the world" — could even lift into the air. And the forty-ton flying boat, laden with eleven crewmen, ten officials, and 4,200 gallons of fuel, raced fifty-eight seconds through the waters of the Patapsco River before rising ponderously aloft. But Captain Harold Gray, a master of ocean flying (and future Pan Am president), was at the helm, a crew that had passed a thousand of Priester's tests was on the job, and Pan Am's long-range direction-finders traced and reported the Clipper's track hundreds of miles out to sea.

Not quite four hours after departure, Captain Gray reported turning Jones's Corner — the invisible crossroads east of Bermuda where early fliers on the southerly great-circle track turned to follow the thirty-fourth parallel eastward toward Europe. At 10:30 P.M., when the plane's radio swung from Baltimore to the Azores for its directional bearings, Gray passed another milestone in the sky — the point of no return. There was now no turning back because the plane was nearer its destination than the place it had started from. At this moment of maximum hazard, the Yankee Clipper, nearing the Azores on its pathbreaking way to Europe, radioed confidently:

1,377 miles out. Making 157 miles per hour. Altitude 8,100 feet. Every star in sky now visible. Utilizing Arcturus, Spica, Polaris, Sirius, Capella for fixes. Horta station came in, reporting our signals clearly.

In May Pan Am won its mail contract and began flying mail across the ocean. In the sense that flying mail was a further preparation for trans-ocean transport of passengers, these were test flights too. Summer's good flying weather followed; with fourteen test flights successfully completed, Pan Am was ready to go. In June the airline published the first Atlantic timetables: departures every Saturday at 7:30 A.M. from Port Washington for Southampton, arrival the next day at 1:00 P.M.; departures every Wednesday at noon for Lisbon and Marseilles, arrival at Marseilles at 3:00 P.M. the next day; fare $375 for either trip one way.

The day came for which Trippe had worked for twelve years. It was the day when the airline opened regular passenger service to Europe. Some five hundred names were on the waiting list. First had been that of Will Rogers, undoubtedly America's number one air traveler in his day. When Rogers was killed flying with Wiley Post in 1935, the name of Bill Eck, whose reservation had been on file since 1931, took top place.

Eck, a Washington railroad executive, arrived at Pan Am's Port Washington base early on June 28, 1939. It was a great day for flying. A fresh breeze blew from the south. There was hardly a cloud in the sky. The Long Island town had declared a holiday, and several thousand flocked to the waterside. The high school band, dressed in blue and white, swung into "Flying Down to Rio." Then after the Presbyterian minister offered a prayer, a bell sounded at the dock. Immaculate in blue, Captain R. O. D. Sullivan and his eleven crewmen marched out, two by two. The ground crew lined up at attention. Bill Eck, the $675 round-trip ticket clutched in his hand, was the first passenger to enter. Then came Mrs. Clara Adams and Julius Rapoport, both of whom had reservations to fly with Imperial Airways from London to Hong Kong, and from there to continue their air trip round the world aboard Pan Am's Pacific Clippers. Among the twenty-two paying passengers were six women in all. As one walked across the landing stage, her little daughter called, "Write me a letter." She laughed and said, "I'll be back before the letter."

The Dixie Clipper's Cyclones exploded into life. The great plane swung slowly and taxied out around the point. Then watchers heard a roar as

Passengers board the Boeing 314 Dixie Clipper at Port Washington, New York, on June 28, 1939, for Pan American's first revenue transatlantic flight.

Captain Sullivan gunned his engines. Moments later they saw the big plane rise over the trees, circle slowly, and point its nose toward the Azores. A new era had begun in travel.

On July 8, the Yankee Clipper, carrying seventeen passengers, began service on the northern great-circle route, and Pan Am stepped up to weekly cross-ocean service — but not for long. Two months later World War II broke out. The U.S. Neutrality Act took effect. Barred from flying to any country at war, Pan Am terminated its scheduled flights at the neutral ports of Foynes and Lisbon.

Of eighty-seven scheduled trips that first winter, Pan Am managed to complete only forty, and many of these reached their destination only after big delays. On westward trips the Boeings ran into the stiff head winds that still bedevil Atlantic travel and were compelled to make an extra fuel stop at Bermuda. But that was nothing compared with the troubles at the Azores. There ocean swells off Horta, the Pan Am base on the eastern side of the principal island, played havoc with scheduled arrivals and departures. Though the harbor lay behind a breakwater, it was small, and flying boats had no choice but to land and take off on the sea outside.

Pilots, and soon passengers, came to dread the swells at Horta. The airline made a rule: if Horta reported swells running "three feet occasionally four," planes could not land there; if the swells ran "two feet occasionally three," planes could not take off from there. The predicament of a Europe-bound pilot approaching his point of no return was especially trying. Again and again captains already six hours in the air heard the lugubrious report from Horta: "Swells three feet occasionally four." Then they had to turn around and fly all the way back to Bermuda. Passengers and crews were held up at Bermuda or Horta as long as two weeks waiting for the waves to let up enough for their Clipper to take off.

In the midst of these vicissitudes occurred the very kind of disaster that Pan Am's flying boat men most dreaded. The accident happened right on the Clipper track to Europe and, so to speak, in full view of the American public. It involved a British seaplane bound for Bermuda, and the impact was great for Pan Am and for the future of flying boats in Atlantic service.

Back in 1937 Pan Am and Imperial Airways had sealed their transatlantic pact by inaugurating a joint passenger service between the United States and Bermuda. Pan Am maintained its weekly schedule with a Sikorsky Clipper; Imperial Airways operated the Empire flying boat *Cavalier*. One day in February 1939, the *Cavalier* was halfway from Port Washington to Bermuda with a load of nine passengers, five of them Americans, when the plane ran into a weather front. After a bit the engines began to labor. The plane gradually lost altitude. Captain Jack Alderson diagnosed the trouble and sent off a radio message: "Difficulties with carburetor icing." Finally, their air intakes clogged, all four engines quit. The crippled *Cavalier*, having radioed its position, nosed down toward the waters of the Gulf Stream.

The *Cavalier*'s seats had inflatable life belts fastened into them, but no seat belts. First Officer Richardson told everybody to hold fast to their seats. When the plane came down, it struck the crest of a big wave. The hull split halfway back, and water rushed in. Disregarding Richardson's order, Gordon Noakes, a Manhattan furrier, had left his seat. He was thrown against a bulkhead, fracturing his skull. Except for William Talbot, a tall young man from Harvard who suffered a broken arm, Noakes was the only casualty. But the door flew open at impact and Donald Miller, a Lincoln, Nebraska, insurance man, leaped out and was never seen again. The plane stayed afloat for ten minutes. For the others there was time to don life belts and scramble up on the *Cavalier*'s wing. Then, as the plane went under, the twelve survivors slipped into the water.

At Captain Alderson's direction they formed a ring, seven men and four women, holding hands. Then Alderson swam round tying the strings on the life belts together so that they were doubly linked. Noakes, unconscious, had floated free of the plane, and Alderson carried him to the middle of the circle where he removed his own life belt and placed it under Noakes to hold him up.

Hours passed. An error in relaying the *Cavalier*'s location slowed rescuers. Alderson, struggling to keep Noakes afloat, grew exhausted. Edna Watson, a big woman, said to herself, "I'm going to save that man," and kept Alderson afloat even after he passed out. Night fell, Noakes died, and in the same desperate hour Steward Bill Bennett gave up. He simply detached himself from the circle and slipped away into the dark.

It was the Standard Oil tanker *Baytown,* answering the SOS, that found the little circle of bobbing heads just before midnight. Its boat put out, and strong arms pulled the survivors from the water. They were just in time. Alderson was unconscious and the rest, having been in the ocean twelve hours, were at the end of their strength. Of course, had Alderson not pushed his plane on far enough to hit the Gulf Stream, they would have gone down in waters thirty degrees colder and none would have survived.

Front-paged in all U.S. newspapers, theirs was a harrowing story. *Life* diagrammed the drama, called them heroes. In Hollywood Sam Goldwyn rushed to make a movie, *Thirteen People Went Flying.* Pan Am mustered enough clout to stop him, but the accident also shocked the airlines. From then on no plane flew to sea without rubber rafts on board. And the crash of the *Cavalier* hastened a moment that was advancing upon American commercial aviation faster than anyone knew.

It is customary to say that World War II, thrusting flight technology forward by decades, sounded the death knell of the magnificent and luxurious flying boats. Actually the bell tolled when the *Cavalier* went under. Looking back, one sees how the crash of the Fokker trimotor with Knute Rockne aboard spelled the end of the wooden commercial planes and ushered in the era of the modern all-metal airliner. In much the same way, following the crash of the *Cavalier,* the flying boat soon lost its primacy in transocean flying. Lindbergh had said, "The use of the flying boat will be temporary," and there was no question that carrying its own emergency landing facilities — that's what the hull really was for — made it heavier than the equivalent landplanes. Airmen already knew that on long-distance journeys the way to triumph over weather was to fly over it, and pressurizing the cabin to make this possible was much easier with a landplane than a flying boat. The most discerning of them also knew that the way to reduce accidents was by increasing the reliability of flight rather than by multiplying facilities for forced landings.

These were arguments Lindbergh made to the airline's management.

And they prevailed. Pan Am, which had pioneered the oceans with flying boats, never ordered another. In his memoirs Lindbergh told how Priester, the great Priester who had overridden his objections four years before, now changed his tune. Within three months after the crash of the *Cavalier*, Pan Am placed a huge order for landplanes with ocean-flying capabilities. On June 11, 1940, Pan Am signed a contract to buy forty four-engine Lockheed Constellations for delivery starting in 1943.

Pearl Harbor compelled cancellation of such contracts, and Pan Am never got its high-flying Connies until after the war. So the airline operated its flying boats, and it says much for the big boats' performance that as long as regularly scheduled airlines continued to use them, there was never another case of a forced landing in the open Atlantic. Indeed, when World War II broke out, only Pan Am's giant Clippers could keep up the ocean service.

The service in no way diminished, although the volume of mail restricted the number of passengers. In the Pacific, travelers like Minister to China Nelson Johnson and journalist Clare Boothe Luce island-hopped regally across the widest of oceans, eating their seven-course dinners and taking their nightly ease in Pan Am's lagoon-side hotels. Nobody had to sit strapped down and crowded between neighbors on a Boeing Clipper. With almost as much room as a jumbo jet, these big planes never carried more than thirty-five passengers. On the Atlantic flights, as on the San Francisco–Honolulu trip, passengers turned in and slept through the night in berths big enough for Wilt Chamberlain. Rising, they found their shoes freshly shined. In the steward's galley was a call-board with bells for every seat; a coffee-maker was kept ever ready. In the dining saloon meals were served at tables set with linen, silver, china, fine crystal, and fresh flowers. Dinner might feature a standing rib roast brought aboard by the purser just before departure. In the next seat you might find Wild Bill Donovan, the New York lawyer traveling to Europe on a supersecret intelligence assignment from the President; William Purvis, chief of the British Purchasing Mission on his way back to report to the cabinet; Del Paine, the *Time* editor flying to take over as the news magazine's London bureau chief; John Lacey, the State Department's marine sergeant courier, with his diplomatic pouch chained to his wrist. That was Atlantic air travel just before Pearl Harbor.

While Pan Am expanded to ocean-spanning dimensions, its Latin American involvements also demanded Trippe's attention. Time changed, and so did Washington policies. Rousing to the Nazi danger, the State Department sought ways to counter growing German influence in Latin America. An early alliance that Trippe had struck with a local airline returned to plague him. The airline was run by Germans. At the time, the deal with Dr. von Bauer's SCADTA had gained Pan Am the indispensable right to fly through Colombia from Panama to the rest of South America. Now, as World War II neared, the diplomacy of Juan Trippe collided with that of the State Department. In the crunch, Trippe very nearly lost control of his company. Pan Am itself was hurt.

According to Trippe's 1931 pact with von Bauer, Pan Am had acquired 80 percent ownership of that company. But the price of that deal was a secret contract agreement that guaranteed the jobs of SCADTA's German staff. Then and later, this seemed highly desirable to Trippe. Inside Colombia SCADTA's Germans, many of whom took Colombian citizenship and wives, were held in high esteem. They had built up an air service that was essential to the country's economy. They had gained lucrative mail contracts. To Trippe it seemed best to let them run the airline and to let the Colombian public see that they did so.

But with the rise of Hitler in Germany, the United States became alarmed at the situation in Colombia. Not only was German commercial aviation expanding all over South America; the State Department had word that the Nazis might be sending in German air force pilots to fly for SCADTA. In Washington, generals and admirals nearly hit the roof at the thought of Luftwaffe types flying daily within a few hundred miles of the Panama Canal. The State Department feared that the German hold on the country's air transport might make Colombia susceptible to hostile influence.

Trippe, busily managing Pan Am's international affairs in his own personal fashion, was caught in the middle. When the terms of the 1931 agreement had to be divulged, his directors became anxious. For the first time the company's biggest stockholder, Cornelius Vanderbilt Whitney, asserted himself. Hitherto holder of the honorific title of board chairman, Whitney announced that he was taking over as Pan Am's chief executive officer. At Chrysler Tower headquarters in Manhattan, he moved into Trippe's big corner office and began holding daily meetings with department chiefs. Trippe took a little office nearby. Executives, at a loss, began reporting to both.

Whitney, at forty, was still best known as a polo player, though his family "felt that Sonny should have a job." For Sonny Whitney, picking up the threads of Trippe's labyrinthine diplomacy was a bit much. Like any airline chief, he visited Washington. But he was embarrassed, he told a State Department man, to have to "come [to you] to find out what the situation is in Colombia."

The United States had installed a strong ambassador at Bogotá named Spruille Braden. To Braden, defense of the Panama Canal was paramount. To forestall any chance of the airline's being used to attack the canal, he wanted the Germans out of SCADTA fast. Accepting a Colombian plan for nationalizing the airline, he insisted that Americans instead of Germans take over and train Colombians to run it — but run it themselves meanwhile. The Americans, he said in no uncertain terms, should be Pan Am. For Pan Am's representatives in Bogotá he had stern words: as owners, Pan Am must pay off and dismiss the Germans, replacing them with its men and equipment. "It will be expensive but Panair will have to pay the bill for past carelessness," he said.

Bound by contractual pledge to the Germans on the one hand and fearful on the other of losing Pan Am's investment in Colombia, Trippe proved so evasive that Braden and his superiors lost all patience. The State Department, demanding action, finally called for a meeting with Pan Am's board of directors. Neither Whitney nor Trippe was up to this. But Thomas Morgan, suddenly made chairman of the company's executive committee, appeared in Washington with Trippe to say that the board had voted to supply funds and equipment for a Pan Am takeover of its Colombian property. "Tearfully" Trippe promised General William V. Strong, assistant army chief of staff, that he would "cooperate fully," and in February von Bauer and other top Germans were removed from SCADTA.

That did it. At the Chrysler Building Sonny Whitney still occupied the big office, but attendance at his daily staff meetings fell off. Executives made their reports and got their orders at the little office nearby. It was evident, as *American Aviation* reported, that Trippe was "back in the saddle." Once Pan Am had bowed to Washington's demand to sever ties with von Bauer, Morgan also faded into the background. The official explanation given out afterward for these gentlemanly and almost wordless scuffles contained no hint that the United States government might have taken a hand. Instead it was said that only Trippe knew the ins and outs of Pan Am's affairs and, though Whitney controlled more shares, Trippe had to be called back to command because he and only he knew enough to run the airline.

On July 11, 1940, Pan Am announced that SCADTA had been merged into a new Colombian airline called AVIANCA. Three days earlier, having flown in its own flight and ground crews by night, Pan Am dismissed the eighteen remaining German pilots of SCADTA in a dawn coup. U.S. warships patrolled offshore. Trippe said later, quite proudly, "We lost only one plane." Braden made sure that the German airmen left in Colombia stayed grounded; later they were interned.

In 1942, when the United States was itself at war, Washington paid Pan Am $1 million to cover the expenses incurred in getting rid of its German partners. When SCADTA was reconstituted as AVIANCA, the Colombian government had an option to buy a majority of the stock. But Pan Am emerged with 64 percent ownership, as well as technical control, and continued to hold a 38 percent interest until 1969.

Since Trippe survived powerful challenges from within and without and contrived at the same time to conserve and extend Pan Am's financial stake in Colombian air transport, the cleansing of SCADTA may be viewed as one of Trippe's most skillful performances. To his quality of audacity could be added another: tenacity. Yet what became painfully manifest during the affair was that the interests of Pan Am and those of the U.S. government simply did not coincide. That was a disclosure Trippe could hardly have desired. For Pan Am the revelation was ominous enough. But the ill will that Trippe incurred within the State Department and even more in the defense establishment by his stubbornly waged defense of his own and his company's position in Colombia cost him dearly later. When the issue of Pan Am's continuing monopoly in the field of U.S. international aviation came up, these forces lined up against Trippe's claim that Pan Am was and should always be the U.S. government's "chosen instrument."

Part IV

Climbing
(The 1940s)

19

Technology and War

If the nineteenth century was the age of sea power, and Britain ruled the waves, the twentieth century has been the age of air power, and America has flown on top. Europe seized the American invention and brought about its spectacular development in World War I. But World War I was only the first part of the great conflict, as we now see, and World War II delivered the final payoff. It made America lord of the air.

Perhaps airplanes did not decide the war of 1939–45. No defender of Stalingrad, no tanker who raced across Europe with Patton would allow us to say air power won. Yet the Second World War was perceived and planned as an air war. Fearing the bomber, Britain saw the Rhine as its new frontier; the Germans, sure that the airplane would avenge 1918's defeat, amassed a mighty Luftwaffe to leap Britain's ocean moat. When the United States joined the fight, President Roosevelt's most electrifying action was to set the arsenal of democracy to producing 50,000 warplanes a year. The nation sent up 303,000, and won. Fittingly, one of those planes, flying its terrible load to Japan, ended the war that Stukas had begun.

Air power was so important in this greatest of all wars because it was delivered — swiftly, suddenly, and from unprecedented distances — through a new element. As the struggle ran its course, the weight, range, and speed at which this new power was wielded rose mightily: it is commonly said that World War II advanced the airplane by fifty years. On this scale of potency, what began with 150-mile-per-hour machines flying seventy-five miles to their Polish targets, culminated in an intercontinental bomber performing its Hiroshima mission by winging across thousands of miles of ocean from a forward base, itself half a world away from the United States.

War is fueled by technological development, and vice versa. Nowhere was this more true than in the air. The demand for supremacy in the air pushed the thrust of engines and the streamlining of wings beyond hitherto known limits. Horsepower, wing-loading, communications, weather, all the technical problems were attacked. Scientists and engineers mobilized to solve them, armies of managers and workers teamed to carry out their answers.

Aircraft manufacture, only recently a cottage industry in America, was turned overnight into mass production. Suddenly Boeing became the biggest employer in Washington — and Kansas. Douglas, Lockheed, North American, and Consolidated spread all over southern California, and thousands of subcontractors mushroomed to make innumerable components the Air Force specified. It was at this time that the wires, pipes, lines, cables, and gunmounts on airplanes multiplied into electrical, fuel, hydraulic, heating, and weapons systems. Under the force of war, aviation got directional systems, navigational aids, electronic gear, and radio-telephone equipment of such staggering intricacy that the airplane became just about the most complicated metal structure ever produced.

The organization of flight was similarly transformed. War demanded oceanic weather forecasting, long-range communications, and a whole chain of costly airfields. Thereupon transports, bombers, and even fighters flew the supposedly impossible Atlantic. Until the Japanese invaded the Aleutians, only reckless bush pilots chanced flying through Alaska's treacherous weather. Engineers brought in Loran — a new kind of radio device for long-range navigation — and soon air transports bored through the thickest fogs, lugged men and supplies to Fortress Alaska, and forged the ties that brought the biggest, coldest state into the union. Similarly, wartime air traffic between San Francisco and Honolulu built up the slender link that Pan Am had pioneered and bound Hawaii to the United States as never before.

In America's air war, research took a back seat, development had a place in the cockpit, but production — sheer production — grabbed the controls. Coming in late as America did, it could hardly have been otherwise. To accomplish the goals that President Roosevelt set, it was necessary to push all thought of improvements aside and concentrate on making the planes the nation had when Pearl Harbor was bombed. This was true of the building of transport planes, and the only major new warplane produced by the United States after it declared war was the P-51 Mustang fighter — and that had a British engine. The two really revolutionary advances of World War II, radar and jets, were British and German creations, and the Americans got in on them only later.

The jet plane was, quite simply, the most important accomplishment in aviation history since the Wrights' first flight. Yet jet power's emergence in World War II, and its failure to influence the war's outcome, only show what a long lead time is required to bring such fateful technological innovations into being. The life-and-death struggle between Germany and Britain forced the breakthrough, but the decisive sequel came much later.

*　*　*

Back in the nineteenth century, when the internal combustion engine was first hitched to wheels, people said at once that there ought to be a simpler way to get power than by pushing pistons. The most alluring alternative was to exploit Newton's third law — for every action there must be an equal and opposite reaction. If you could just channel the force of explosions to thrust rearward, then you could propel a machine forward in about the same way that a cuttlefish, expelling jets behind it, advances through the water. What simpler prime mover could be imagined? The idea engaged the attention of inventors, almost to the same extent as the idea of perpetual motion, and with about the same results. Propulsion by the exhaust of a gas turbine, the best instrument available, fell wildly short of even the most workaday piston engine in efficiency.

But there was a special application for the gas turbine in the sky. At higher altitudes there is less air. Then the engine of an airplane, like a human being gasping for oxygen, gets out of breath. It has to gulp more air to keep working as efficiently as it would at sea level. In 1906, only three years after the Wrights' first flight, Alfred Buechi of the Swiss engineering firm of Brown-Boveri proposed attaching a little turbine to an airplane's engine so as to force the extra air needed at high altitude into the cylinders.

The first such turbosupercharger — so called because the turbine "charged" the cylinders with a richer air-gas mixture — was tested in 1911. When pilots learned the importance of gaining superior altitude in aerial combat, Professor Auguste Rateau built a turbosupercharger that tested well over France and was produced in quantity by the end of World War I. A sample was sent to the United States and placed in the hands of Dr. Sanford Moss, pioneer turbine man at General Electric. With the first Moss-designed turbosupercharger providing the extra lift, U.S. Air Corps Lieutenant Schroeder soared to a record 33,000 feet in a Packard Le Père biplane over Dayton, Ohio, in 1920. Boeing's 1939 Stratoliner, first high-flying commercial plane, topped clouds and peaks by dint of Moss's turbosupercharger.

But it was one thing to rig a small turbine auxiliary to supply bursts of power to a prop-driven plane, and quite another to create a turbine that would propel an airplane through the air by itself. And in the years after World War I, investigation into radical ideas lagged. In 1928 Dr. A. A. Griffith of the Royal Aeronautical Establishment proposed design of a gas turbine that would drive a conventional propeller, but the authorities declined to take up his suggestion. In 1930 Frank Whittle, a twenty-three-year-old Royal Air Force flier, patented a jet engine design, complete with a turbine, compressor, and combustion chamber. But his superiors brushed him off.

Then Hitler came to power and immediately organized an Air Ministry. Europe stirred to rearm. The first efforts were by private individuals. Friends of Whittle organized Power Jets, raised some funds, and gave him the chance to develop his revolutionary design. The Germans resorted to private projects because the Allied ban on public arms manufacture still prevailed. The universities took the lead. Since 1918 there had been little improvement in basic airplane design — the only substantial gains in speed had been obtained by boosting piston-engine power. While Professor Willy Messerschmitt's firm in Augsburg undertook the design of the world's fastest conventional fighter, the Bf-109, at Göttingen the Aerodynamic Research Establishment assembled leading theorists to design something still faster.

Early in 1935 a young student at Göttingen named Hans von Ohain began work on a power plant that, unbeknownst to either, was very like Whittle's. The Göttingen professors sent him to aircraft manufacturer Ernst Heinkel, who provided such prompt support that von Ohain became the first to install a jet engine in a plane — the Heinkel 178. It was not a very inspiring or successful plane, but it was the first jet plane to fly, four days before war broke out, on August 27, 1939.

The Göttingen theorists' studies had already convinced them that there was no way for a prop-driven airplane's speed to rise much beyond that of a souped-up Messerschmitt Bf-109. The mysterious phenomenon called "compressibility" — the same aerodynamic barrier that Kelly Johnson bumped into at Lockheed when he tried to beef up his P-38 Lightning — set the limit, which they calculated was 623 miles per hour. With a propeller even the highest-flying plane could not get much over 500 miles per hour, and any engine built to lift a prop plane high enough for such speeds would be too heavy to be practical. In sum, the time was ripe to turn to the turbojet, which did not appear to be prone to such limitations in forward speed. The Göttingen studies showed that compressors could be designed for jets that would be efficient enough to yield the desired speeds.

Flouting the arms ban, the Germans went all-out, producing conventional fighters and blitz bombers while setting three of their four biggest engine firms to work on turbojets. The British, as unaware as the Germans of what their foe was up to, did the same. The unearthly whine of the jet was heard first at Rugby in 1938 when Whittle's turbine "ran away" in early tests. Everybody made a dash for cover, but an explosion was averted. Later, clouds of fuel vapor rose from leaky engine joints. When the spilling fuel caught fire on the red-hot exterior of the combustion chamber, Whittle and his men leaped like demons amid the spurting flames. Like the early German model, Whittle's turbojet was really a glorified turbosupercharger with a combustion chamber welded on to it. Its compressor sucked air like a vacuum cleaner, and the turbine expelled the hot gases with the force of a thousand fire hoses. The engine worked, and Rolls-Royce, adapting and installing it in the Gloster Meteor, quickly built up the proficiency in aircraft gas turbines, as they had previously done for piston engines.

The Germans, following up the findings of the Göttingen theorists, produced the most arresting jetplane of the war. The Messerschmitt Me-262 had the sweptback wings that overcame the "compressibility" phenomenon. It was powered by a pair of long, thin turbojets of a new type. Presenting a much smaller frontal area, they passed the air straight through to the combustion chamber by a succession of compressor stages. Turbine blades of finest, heat-resisting alloys, hollow and internally cooled, turned up to 8,700 revolutions per minute. The Germans had their troubles too. Turbines jammed, fires broke out. The plane's original design with an old-fashioned tailwheel proved a costly mistake. Visibility from the cockpit was bad, and shortly after it first flew with the new jets at Rechlin on July 18, 1942, the test pilot failed to see a fence, taxied through it, and smashed the prototype on a trash heap.

Redesigned with tricycle landing gear and other improvements, the Me-262 V4 was ready to go the following spring. On May 22, 1943, General Adolf Galland, the Luftwaffe's fighter chief, flew it, and rushed to tell Goering that it clocked speeds fully a hundred miles per hour faster than anything seen in the war. Even so, production had to wait for Hitler's OK, and it was not until November that the Führer saw a demonstration of the plane that his fighter commander said could regain command of the air. To everybody's astonishment Hitler declared that the Me-262 was the long-awaited blitz bomber that should carry destruction to the enemy. For the next six months, the world's fastest airplane was fitted with bomb racks and assigned to bomber squadrons while Germany lost ground on all fronts. Not until late in 1944 did the Me-262 jets begin to be used against Allied bomber formations. They struck like lightning bolts, but they were too late and too few, and they suffered many accidents besides.

But the Me-262, direct ancestor of the Russian MiG-17 and the American F-104, pointed the way. During the war General Hap Arnold, U.S. Air Force chief, got wind of the jet breakthrough, and in October 1942 — a terrific piece of reverse lend-lease — an Air Force B-24 flew one of Whittle's first engines to Washington. General Electric, because of its long experience with turbines, was assigned to build one like it, and in a year's time it flew in America's first jet plane, the XP-59 Airacomet built by Bell Aircraft. Before the war ended, Lockheed built the second U.S. jet fighter around another British engine and flew it on January 8, 1944. Too late to see enemy action, the American jets were plagued by bugs and consumed fuel at twice the rate of conventional aircraft. But in November 1945, the British Gloster Meteor set a world speed record of 603 miles per hour, and Sir Frank Whittle announced with quiet confidence, "In ten to fifteen years' time we shall see the piston engines displaced in all new aircraft except light aeroplanes."

* * *

The second great technological leap of the war produced radar. Desperately fearful of their new vulnerability to attack through the air, the British bent every effort to gain a means of early warning against approaching aircraft. They succeeded. Even before the war began, a line of huge towers rose along the North Sea coast whose receptors could detect and track planes as far as two hundred miles away. By the time of the 1940 Battle of Britain, radars in England were tracking German warplanes from the moment they took off from their fields across the Channel, and men seated before radar scopes on the ground in Kent were directing Spitfire pilots circling high overhead to their tallyhos.

That was radar's finest hour, but it went on to many other wartime triumphs. The beauty of this marvelous instrument was its versatility. The acronym the U.S. Navy gave it, instantly adopted, stood for "Radio Detection And Ranging." But Sir Robert Watson-Watt, who had led in its development since 1916, said, "Radio Direction-Finding and Ranging" would have been a better, broader definition.

As long ago as 1887 Heinrich Hertz, the discoverer of radio waves, had shown that they were reflected from solid objects. He also demonstrated that if a square of wire were turned perpendicular to the source of waves they were then received most strongly. Exploiting these potent clues, radiomen in World War I pointed their aerials at aircraft transmissions and plotted the bearings to locate the planes. By using its network of shore stations as a radio direction-finding system, Pan Am tracked and guided its transocean planes in the 1920s and 1930s.

A simpler, stronger, and much more accurate way was foreshadowed by scientists at the Carnegie Institution in Washington in 1925. They wanted to measure the distance to the layer of ionized particles that encircles the globe at great altitudes and acts as a mirror reflecting radio waves back to earth. They did so by training a beam of radio pulses upward and measuring the echo.

The British took it from there. They devised units that permitted use of the same transmitter for sending and receiving. They perfected the cathode-ray oscilloscope. Painted inside with fluorescent material just like a TV screen, this phenomenal gadget presented a maplike outline of the area scanned by the turning radar antenna on which the approaching aircraft, singled out by the returning pulses, appeared as a distinctly moving "blip." On top of all this they invented a tube that harnessed microwaves, thereby increasing enormously radar's power and image resolution. When Sir Henry Tizard, Britain's boss scientist, carried one to the United States in his briefcase in 1940, it was called "the most valuable cargo ever brought to our shores" — yet another priceless item of reverse lend-lease.

As the tide of war swung, the Americans seized upon radar for offensive purposes. At the newly formed Radiation Laboratory in Cambridge, Massachusètts, scientists used its heightened powers of measurement to train guns and land planes. With their electronic stopwatches, calculations could be in microseconds. If a pulse was reflected back to the transmitter in a microsecond, the speed of radio waves being equal to the speed of light, then the distance to a target was 0.093 miles, or 164 yards. By May 1941 the U.S. fleet began to get gun-laying radars that could determine the distance to a Japanese cruiser some twenty miles over the horizon with an accuracy of one-thirtieth of 164 yards, or a little over five yards. The Radiation Lab designed radar scopes that outlined the shape of ships in their own formation around them — invaluable for station keeping. And they perfected search radars that pictured the blips of planes up to one hundred miles away.

But how could radar operators tell which blips echoed back from enemy planes and which from friendly planes? That problem was solved early. Their own planes were equipped with transponders, known by the code name of IFF — Identification of Friend or Foe. When turned on, these devices caused friendly planes to return a distinctively different signal, which appeared in a different-shaped blip on the radarman's screen. Appropriately miniaturized, radar was then placed aboard the airplane itself. It was first used to locate enemy ships and then, in specially designed fighters, to track and shoot down hostile planes in the dark of night.

So versatile was radar that scientists believed it could be used to solve aviation's most intractable problem — landing planes in zero-zero weather conditions. .Reports from the front told of Air Force fighters hampered or grounded by adverse weather. For victory, combat squadrons needed a dependable system to get their planes back down no matter how low the ceiling and visibility.

The physicist Luis Alvarez brought radar to the rescue. At the Radiation Lab he conceived what came to be called Ground Controlled Approach (GCA), a method of bringing down to a blind landing any plane equipped with standard two-way radio gear. In Britain, men called fighter directors in dugouts were already peering into radar scopes to guide planes overhead toward enemy planes. In Alvarez's scheme, men on the ground would use the new precision radars to "talk down" pilots when they returned to land.

To provide the ground controller with the vital information, his GCA setup called for two receiving systems — one radar for search and surveillance and a second, high-resolution radar for guiding the plane to its final touchdown. Both were installed in a truck alongside the runway, and the controller used the first radar to instruct the pilot how to find his way into the landing circle over the field and the second for the final "talk down" to the runway.

"Our technique," Alvarez's aide Larry Johnston said, "was simply to keep reading the data to the pilot — 'You are two miles from the airport, go up ten yards, go twenty yards to your left.' Then, as the pilot started the azimuth correction, 'Go left fifteen yards, ten, six, straighten out, five, you're on the path.' " On December 22, 1942, Ensign Bruce Griffin soaped the windows of a North American SNJ trainer at the Navy's Quonset Point, Rhode Island, air base, and made the first completely blind landing. Early in 1943, Alvarez as controller guided Colonel Stuart Wright through snow and low clouds to the first completely blind foul weather GCA landing. That was at Boston airport, and a lot of back-slapping followed.

The Air Corps ordered GCA sets into production, and Alvarez went off to Berkeley to make the discoveries in elementary particle physics that won him a Nobel prize. But his Radiation Lab colleague Chuck Fowler was on hand when the Ninth Air Force put GCA into use at Verdun in December 1944. Wrote Fowler:

A couple of weeks ago two P-61s returning from a nightfighter mission were caught up in the soup. The ceiling was essentially zero and so was the visibility. The first P-61 came in right on the button the first time. After he was down they asked the pilot when he saw the runway and his now famous answer was: "I didn't, I just felt a bump."

The second P-61 was off in azimuth on the first run so they told him to go around again. He came back with, "You'll have to make it quick, or I won't have enough gas to go up high to bail out." They brought him around in a very tight pattern and this time he came in all right. Visibility was so bad that he couldn't see more than one runway light at a time. When he finally stopped he had 20 gallons of gas left — which for the 2000-hp engines in a P-61 is just about one good cough.

Of course the crucial part of GCA was that the pilot had to place himself and his machine in the hands of someone on the ground. Fowler said:

One colonel couldn't see how a groundling could instruct *him* on how to land *his* plane. The crew waited for a day when the colonel got caught up in a heavy fog, and had no choice but to use GCA. He was brought in to a perfect landing, though the field was so thoroughly closed in that a jeep had to be sent out to guide the plane to its dispersal point; the jeep drove off the perimeter track in the fog. The colonel became a staunch supporter for GCA from then on.

It was said that the entire cost of wartime GCA was more than paid for just by the number of B-29 Superfortresses it brought down safely at Iwo Jima after they bombed Japan. GCA is still the Air Force's answer to bad weather landings — in use at all military fields.

20

The Airlines Go to War

A*irlift* was one of the great military achievements of World War II. It was a new and unlooked-for use of the airplane — not trumpeted in advance by the apostles of air power, not even anticipated when President Roosevelt proclaimed production of 50,000 planes a year as America's national goal.

When the fighting started, commercial air transport was thought of as a mere handmaiden of the forces of aerial combat. For the Air Corps chieftains, schooled in the doctrines of Billy Mitchell, bombers and fighters were the machines that won wars; in order to produce 50,000 of them a year the building of civilian planes must be stopped altogether. Even after Donald Douglas convinced Secretary of War Henry Stimson that transport planes would be needed, and the order was reversed, the Air Corps ordered its DC-3s primarily as troop carriers, to fly parachute drops and other combat missions. It took time until the potential of airlift was seen, but it became, in this biggest of wars, the military's longest arm.

The story of World War II's airlift began well before Pearl Harbor, when the British stood backs to the wall. President Roosevelt, driving to help them with aid "short of war," turned to Pan Am, America's only airline then operating overseas. When he traded fifty old destroyers to the British for West Indian bases, he had Pan Am build airfields on these islands. Trinidad was the most important. Long a stop-off on the airline's flying-boat track to South America, Trinidad now loomed as a stepping-stone on a new kind of supply line to the British.

For a starter Pan Am was called upon to open up a row of airfields across the heart of Africa so that planes flying down from Britain along the African west coast could carry emergency aid to forces in Egypt that would otherwise have to be transported in ships around the Cape of Good Hope. When Pan Am's engineers readied eight airports in just sixty-one days, Washington signed up Pan Am to begin ferrying lend-lease bombers across Africa to Khartoum. This called for the next step: to build a chain of airfields on the New World side that would link the United States across the Caribbean with a jumping-off point at Natal on the Brazilian side of the Atlantic. Declaring the importance of the South Atlantic crossing "cannot be over-estimated," Roosevelt gave the mission of building the airports to Pan Am for two reasons. First, the program was centered in Latin America, Pan Am's old stamping grounds. Second, since the United States was not yet at war, the government needed just such a private company as an unofficial go-between to get the fields built on foreign soil.

Pan Am proceeded to build forty airfields and bases in twenty American republics at a cost to the United States of $90 million. The new kind of aerial lifeline thus created sustained British and soon American forces in distant theaters of war. But these same bases also provided the airfields along Pan Am's South American and Caribbean routes for lack of which the airline had developed its flying boats. By the time the program was completed in 1943, Pan Am was shifting from its old nautical traditions and starting to fly landplanes like any other American airline. And because of the way the original contracts were drawn, Pan Am or its local subsidiaries held title to these new airports paid for by Uncle Sam. In the postwar round of contract renegotiations, the company relinquished that title, but not before using its special position to try to keep rivals from using the fields.

After France fell in 1940, desperation drove the British to an airlift across the North Atlantic — in the winter at that. A year before, they had placed an order with Lockheed (the biggest order received by an American plane maker until then) for 177 twin-engine Hudson bombers. These were modifications of the familiar twin-finned Lockheed Electra 14, the airliner in which Howard Hughes had circled the world and in which passengers traveled on such lines as Northwest, Braniff, and Delta. In the summer of 1940 the British contrived to fly these planes to Halifax for shipment across. But even that was too slow. Britain's Canadian-born Minister of Aircraft Production, Lord Beaverbrook, called on his business friends in Canada to get them flown over. They did.

Near the tip of Newfoundland, a thousand miles out into the Atlantic, a crude runway had just been cleared for the Gander airfield projected five years earlier. Dr. Andrew McTaggart-Cowan, Canadian member of the North Atlantic planning committee, had already set up a weather-forecasting station amid the Gander ice and was receiving coded data from Britain, Iceland, and Greenland. BOAC crews headed by Master Pilot Don Bennett, who had flown the route in 1938, would lead the way. Beaverbrook's posse of Canadian businessmen had snagged fliers — ex-airline pilots, barnstormers, bush pilots, skywriters, stuntmen — by offering them $1,000 a trip.

Bennett decided to lead the first seven Hudsons over in formation at night. At Montreal they filled the bomb bays with extra gas tanks. On November 10, 1940, the crews — nine Americans, six Britons, six Canadians, and one Australian — stayed overnight at Gander in three railway cars on a snowy siding. McTaggart-Cowan, grease-penciling bold lines across the pilots' charts and writing out separate analyses for each five-degree segment of the crossing, gave the first of his famous two-hour meteorological forecasts. Joe Mackey, ex–U.S. government test pilot and later founder of Mackey International Airlines, recalled afterwards:

The greatest weather briefing I have ever known. The weather ships out there had all been sunk. All he had to predict by was the weather that had already gone by Newfoundland. He gave us the latitude and longitude of where we would find a front; we could almost navigate by it.

On the strength of McTaggart-Cowan's forecast of a twenty-three-mile-per-hour tailwind, a full moon, and little cloud above fifteen thousand feet, Bennett gave his pilots a flight plan of nine hours, thirty-two minutes to Aldergrove in northern Ireland. For a wonder, they stuck together halfway across. Then they hit the front, and the formation broke up. Bennett continued to give directions on his radio. But some climbed as high as eighteen thousand feet trying to get above clouds. When their twenty-minute supply of oxygen ran out, some crew members lost consciousness. Some pilots flew the last part in a dream haze sipping the last driblets of oxygen. At Aldergrove Bennett was first down. Two others arrived within ten minutes. Bennett watched by the window till the seventh Hudson hit the runway. Then he signaled Beaverbrook: "All arrived safely, without incident."

Only forty planes flew over that winter. But Beaverbrook's improvisation showed that it could be done, and that ferrying was the only way for warplanes to cross. First the British military joined in, using the new four-engine B-24 Liberators lend-leased by the United States to shuttle ferry pilots back to Canada. Then, well before Pearl Harbor, the U.S. Army Air Corps formed a ferrying service to rush bombers to Britain. By then TWA, the only domestic airline with four-engine planes, got into the act, setting up the Eagle's Nest flight training center in Albuquerque to show British and American pilots how to operate the multi-engined planes. The B-24s were true warplanes, big, ugly, and no better than they needed to be. Henry Ford later turned them out at a rate of 500 a month at his half-mile-long-by-quarter-mile-wide Willow Run plant near Detroit. "Throw a rock at one and you could set it afire," Otis Bryan, Eagle's Nest boss, said later. "They had sight gauges for fuel on the instrument panel," recalled Ray Dunn, later TWA vice president. "A bullet, a kick, or a knock from a flight kit could break the glass and spray gasoline all over the cockpit."

*   *   *

When at last the United States went to war, all the airlines went to war. By prior understanding with the military, they kept their identity and profit-making function. The armed forces took over 200 of the industry's domestic fleet of 360 planes, then contracted with them to fly various missions, often in the very planes that had been taken over.

British historian Sir Denis Brogan said, "To Americans, war is a business, not an art. They approached it with business resourcefulness and efficiency as they approached the harnessing of the Tennessee Valley or the exploiting of the Mesabi iron range." Soon after the United States joined the fight, Air Force chief Hap Arnold took decisive action. He formed one big Air Transport Command (ATC), separate from supply and other units and separate also from the dozen or more air forces under Arnold's command. To the ATC he gave the job of ferrying planes, lugging supplies, and carrying men for all. General Harold George was put in charge; C. R. Smith, most formidable of airline executives, moved to Washington as his second in command. Smith lost no time tapping the airlines for more good men. Soon, paralleling the pipelines of tanks, trucks, jeeps, and Liberty ships flowing to the fronts, a vast new network of fast-moving air communication sprang up.

The airlines were central to this military transformation. On the home front, they used the limited number of airplanes the armed forces let them keep to maintain swift and reliable transport between the centers of war production. They sped the blueprints, the contracts, the crucial spare parts. When factories faltered, they delivered the troubleshooters from Washington. Only heaven knows how much victory was hastened by fast air communications inside the arsenal of democracy. On the war front, the airlines flew regular contract trips to all the theaters of war. For a time they provided all the airlift there was. Then as the ATC and the Naval Air Transport Service grew to great size, they set up the Airlines War Training Institute and trained 43,000 crewmen and mechanics for the services. At all times the airlines functioned as the old pros, delivering the most important passengers and cargos throughout the worldwide system.

On the day the United States declared war, Western Airlines DC-3s pulled off their routes to fly ammunition to the West Coast. On December 14, 1941, fifteen American Airlines DC-3s wheeled at orders in midflight to land at the nearest airfield and discharge the passengers. Flying to Fort Lauderdale, Florida, the DC-3s took aboard 300 Signal Corps specialists. Twenty hours later the troops stepped out at Natal to take over communications for the transatlantic ferry in still nominally neutral Brazil.

The first domestic airline to begin regular flights overseas was Northeast. Until then Northeast had operated a grand total of six planes between New York and points as far north as Portland, Maine. On February 13, 1942, a Northeast DC-3, designated a C-39 by the Army because it was powered by the original Wright Cyclone engines, began flying passengers and cargo from Presque Isle, Maine, across maritime Canada to Gander in Newfoundland. By April Northeast's flights were extended to Greenland, where the United States won rights to build airfields, and in May to Reykjavík, Iceland. Recommendation by Northeast pilot Milt Anderson that the Douglases could use more power on the Arctic crossings hit Washington about the time General Arnold decided he would need transport planes everywhere. The result was that the ten thousand DC-3s Douglas proceeded to build for the military got newer, 1,200-horsepower Pratt & Whitney engines instead of Wright Cyclones, and were designated C-47s.

Because it had so much experience with snowy airfields between Chicago and Seattle, Northwest was picked to fly to Alaska. On the first survey flight after Pearl Harbor, Captain Lee Smith's DC-3 broke a new airline trail — north from Great Falls, Montana, through Edmonton and the Yukon to Fairbanks, north of the Arctic Circle. It was so cold that rubber hoses shattered like glass. The northern lights played weird tricks with radio communications. Making an instrument landing at Whitehorse, Smith saw a Pan Am plane that had flown in on charter. It was a first encounter, and it soon became apparent that domestics flew a different style. Four times the Northwest party flew in and out of Whitehorse. Each time the Pan Am plane came on the radio and asked how the weather was. Smith said later, "They told us if they'd been caught flying on instruments there they'd have been fired."

Flying by instruments day and night, Northwest was soon supplying the military all the way out to the end of the Aleutians. Pilots like Vince Doyle learned about Alaskan icing. Forming on the wings, the ice drove their C-47s down even when they were flying at full power. Finally they found themselves flying close to the water, so close they could sometimes see salt spray spattering their plane. At that level, friction with the air melted the ice off the wings. But, Doyle said later, "If I climbed up again, more ice quickly formed — and then I was back down again just above the water." In 1941 Northwest had asked the CAB for the right to fly the North Pacific route to the Orient. Its wartime performance gave Northwest the basis for the postwar grant of routes to Japan and China that turned it into a major airline.

Six Eastern Airlines C-47s started flying for the Army to Puerto Rico, Trinidad, and Brazil. It was another incursion into Pan Am territory. Pan Am's flying boats had always flown by day, and below the clouds or not at all. Eastern's Vice President Paul Brattain said, "Fly at night? We do it all the time. Over water? Hell, flying is flying." And when American took a contract to fly to South America, it brought along its stateside ways. With the help of a couple of Civil Aeronautics Authority engineers, American Captain Ham Smith took a month to put in sixteen radio ranges, paid for of course by the U.S. government. Thereafter everybody including Pan Am flew on Smith's beams from Surinam on the Venezuelan coast all the way down to Montevideo.

The only planes in domestic service capable of flying the Atlantic were TWA's Hollywood High Flyers — the five four-engine Boeing Stratoliners. After Pearl Harbor the Air Corps commandeered them, stripped out the fancy furnishings and the pressurization system, and installed extra gas tanks in the front half of the cabin. With that, the United States had its first landplanes with transatlantic range. Signing a contract to fly the stripped-down Stratoliners to North Africa, TWA became an overseas airline overnight. TWA Chief Pilot Otis Bryan started the airline's intercontinental division, based inside a square of file cases on the second floor of the Washington National Airport terminal building.

TWA's first assignment was to deliver antitank shell fuzes to the British, trying to hold off Rommel in Libya. For a mission of such urgency, General Arnold asked for a temporary certificate authorizing the airline to fly to Egypt, and the White House so ordered. The Civil Aeronautics Board complied, with history-making consequences. Chairman Welch Pogue said later, "At the time I thought it was pretty bizarre — a civilian certificate for such a purpose. But it was only temporary, and it was to win the war."

Before takeoff TWA President Jack Frye invited all five CAB members to the airport to view the preparations. Bryan, whose managerial maxim was, "Never ask a pilot to do what I wouldn't do myself," had put himself down to make the first crossing. A board member asked him, "Have you ever been out of the U.S.?" Bystanders said that Frye's jaw dropped a foot when Bryan replied, "One week in Mexicali."

Without pressurization, Bryan's Stratoliner bumbled along at four or five thousand feet. But he had no trouble boring through the towering intertropical front north of Brazil. Crossing the South Atlantic, navigator Pete Redpath took star sights. And when it was time for the African landfall, he deliberately called on Bryan to veer a bit north of the charted track. That way, when they hit the coast, Redpath knew for sure that the airfield must lie to the south. Following this plan, Bryan turned south and followed the coast safely into Roberts Field at Monrovia. Pausing to refuel, they roared on to Accra, El Fasher in the Sahara, Khartoum, and Cairo; theirs was the first American commercial airliner ever to appear in the Middle East. Four days later Bryan was back in his file-case corral at the Washington airport, and General Arnold sent over a "well done."

What Bryan started turned out to be anything but temporary. When Portugal gave permission to land on its field at Lages in the Azores, TWA's converted Stratoliners found a shorter transocean crossing. They also started flying the North Atlantic. On the first trip Floyd Hall, later president of Eastern, flew copilot. He said later:

All we knew was that Bernt Balchen was up in Greenland building an airstrip, and had a homing beacon. It was 25 below when we landed, and there was Bernt with shirt wide open greeting us. He made us take a Finnish bath that night.

The next day the Stratoliner landed at Prestwick in Scotland, more proof that the North Atlantic could be flown in winter.

Because TWA flew landplanes across the Atlantic, its men got first crack at piloting the next big airliner when it came along in mid-1942. This was the four-engine Douglas DC-4, which had been designed and built for United and American to use on their domestic routes as the successor to the fabulously successful DC-3. But for the war, they would have been delivered in 1942 to the airlines that were TWA's biggest rivals. Taken over instead by the military as the C-54 Skymaster, this splendid plane was handed to TWA for its transatlantic war route.

The C-54 was fifty miles an hour faster than the Stratoliner, had more power, was bigger and more comfortable, was not so noisy in back, and had modern features such as tricycle landing gear. Although unpressurized, it was perfectly capable of flying the ocean nonstop, at least in the easterly direction. Just in time, the plane gave the U.S. airlift global capability; the ATC bought 1,163 of them. The smaller C-47s, ten times as numerous, flew with the combat forces, shuttling in and out of the muck and grit of front line fields. But the C-54s could haul payloads twenty-five hundred miles.

Eighteen of the first twenty C-54s went to TWA's intercontinental division. Bryan had taken the first one across in September 1942. Before long, TWA was flying five trips a day across the North Atlantic to Prestwick, five more by the southern route to North Africa. Bryan himself made fifty crossings. He also flew President Roosevelt to the Casablanca, Teheran, and Yalta conferences. "We ran a complete airline," said Bryan's right-hand man, Captain Hal Blackburn. "We carried supplies but mostly we carried VIPs. We flew them all — President Roosevelt, the joint chiefs, Ike, Bradley, Chiang Kai-shek, Ambassadors Harriman and Davies, Jimmy Doolittle. You name 'em, we flew 'em."

The second big airline to acquire the Atlantic habit in World War II was American. In the regional carve-up of assignments, the military sent Northwest to the northwest, United to the west, Eastern to the southeast, Braniff to the southwest, TWA across the Atlantic — and American everywhere. Already the biggest airline, American in midwar made a significant acquisition. American Export Airlines, having dented Pan Am's overseas monopoly by winning a certificate to operate its flying boats across the Atlantic in 1942, experienced vicissitudes. With plenty of encouragement from Pan Am, the Post Office Department dragged its feet in awarding the new line a mail contract; then the Senate Appropriations Committee voted to withhold mail pay for the new airline. John Slater, Export's president, fought valiantly on, and was rewarded when the

The wartime DC-4, prototype of the later Douglas planes, was the first with tricycle landing gear.

Navy, after Pearl Harbor, took over its four-engine Sikorsky flying boats and contracted with it to fly to Portugal and Britain. In spite of all Juan Trippe's efforts to prevent it, Amex flying boats operated side-by-side with Pan Am's out of La Guardia's Marine Terminal. Then the CAB ruled that it was unlawful for a company in a competing form of transportation to control an airline. Forced either to get out or find a partner, American Export Steamship Lines sold 60 percent of the airline's shares to American. From that point on, American was in the Atlantic in a big way, flying scheduled services daily and contract flights even more often.

If TWA flew ten thousand crossings in the war, and American upwards of five thousand, Pan Am's transocean flights topped all the contract fliers. Pan Am seaplanes and landplanes flew the Pacific for the Navy, the Atlantic for the Army. Fifteen thousand wartime trips were made across the Atlantic alone. Its huge Boeing flying boats, operated for the military, never ceased to make their stately way to Europe and back from the day Hitler bombed Warsaw until the fighting ended in Japan. It had been Andy Priester's practice to pull them out for overhaul after every crossing. But when the North African front crumbled in 1942, four of the big boats shuttled back and forth between Natal and Monrovia, lifting top priority cargos across the ocean for lesser landplanes to rush the rest of the way to Cairo. One Clipper made six crossings in ten days, a record without rival till the day of the jets. Pan Am Skipper J. H. Hart flew back and forth across the Atlantic twelve times in thirteen days, a record that has never been equaled.

For the big boats daytime-only flying ended the day the war began. At La Guardia's Marine Terminal the Clippers took off at 4:00 A.M., before first light, so as to get to their overseas destination before dark. "We didn't want to have them see who or what we were carrying," said Captain Tom Flanagan. With its nautical training and discipline, Pan Am observed wartime radio silence and steered strictly by the stars. To facilitate Andy Priester's rule that sights must be taken every hour, the Boeing 314 was equipped amidships with the world's first astrodome. There the second officer peered through his sextant to record the angle of the stars. Captain Hack Gulbransen recalled later, "It was fun. Navigation was what made flying interesting, and the Boeing was steady as a bridge." Even so, it was customary to pass word to passengers to remain quiet in their seats "while the navigator takes his sights."

Solid as they were aloft, the big boats could be quite unmanageable when they were on the water. Captain Howard Cone, who piloted President Roosevelt's first crossing in early 1943, wrote a manual on the perils, "How to Sail a Boeing 314." Taxiing up to a buoy in an African night squall was a fine art. On the notoriously still waters at Bolama, a regular stop in Portuguese Guinea, it often took three thundering three-mile runs to get airborne. In treacherous crosswinds prevailing at places like Port Lyautey, taxiing to take off could be hairy. The Boeing had no wing floats at all, and the huge wing was apt to tip steeply toward the water. Then the captain roared, "Eight men in Number Four nacelle," and everybody scrambled out through the passageway in the opposite wing to help the skipper right the tilting giant.

The boats' worst problems on the water were at Horta in the Azores. The offshore swells that prevented takeoffs were responsible for delaying 59 percent of Pan Am's mid-Atlantic crossings an average of three days in 1939. Two years later, planes trapped at Horta by seas too high to permit takeoffs suffered delays averaging four days. So bad were the holdups that the airline had no choice but to reduce payloads and overfly the Azores.

When the Boeings were obliged to land at the Azores, Pan Am went to great lengths to get them out. Once winds suddenly changed and two

Boeings, unable to come down on the ocean at Horta, were told to land instead at Pym's Bay, a shelter on the opposite side of the island. Before they could tie up, the station manager heard a forecast of another wind change and ordered them out. Captain Ed Sommers's Clipper was stripped bare — all passengers, mail cargo, and even extra crew were taken off. Tom Flanagan, who watched from the other boat, said later, "Never forget, water is a solid. They simply had to be airborne before reaching open water. Hitting those waves would have been like running into a brick wall." Sommers's Boeing made its getaway, but several windows were smashed, and when the Clipper reached Lisbon a ripple was found in the hull. Flanagan's boat was ordered to taxi around the island to berth behind the Horta breakwater. "That was a 30-mile ride I'll never forget — at one moment high atop a wave and looking for miles in every direction, the next plunged down into a trough and unable to see anything but green water on either side." The Clipper made it to Horta — then waited two weeks for the swells to subside enough for him to fly back to New York.

At Lisbon the station manager would steal out in his launch and lay a string of landing lights on the river Tagus for the Boeings' arrival at Cabo Rivo. "And always," Flanagan said later, "when you taxied up in the night, there was the German consul down at the dock to welcome you." Neutral Lisbon, frequented by Allies and Axis alike, was a center of espionage and intrigue. Pan Am, KLM, BOAC, and black-painted Lufthansa airliners came and went. It was mostly live and let live. Then suddenly in June 1943 a KLM DC-3 flying to Britain was shot down by German fighters with thirteen people aboard, including the British film star Leslie Howard. In his memoirs Winston Churchill blamed the incident on German agents who, seeing a "thickset man smoking a cigar" board the plane, "signaled that I was aboard." The Germans knew he had just left a big powwow at Algiers, Churchill said, and jumped at what they thought was a fine chance to kill him.

Pan Am Captain Jim Waugh, whose Clipper was two hours away on its flight to Foynes that night, was ordered without explanation to return to Lisbon. He always wondered if the Germans could have made so silly a mistake. Churchill himself thought it "difficult how anyone could imagine that with all the resources of Great Britain at my disposal I should have booked passage in an unarmed and unescorted plane." During a flight to Rio two years after the war, Waugh found out what really prompted the German action. A passenger who had worked for the Office of Strategic Services told him this story. The OSS had managed to tie up Spanish wolframite that had been going to German warplants. The deal was to supply black gunpowder to the Spaniards. An OSS man had to go to London to get gold that was part of the payoff. He was on the KLM plane. By shooting down the plane, the Germans eliminated him — and his fellow pas-

sengers. The principals in the deal were also assassinated. Spanish wolframite was the Germans' only source of steel-hardening tungsten, vital
to their war effort. When the Americans tried to cut off their supply by
their preclusive-buying ploy, the Germans struck back viciously to protect it.

Lisbon was also the scene of Pan Am's only fatal Boeing flying-boat
crash. On the night of February 22, 1943, the original Yankee Clipper, pioneer of commercial flying on the Atlantic, came in for a landing with
forty passengers and crew aboard, including a party of USO entertainers.
At the wheel was one of Pan Am's most senior pilots, Captain R. O. D.
Sullivan, who had been Musick's copilot and had led inaugural flights in
the Pacific and Atlantic. The evening was clear, visibility was seven miles,
and the station manager in his launch had laid his string of lights along the
Tagus to guide the landing.

Flying over the city parallel to the lights, Sullivan made his turn at five
hundred feet on his way to the final touchdown. Then unaccountably the
plane's left wing dipped into the water and the huge flying boat plunged
heavily into the Tagus. Twenty-four aboard were killed, including nineteen passengers. One of the fifteen saved was Broadway singer Jane Froman. Though badly injured, she fought to the surface. She said later, "I
saw the head of somebody. I called, 'Who's there?' A voice replied, 'It's
your old first mate.' " It was Fourth Officer John Burns, and he held her
up until a boat rescued them.

By all accounts the plane made a normal and smooth approach, and
Sullivan, who survived, seemed unable to explain the crash. The CAB investigation decided that he had "erred in his judgment of the position of
the aircraft in relation to the water," and it seemed final. But Tom Flanagan, who was in Lisbon at the time, came to believe that Sullivan, a master pilot, was probably the victim of what later became identified as the
"black hole" effect. This is a phenomenon that passengers as well as pilots
have noticed. Approaching a city like New York at night, one can see the
carpet of city lights below and to one side. But when the plane turns away
toward the Atlantic, there may be only one or two lights out there in the
blackness, and one becomes disoriented. One has no idea how far or near
the lights are. Pilots call this black hole. It overtook the pilot of an Eastern Lockheed 1011 jet that piled up at Miami in 1965 at the same point
when Sullivan apparently had become confused. On his final approach to
the landing, thinking the runway lights were much nearer than they were,
the pilot tilted a wing too low and slammed into the ground. The Lisbon
crash had one happy sequel. Jane Froman, who had been on her way to
sing for the troops in Britain, recovered use of her crippled legs and married Fourth Officer Burns.

Pan Am's big flying boats were the only airliners that flew throughout
the war with all their luxurious peacetime appurtenances. Yet when the

war ended, the airline was quite content to let the Navy keep them. As early as 1940 Pan Am had hedged on its future by placing an order for the four-engine landplane that became the Lockheed Constellation. The war completed the change. When the long-legged C-54 became available in quantity, Pan Am flew it on its overseas contract flights just as TWA and American did. It was no longer necessary to shift passengers and cargo at Natal for the hop across the Atlantic. Pan Am's Cannonball service to the Persian Gulf and India became an eighty-five-hour landplane operation all the way from Miami to Karachi. And when the Germans were finally driven out of Africa in 1944, Pan Am's sturdy C-54s shifted to a more direct transatlantic route and flew as many as seven flights daily to Morocco. While the Boeing flying boats sweated out the swells at Horta, these planes refueled briefly at the land base nearby and roared away.

In three years Pan Am's Africa–Orient operation, in every way a full-fledged airline, completed two thousand Atlantic crossings for the military. It flew seven times as many miles as the entire Pan Am system achieved before the war. By war's end Pan Am's many different services — Caribbean, Alaskan, Pacific, North Atlantic, Africa–Orient — had racked up half the total contract miles flown by U.S. airlines for the military in World War II. Most of these were amassed flying landplanes.

Thus war wrote an end to one of commercial flying's most splendid traditions. Though the British made a half-hearted effort to restore flying-boat airliners after V-J Day, they soon had to give up. Once Juan Trippe had talked of seaplanes that would carry hundreds from continent to continent. But his company never ordered another. Two more big boats were built in wartime for the Navy — Consolidated's PB2Y and Glenn Martin's eighty-ton Mars — and Pan Am flew them on contract. After that, even the Navy saw little use for the giants. It is true that Howard Hughes won a contract in 1942 to build a flying boat eight times larger than the Boeing 314s, with a wingspan a third larger than that of the jumbo jets flying in the 1970s. But that monster was born to meet a short-lived wartime logistical problem. At the time, submarines were sinking Allied ships at such a calamitous rate that it seemed possible that men and freight might have to be transported instead by air. However, planes — small planes equipped with radar — quelled the submarine threat, and the convoys delivered the goods. Built entirely of plywood, the eight-engine Spruce Goose, as it was called, was an instant relic when completed in 1947. Hughes himself took it into the air once to fulfill a contract requirement, reaching an altitude of about seventy feet. It never flew again, although Hughes kept it the rest of his life in a hangar in Long Beach with a security force standing guard over it.

Thus flying boats, so spacious and so comfortable, faded out just like the zeppelins, with all their luxurious appointments, a decade before them. The boats were terribly expensive to maintain, difficult to service and repair in the water, and subject to the sea's corrosion to boot. These disadvantages were acceptable when there were no other places but harbors for airliners to alight. But World War II planted airports everywhere; the boats were doomed. They were more comfortable, but they were no safer. Weighed down by their huge hulls, they were inevitably outclassed in the one aspect of flying that men prized most — speed.

Though the airlines performed invaluable service, they were dwarfed by the immense organization built up by the nation for its World War II airlift. The great bulk of wartime transport flying was done by members of the armed forces called to the job as their part in the great mobilization of the nation's manpower and resources. The performance of these young men bespeaks, perhaps better than anything else, the measure of air transport's advance between 1941 and 1945. The 25,000 pilots of the ATC and NATS delivered the bulk of the 170,000 planes built in those years to the fifteen theaters of war. In the 11,000 transport planes turned out by Douglas, Lockheed, and others in those years, the aircrews of the ATC and NATS flew 28 million miles, lifting supplies and men over oceans that had been thought impossible to cross.

Beginning in 1942 not just grizzled airline pilots but also downy-cheeked second looeys with a couple of hundred flying hours began getting McTaggart-Cowan's celebrated briefings at Gander and then disappearing into the night in the direction of ATC's terminus at Prestwick. Before long not only Gander but also Goose Bay on the tip of Labrador, and a half-dozen fjord-side strips with strange names in Greenland, heard the roar of bombers and even fighters winging across the Atlantic to Britain. And the losses along the way turned out to be fantastically light. To be sure, these were warriors, not civilians, and even the passengers who caught rides with them wore parachutes — as nobody on a scheduled airliner ever did. But it was by this massed military airlift that the challenge of the North Atlantic was met and mastered. Without these exertions, five years of careful experimenting would have gone by before regular commercial flying with landplanes would have been hazarded. Credit this gain to the war.

Possibly the outstanding job of the airlift in World War II was keeping China in the war. This involved flying supplies from the northeast corner of India over the Himalayas to Kunming in the Chinese interior. Pan Am's subsidiary, China National Airways Corporation (CNAC), opened the line in 1941 just as the Japanese grabbed Burma. One of CNAC's DC-3s set a record carrying seventy-four people out. The pilot of that plane went back to use the lavatory; he opened the door and six Chinese children spilled out.

The chief beast of burden on this route was a transport plane that never saw service on a peacetime scheduled airline. The twin-engine C-46 Curtiss Commando, touted because it could carry twice the load of a C-47, was another war baby. Eastern Airlines had time to take out only the worst bugs before delivering the C-46s to the ATC's pilots in India. These young fliers, picking up their loads nine thousand miles from the United States, were at the end of the line in almost every way. Their job was to lift the stuff over the world's highest mountains and the world's steepest gorges some six hundred miles to Kunming, an air base still held by Chiang Kai-shek in the Chinese hinterland. The drill was to get over the Hump, as they called it, and back the same day, kicking out their cargo on the other side and returning to their Indian jungle base — all on the same tank of gas.

On a clear day you could find your way to Kunming by following the wrecks strewn along the route. An airline pilot, after flying over and back in 1943, called the operation "just plain mass suicide." Eric Sevareid wrote, "It is not often that one sees fear in the face of fliers, but I saw it there." C. R. Smith, who had advised that "the India–China Ferry must be conducted on the best standards of transportation," said later:

Here was this bunch of kids flying it. They had to fly to 28 or 29 thousand feet to get over the mountains. They had no navigation gear. And we had this lousy C-46.

We lost 300 planes on the goddam Hump. So many had to jump out over upper Burma that we put in camps to rescue fallen pilots. That was eat 'em up country. And that was the country where all the missionaries went that my mother used to support.

One of these pilots was walking along this trail with his rescuers. They met some Kachins led by a guy with a wispy gray beard. Both parties stopped. The guy with the beard comes across to the pilot, taps him on the chest and says, "You Baptist?" That boy was a Texan and a Baptist. So they all stood on that trail and sang "Jesus Wants Me For a Sunbeam."

What an operation — 125 mph crosswinds, 56 distress signals in one day, and 1,000 round-trips some days.

Captain Alexis Klotz, TWA skipper who flew VIPs around in the war, has told how the B-24 in which he was chauffeuring ATC boss Harold George got caught in the Hump traffic one rainy day over the Kunming terminal:

By a rough estimate of the tower's ability to ladder the planes down I figured it would be two hours until our turn to land. The ceiling below was getting worse. Every once in a while we would catch a ghostly glimpse of a passing plane also shuttling at our altitude.

We heard a pilot in a C-46 just ahead of us say that one engine had quit out of gas, and they would bail out as soon as the other quit. A minute later the sickening word came, "Bailing out." The tower said, "Good luck, fellows"; and we got his place in the stack, hoping his abandoned plane was not in our path or we in his.

We gradually worked our way down. Finally I was let off the hook, put the gear down, and the field came in sight ahead as a red gash in the rain. I landed in about two feet of water and taxied up to where General Claire Chennault [commander, Fourteenth Air Force] was waiting for us.

That night we slept in the hotel for transient crews. They had four double-deck beds sleeping eight people in each small room. A C-46 sergeant turned in at midnight and was sick in the night, something warm dripping down through the ropes on me.

General George said the show must go on, and it did for three years and 150,000 trips by V-J Day.

The nature of an airlift in war is getting the stuff to the front, and that was the ATC's big job. But its planes did not fly back empty. Millions of American servicemen got their first taste of air travel catching a lift to a rest area, or even home leave, in the bucket seats of a dun-colored C-54. Men at war, C. R. Smith said later, had good reason to travel. Masses of them moved by air, and he was sure that accounted for the big change in attitudes toward air transportation afterward.

The ATC maintained a whole fleet of hospital planes. These were the litter ships. Some lifted the wounded from the front, and some carried them all the way back to the United States. In combat areas it was a scramble. Once First Lieutenant George Jones's C-47, with twenty-four litters in the cabin, crash-landed out of gas in the Solomons. The pilot did a good job of coming down in a hundred-and-fifty-foot clearing. But a piece of a propeller blade flew in, struck a litter, and cut a wounded man's trachea. Second Lieutenant Mary Hawkins, improvising suction tubes from a syringe, a rectal tube, and finally the inhaling tubes of Mae West jackets, kept his throat cleared of blood till a destroyer took them all off next day. She also calmed a hysteric, and nursed another man whose malaria returned through the night.

In Europe troop-carrier C-47s flew the wounded out to Britain, where, by 1944, specially equipped ATC C-54s lifted them almost instantly across the Atlantic. The men were carried aboard in litters that were then hung in tiers of four along both sides of the plane's long cabin. As a rule there were twenty-four litters. The planes stank of wounds and bloody bandages. Clamped to the supporting poles by each man's head were the bottles from which fluids dripped into their veins as they flew. From time to time a nurse walked down the narrow aisle and knelt beside a stretcher. The plane stopped at Iceland so the men could have a hot meal. But there was little talk as the plane bored through the night. Even at Gander the long, dark cabin with its burden of wounded was strangely quiet. These were the most somber of Atlantic crossings. A fifth of all wounded returned to the United States in this way.

21

Air Travel in the War

After the United States went to war, young Henry Gunderson, the Minneapolis flour salesman who flew every week in the late 1930s, stopped traveling on the airlines. He couldn't get a seat. All of a sudden, the nation's domestic airlines found themselves with just 165 transport planes — the military had taken the other 200 for its needs. And when the airlines tried to keep up their peacetime rates and schedules with these few planes, Washington stepped in and said only those with "priority" could fly. A flour salesman, no matter how cordial his welcome in the Admiral's Clubs, no matter how firm his friendship with airline passenger clerks, was out of luck. Like thousands of other civilians who traveled a lot, Henry·Gunderson had to take the train.

Priority One was reserved for those with orders from the President. In fact, anyone holding it was said to be traveling on White House priority — on some errand for the President, or later, the Secretary of War or Navy. Priority Two was for military pilots on their way to ferry planes to the front. Priority Three was for military people or civilians traveling on business essential to the war effort. Among uniformed personnel, some carried orders that authorized air travel: these individuals, including servicemen on leave, might be asked to wait for a later flight. Others' orders *directed* travel by air, and these people the airline clerks put on the next flight out. Finally, there was Priority Four — for military cargo.

After the priority passengers had taken their seats, there were few places left for the private citizen summoned by family illness; those who wanted a Florida vacation dismissed the very thought of flying. At times the airlines had as much as half the space free for nonpriority passengers, but this was usually on out-of-the-way routes, at off-peak times, and on flights that stopped in a lot of places. A new set of words entered the language. Passengers, including servicemen holding Priority Three, were sold tickets and permitted to appear at the airport as "standbys," on the chance that when it came time for takeoff, some of those who had booked seats would fail to appear. Then the standbys, hovering by the agent's elbow, might go aboard in place of the "no shows." The most baleful of the new terms was "bumped." Any passenger without a priority could be "bumped" from his seat at any stop by another passenger who had one. Newspapers gleefully reported how Eddie Rickenbacker, president of Eastern Airlines, was bumped from one of his own planes in 1944 by a GI with Priority Three papers.

One result was that the proportion of women air passengers dropped — from 25 to about 15 percent. Another was that every flight flew with maximum load. The highest load factor — the term airlines used for the proportion of seats occupied — in any war year was 89 percent. Any traveler at the time would have sworn it was 100 percent. But no, the necessity for rotating planes, pulling them out for five-hundred-hour overhauls and the like, held the percentage down to that figure. At that, all sixteen trunk lines began to make money for the first time.

Compared to those in Europe, where commercial flying came to a dead stop, Americans who flew continued to get quite a good shake from the airlines. Although practically all their men were reservists, flight crews stayed on because the Pentagon ruled theirs was essential work. Passengers still got hot meals — bouillon soup, fillet of beef, fresh beans, hot potatoes, salad, rolls, lemon meringue pie, mints, and coffee were on one 1942 menu on United. On Friday there was fish as well as meat.

At the same time paperboard replaced the plastic trays of peacetime. That saved a hundred pounds per trip, and the trays were disposable besides. Such changes helped the stewardesses, who had to jump when every twin-engine DC-3 flew with full twenty-one-passenger loads. This was the time when the airlines dropped the requirement that each stewardess be a registered nurse. Passengers noted that this was also the time when the airlines ended discounts for Air Travel Card holders and stopped selling tickets through travel agents.

Suddenly, having tried by every advertising, public relations, and promotional device they could think of to lure people into flying, the airlines did not even have to give a thought to passengers' opinions or needs. And of course people were bumped, standbys disappointed, arrivals late, terminals jammed, bags lost, and a fantastic proportion of flights seemed to involve flying all night through frightful weather in ill-heated planes. In fact, conditions were miserable. But when a man complained on the airline bus from Midway Airport into Chicago, the other passengers jumped on him and said, "You're lucky to be riding at all."

In the early crunch the only passengers who got to fly were — another coinage of wartime flying — VIPs, Very Important People. Ambassadors, bureaucrats, staff officers, war correspondents, scientists — they were the first wartime intercontinental voyagers. The best rides anywhere were on Pan Am's big Boeing flying boats, which kept their regular transatlantic schedules and style from first to last. "My God, what luxury," recalled

Roy Alexander, an editor who flew to London early in the war. "The bunks were so big that only a seven-foot basketball player could have touched the bottom." There were only five or six passengers, but a brace of stewards cooked their meals on board and served them on linen table-cloths. By custom everybody stayed in his seat until the captain got off. Each time he stepped through the door, the lanky skipper would hit his head on the coaming; each time he would turn and say to those behind, "Look out for your heads." When swells held them overnight at Horta, the captain led them to Pan Am's hilltop guesthouse. "Want a drink?" he asked, and invited the passengers to reach into the well-stocked icebox for whatever they pleased. One of the passengers was later identified as Sir William Wiseman, chief of British intelligence. Another was a sinewy, sport-jacketed young fellow, a former Marine, who carried the State De-partment diplomatic pouch and slept with it handcuffed to him. At Lis-bon he went off in the back of a truck, sitting on a pile of mailbags and holding a shotgun, with the diplomatic pouch still chained to one wrist.

On its historic mid-Pacific route, Pan Am lost one Clipper, strafed on the water at Hong Kong on Pearl Harbor day. Approaching Honolulu that morning, Captain Lemoyne Turner saved his Clipper by ducking two hundred miles south to Hilo, and Captain John Hamilton saved a second by racing to take off at Wake Island through a salvo of Japanese bombs. The airline had a second transocean route that led from Hawaii to the Southwest Pacific that at once became a highroad of war. Clipper crews flew VIPs along this track using four-engine Navy Consolidated PB2Y fly-ing boats almost as stately as their Boeings. One such scheduled flight in early 1943 stopped at Nouméa to carry twenty uniformed passengers back to Honolulu. Vice Admiral Kelly Turner, commander of U.S. landing forces in the Solomons, was fighting off malaria and occupied the plane's bedroom. Inured to the mud, dust, and canned rations of the forward areas, the other nineteen passengers sank into big, leather-covered chairs only to be invited to lunch at tables set up by the stewards. "Everything is so clean," a Marine captain said. "It's like a West Indies cruise," said a Navy airman whose carrier had been torpedoed. Taking off each morning at 9:00, the big plane flew until about 3:00. The first afternoon it alighted at Suva, capital of the Fiji islands, where the Pan Am manager ordered drinks and dinner served to his passengers on the moonlit balcony of the Grand Pacific Hotel. The next day's flight ended at Canton Island, where the travelers feasted at Pan Am's only surviving midocean sandspit hotel. Shortly after midnight there was a flurry of excitement when a lone Japa-nese bomber appeared overhead from Majuro, another sandspit fifteen hundred miles away. Searchlights caught the plane in crisscrossing beams and the plane dropped its bombs in the lagoon. An intelligence officer snorted, "Never in all their glory had the Solomons a raid like one of these," and went back to his screened bedroom. Next day's trip was the longest, so long that it extended into the night. At 7:30 P.M. the flying boat swung low over the crowded Pearl Harbor anchorage, found its row of landing lights, and alighted in the dark water. After three days and 5,000 unhurried miles, the twenty voyagers went ashore on Oahu in Pan Am's launch.

In contrast with the Clippers' cruiselike comforts were the rigors of crossing the Atlantic on bomber-ferrying flights. On one four-engine B-24 flying from Canada before Pearl Harbor, five British civil servants went along as passengers. It was of course a night flight. The travelers entered the plane through a small opening in the bottom of the fuselage, through a door opening downward and bearing steps on its inner side. Inside was pitch darkness. Groping up to the radio operator's cabin, they were told the captain preferred passengers to be in the bomb bay when the plane took off. One gripped a couple of struts in the cavernous darkness, braced his feet against a bag opposite, and amid whines, scrapes, and the sound of pumps, felt the rush of air and, peering along the passage, saw the tail come up — takeoff. For the flight above clouds, oxygen masks were handed round — rubber nosebags with head straps. By flashlight the travelers located a couple of oxygen tanks lashed to the bomb bay wall. From these, metal tubing led along the walls of the fuselage to a dozen or more stations where long rubber hoses were attached. The passengers found they could attach the hose to their masks at any of these stations, and thus had some freedom to move about the rear of the plane.

Though it had been a summer's day in Canada, the crew donned bearskin flying suits and fleece-lined boots as the plane climbed past ten thousand feet. One foresighted passenger had brought along fleece-lined flying trousers that were enclosed at the bottom. Before the night was over, he needed not only these but also a hood pulled over his face, and his raincoat draped on top. He thrust his gloved hands into his trousers to warm them. Another passenger slept in the bomb bay in his overcoat under two blankets and yet another overcoat flung on top. A third who had brought a couple of blankets slept only long enough to see the red orb of the sun rise between the two engines on the port wing at 11:00 P.M. — his watch was still on Canadian eastern standard time. Lying prone near the radio operator's perch while his companions eked out the night in the tail, in the bomb bay, and on various ledges, he raised on his elbows three hours later to see mountains through gaps in the clouds. The plane's mechanic scurried past to the tail once, then came back again. As he rushed past the second time he said the landing wheel on the right side was sticking and would not come down. It happened so quickly that the passengers "hadn't time to remember to be nervous." Somehow the wheel was lowered, and the plane, cutting through the clouds, put down at Prestwick, Scotland.

The man with the watch noted that it had taken just nine hours, fifty-five minutes, to hop the ocean, and with no fuss about submarines along the way.

Most who flew the oceans were passengers in uniform. It was for them that the commandeered airliners were fitted with bucket seats — aluminum seats along the walls of the fuselage. Paratroopers hunched on these seats on their way to the jump. When the plane carried cargo only, the seats could be folded back against the wall. And they could be put down for casual passengers. But when the ride lasted more than an hour or two, bucket seats were hard on the rear end. Since all insulation and upholstery had been pulled out, the cabin was also very noisy. The individual carried his own comforts — a box of K-rations for lunch, a swatch of toilet paper for the john if there was one, a blanket for the night. On long flights the passengers wrapped up in sleeping bags or blankets and threw themselves on the bare metal floor, or draped themselves over boxes and crates. If there was room, some laid pads down and stretched across the bucket seats. Even in summer it was cold at the altitudes at which planes flew. In winter, if a man on his way across the North Atlantic as much as removed his gloves and dozed off, he suffered frostbite.

VIPs too had to take the wartime airlift as it came. Vice President Wallace flew all the way to China wearing two wrist-watches — one to tell the time where he was, the other to tell the time back home. Wendell Willkie flew all the way to China, and won everybody's startled nod by saying such trips meant that now we all inhabited "one world." Ambassador Averell Harriman seemed to be forever shuttling between Washington and Moscow, two capitals that until the war had seemed impossibly far apart. Once he crossed on a converted B-24 that, after a stop in London, flew all the way around northern Norway to land at the Soviet port of Archangel. But Harriman thought the most hazardous leg of the journey was his return flight from Moscow to Archangel in a Soviet-built copy of the DC-3. The clouds hung so low that the plane flew at tree-top level. Then when trigger-happy Russian antiaircraft gunners fired at it, the plane flew *between* the trees.

On one of the first bomber-ferrying flights across the North Atlantic the passenger was Sir Frederick Banting, the famous Canadian scientist who had discovered insulin and whose presence was urgently desired in British war laboratories. The flight was the second piloted by Captain Joe Mackey, who had flown one of the first seven Hudson bombers across a few weeks before. This time one of the plane's oil coolers sprang a leak after takeoff. Circling back to land, Mackey gave orders for crew and passenger to take to their parachutes, then he plunged through the night to a crash landing in the Newfoundland wilderness. When Mackey came to, he saw through the blood streaming down his face that nobody had jumped. The two crewmen lay dead, Banting unconscious. He dragged Banting into a bunk in the smashed cabin. Delirious, the great scientist dictated memos all night to the dazed Mackey, then died at midday. Three days later trappers found them. Mackey was nearly dead from cold and exhaustion.

The Danish physicist Niels Bohr, bearing the secret of the A-bomb to the West, made his way to Sweden. There he boarded one of the world's chanciest airlines. The British found it so important to keep contact with the neutral Swedes that they kept up a regular fortnightly service from Scotland throughout the war. It was called the "ball-bearing airline" because the planes brought back highly prized Swedish ball bearings for British war plants. When the planes — small, fast, twin-engined Mosquito bombers constructed entirely of wood — flashed across Nazi-occupied Norway and Denmark, the Germans always tried to intercept them. In four years they managed to shoot down four. On the night of October 7, 1943, the plane flew so high that Bohr, lying prone in the bomb bay, lost consciousness when his oxygen tube slipped from his mouth. He narrowly escaped death.

Every passenger on the ball-bearing flight took his life in his hands. TWA's famous navigator Pete Redpath, sent to Stockholm on a special assignment, found this out in a 1944 crossing:

It was just the pilot and I — I in a heavy flying suit, sitting against the bulkhead in the bomb bay. The pilot said, "We've got an intercom. You're an old pilot. I'll tell you when to put on your oxygen." It was very noisy. He said, "Now we're at 6,000 feet. I'll call you at 15,000, because we're going up to 27,000."

I waited for the call but I heard nothing. I called him, nothing. I slapped on my oxygen mask. Then one of the bomb bay doors started to bleed open. Soon it was half way open. I scrunched up. I was holding out pretty well against the cold — I had mufflers around my face. I was furious.

That bird didn't say a word until we were coming down, 20 minutes out of Stockholm. Then he said, "Sorry, I forgot." I said, "Close the bomb bay doors or you'll have a corpse on your hands."

Then there was General Wladyslaw Sikorski, prime minister of the anti-Communist Polish government in exile, whose flying mishaps were all too suspiciously frequent. One plane carrying him crash-landed with engine failure, a second just missed crashing when an incendiary bomb was found smoking in the toilet compartment. In a third crash, on July 4, 1943, the general finally lost his life. A converted B-24 bomber piloted by Flight Lieutenant Edward Prchal, a Czech flying for the RAF, took off from Gibraltar, climbed to two hundred feet, then plunged into the Mediterranean a half mile from the end of the runway. Prchal, the only survivor, said the controls locked when he put the nose down to gain more speed. It seems a fair presumption, thirty years later, that someone wanted General Sikorski out of the way.

Such wartime air travelers were marked men. When Admiral Yamamoto, architect of Japan's early war victories, flew as a passenger to inspect front positions in the Solomons, U.S. forces in the Pacific used their ability to crack coded messages to intercept and kill him. Winston Churchill said later, "These airplane journeys had to be taken as a matter of course during the war. Nonetheless I always regarded them as dangerous excursions."

Churchill first crossed the Atlantic by air after his first wartime confab with President Roosevelt at Washington in January 1942. He traveled in one of the three Boeing 314 flying boats the hard-pressed British had won permission to buy from Pan Am. His original plan was to fly only to Bermuda, and go the rest of the way aboard the battleship *Duke of York*. But his pilot, BOAC Captain J. C. Kelly-Rogers, assured him winds were favorable, and Churchill was in a hurry to get back. Tanking up with 5,000 gallons of gasoline, Kelly-Rogers took off eastward next morning on a flight attempted only once before. "Nothing was lacking in food and drink," Churchill wrote later. "The motion was smooth, the vibration not unpleasant, and we passed an agreeable afternoon and had a merry dinner." Then, while the prime minister slept in the bridal suite at the stern, the plane flew into thickening weather. Toward morning Kelly-Rogers decided to switch his destination from Pembroke in Wales to Plymouth on the south coast. This required changing course toward the south but not, as Churchill wrote later, to the point where they were almost "over the German guns at Brest." That must be put down to a prime ministerial sense of drama. Although Churchill occupied the copilot's seat for an hour that morning and said later, "we did not know where we were," he was not flying the plane. Having radioed his estimated time of arrival at 9:00 A.M., Captain Kelly-Rogers brought the big plane across the breakwater at fifty feet and alighted on Plymouth Sound at 8:59 — hardly the performance of a pilot who did not know where he was. Home safe after a 3,365-mile flight of seventeen hours, fifty-five minutes, Churchill cabled Roosevelt: "We got here with a good hop from Bermuda and a thirty-mile wind."

For greater speed Churchill turned thereafter to the converted B-24. "I flew with Churchill in his B-24," Averell Harriman said later. "It was converted for passengers in the most primitive manner, without insulation and with two rows of hard benches facing each other. The noise was so great it made conversation impossible [except] by passing notes." On the first flight to Gibraltar and Cairo, Churchill bustled into the cockpit and addressed his American pilot, Bill Vandercloot, who had joined the British after flying for TWA. Churchill said, "You know what Hitler would do to me if he ever got his hands on me, don't you, Vandercloot?" "Yes, sir," the pilot said. "But you are not going to let him do it, are you, Captain?" Churchill said.

"Fortified by a good sleeping cachet," Churchill then retired to the after cabin "where two mattresses had been dumped," and slept soundly through the flight over German-occupied North Africa. After they landed in Cairo, a plane bearing the man Churchill had chosen to command the Eighth Army against Rommel was shot down by a prowling fighter of the Luftwaffe "in almost the very air space through which I flew." Appointed Eighth Army commander as a result: General Bernard Montgomery.

Allied summit conferences occasioned the most famous VIP flights of the war. The first, in January 1943, was held at Casablanca, a city captured only two months before by American troops. In the dead of winter, Vandercloot's B-24 bore Churchill southward from an airfield near Oxford. This time there was a bunk for Churchill, although his chiefs of staff had to sleep in chairs. In view of the chill, the RAF had put in a sort of jury-rig heating system. At 2:00 A.M. Churchill woke to find heat from an outlet "burning my toes." Jumping out of bed, he caught a whiff of fumes from the gasoline-fueled heating unit. He ordered it turned off as dangerous. At once the air grew icy. Churchill crawled about pushing blankets against walls against the piercing drafts. "The prime minister is at a disadvantage in this kind of travel," his physician noted in his diary, "since he never wears anything at night but a silk undershirt. On his hands and knees, he cuts a quaint figure with his big, bare, white bottom."

To get to the Casablanca conference, Franklin Roosevelt took to the air, the first time a President of the United States had done so while in office. Harry Hopkins offered his explanation: "They are always telling him that the president must not fly — too dangerous. This is his answer. He loves the drama of a journey like this."

For his first such jaunt Roosevelt made what appeared to be a wise decision: he called upon Pan Am to carry him across the ocean; for that part of his journey where he traveled over land, he would change to a multi-engined landplane. The landplane would be a C-54 that belonged to the Army Air Forces but was flown by TWA.

When his train arrived at Miami he boarded a Boeing flying boat piloted by Captain Howard Cone. A second Boeing Clipper, Captain John Vidal commanding, revved up to fly along — the backup plane. High overhead the Air Force had thirty-six fighters circling. On the first day the plane flew to Trinidad, where Roosevelt took time out to inspect lend-lease bases. On the second day the President flew to Belém at the mouth of the Amazon, where he spent the night on a Navy seaplane tender. On the third day the Clipper flew nonstop across the South Atlantic narrows and landed splendidly at the British West African port of Bathurst.

Waiting at Bathurst was the C-54 piloted by TWA's Otis Bryan. The President was buoyant, in high spirits. He sat attentively through the six-hour flight northward — past Dakar to the coastal port of St. Louis, then inland over the tawny shadows of the Sahara, finally skirting the snow-topped Atlas Mountains. Two other C-54s flew just behind the presidential plane — one the spare, the other carrying secret service men. For the last three hundred miles "within the combat zone," two lookouts, one at the plane's astral hatch and the other at a cabin window, peered out for hostile aircraft while thirty-six more fighters orbited overhead. Although visibility was perfect, no Messerschmitts were seen. The Mediouna airport was deep in mire. But Otis Bryan made a faultless landing on the steel-mat runway and taxied the C-54 alongside Vandercloot's B-24, which had arrived with Churchill two days earlier.

The summit over, Churchill flew to the Middle East. On the way back the No. 4 engine refused to start at Algiers. But the crew chief hitched a rope to a propeller tip, hooked it to a tractor, and raced off. It was exactly like a kid spinning a top with a piece of string, and it worked. The No. 4 engine gave a mighty cough and then settled down to a purr — all the way back to London.

When Roosevelt boarded Otis Bryan's C-54 for the flight back to Bathurst, the Air Force tried to move a bed into the plane for the President. He refused it, and sat up sleepless for eight hours like the rest. After a night at Bathurst, Roosevelt rose early and flew with Bryan to Roberts Field, Liberia, for lunch with President Edwin Barclay at a camp outside Monrovia. Returning to Bathurst, he went aboard Cone's flying boat shortly before midnight. While the Boeing droned across the Atlantic, the President slept in the bridal suite at the rear of the plane.

The next morning President Vargas of Brazil was waiting for him at Natal. The two men talked all day, then spent the night on a U.S. destroyer in the harbor. One of the advantages of the Clippers was that the door was scarcely a step above the landing barge, and the President could be wheeled on in his wheelchair. On the morning of the twenty-ninth Captain Cone and his eleven crew members stood at attention as Roosevelt came aboard for the flight to Trinidad. The next day was the President's birthday, and Cone and his men gave him a sixty-first-birthday party with a cake crowned by seven candles. At the time the Clipper was flying over Haiti. Back in his days as assistant secretary of the Navy, Roosevelt told Cone, he had written that country's constitution. Altogether the President's trip took twenty-two days and covered 18,965 miles.

Roosevelt never crossed the ocean again by air. On his second trip abroad, he rode the cruiser *Memphis* all the way to Oran, in western Algeria. There Captain Otis Bryan picked him up in another C-54 and lifted him to General Eisenhower's headquarters at Carthage. Then they hopped to Cairo, where Roosevelt and Churchill held their meetings with Chiang Kai-shek before flying on to the summit with Stalin at Teheran. Afterward Roosevelt flew in the C-54 to Cairo, to Carthage, and finally to Dakar, where he took a ship home.

For his third and last transatlantic trip, the President traveled all the way to Malta by sea. For the trip onward to Yalta in southern Russia the Air Corps had provided a special Douglas transport plane, the one that came to be known as the *Sacred Cow*. On this plane, designated a C-54D — the only one ever built — the ATC had fixed up a stateroom for the President. But the plane's distinctive feature was a built-in elevator.

On earlier trips a long ramp had been trotted out at each stopping place for Roosevelt's wheelchair. That told everyone who saw it that the President was either present or expected. "There could be no security when you had to fly a ramp into places like Dakar and Cairo," Bryan said. It was for this, more than for comfort, that the new plane carried an elevator. The elevator was a weighty apparatus, heavy enough to make an appreciable difference in the total weight of the plane. Located just aft of the plane's rear door, it was operated by means of a gear on a long shaft. It worked vertically, not on a slant — "something like a landing gear," said Bryan.

There was just room enough in it for the president in his wheelchair — and his valet. The valet would get in with him, you would close the door and press the button, and the elevator would lift him into the plane.

It was always an anxious moment when that door clanged shut on the President of the United States. But the elevator always did the job. It was a six-foot lift, and "once it was up, you would never know from the outside there was an elevator there, it fit so smoothly."

The aerial logistics for the Malta–Yalta airlift were the most elaborate of the war: fourteen C-54s for the U.S. summit party, 800 pounds of deluxe rations for each plane, all planes flying through the night to give Cyprus-based German fighters the slip.

Bryan said later there was a huge difference between Roosevelt's first flight and the last. On the trip to Casablanca the President was buoyant, in high spirits, a man who had escaped from his desk and was enjoying himself. On the journey to Yalta he was a sick man. By this time Bryan knew the President pretty well. Roosevelt called him back to the cabin to tell him about the bargain he'd struck with Stalin to get Russia into the Pacific war. Roosevelt listed his four points — a warm-water port in Manchuria, the cession of southern Sakhalin, a boundary rectification in Siberia, the control of the Manchurian Railroad. "Little enough, eh?" Roosevelt said. At Cairo Roosevelt wanted to get President Inonu of Turkey to come and see him. Turkey was still neutral; Inonu said he was too busy. Roosevelt replied, "If you come I'll send my personal plane for you."

To avoid German interceptors, Bryan picked up Inonu in eastern Turkey. As he sat in the Adana officers club drinking coffee, the club radio blared a German Turkish-language broadcast: "Colonel Otis Bryan has just landed at Adana and is going to fly the president of Turkey to Cairo in the morning." His host, the base commandant, went white. In the morning Bryan took care to fly well inland. No Messerschmitts appeared. Invited to the cockpit, Inonu sat down in the copilot's seat and began identifying place after place in Jordan and Sinai where he had fought the British in the First World War. Inonu stayed up front to the end. Later he sent Bryan a gold Swiss watch for a present.

After his one ocean flight, Roosevelt said to reporters, "Amazing — Wednesday in Liberia, Thursday in Brazil. And I don't like flying, not one bit. The more I do of it, the less I like it." He exhibited his short-snorter bill autographed by Captains Cone and Bryan.

The short-snorter bill arose as the inevitable certificate of wartime transocean travel. Joe Crosson, the famous old bush pilot, is said to have bestowed the first one on a brave soul who risked crossing a corner of the Gulf of Alaska with him in 1929. The badge was a dollar bill (or, later, a ten-shilling note) carrying the signature of a prior crosser, awarded after standing him to a short snort at the bar. In World War II some short-snorters traveled so much and initiated so many members that they had to fasten a whole string of bills to the original one to hold all the signatures they collected. One passenger from the Pentagon was said to carry a chain of short-snorter bills thirty feet long.

22

Taking the Fifth

The surge of American might that propelled the nation's planes to the farthest parts of the earth raised questions about the role of American aviation after the war. Would American planes in overwhelming numbers fly overseas in peace as they did in war? And would these peacetime planes fly as they did before the war, under the management of a single American international airline?

Anticipating the American Century, Republican Congresswoman Clare Luce said, "We want to fly everywhere. Period." Having learned to fly the oceans alongside Pan Am, the domestic airlines said, "We want to compete for the overseas business."

Freedom of the air — was it not a natural extension of freedom of the seas? Back in the seventeenth century the Dutch had invented the idea of freedom of the seas, and thereafter Dutch ships made the transportation of other people's goods a way of living. Now Vice President Henry Wallace, no doubt a bit of a dreamer, said the postwar United Nations organization ought to open the skies and police them. Americans like Wallace quoted a new Dutch precept: "The air ocean unites all people."

The British, fearing that the Americans with all their planes would grab all the business, wanted no part of the open skies notion. To win the war, American transports might land in Britain at the rate of a dozen a day. But in peacetime limits must be imposed on such traffic. British airlines, pushed to the rear in transport flying because Britain could build only warplanes, must have their chance. Yes, the Americans had won lend-lease airfields in the British West Indies. And yes, the Americans had built many more across British possessions in Africa. But the right to land on British soil and the right to fly through British air space were privileges granted only by the British government, and granted only temporarily. Air Minister Harold Balfour rose in the Commons to say gruffly, "After the war, all bets are off."

Suddenly Americans were told that freedom of the air was nothing like freedom of the seas. You did not talk about freedom of the air at all.

Rather, you talked about freedoms of the air — and, Americans were told, there were several freedoms, five, no less. One by one, these separate freedoms would have to be negotiated by the sovereign nations of the planet.

The first one the British mentioned was right of transit. That got you freedom to cross through the air space above a country without stopping, and no more. Second, was the right of technical stop. That allowed you to come down at another country's airport for fuel and servicing, and no more.

But this was nothing. If these two rights constituted freedom of the air, then the open skies concept that Americans demanded was absurd. Without such elementary rights, U.S. airliners might as well stay home. The third and fourth freedoms identified by the British promised slightly more sky — but hardly the wild blue yonder. These were the right to discharge passengers at another country's airport, and the right to pick up passengers there and bring them home. What was commercial flying without these two rights?

Finally, there was the fifth freedom, and the way the British talked about this made it sound so rare and remote that no greater concession could ever be bestowed by a sovereign nation. This was the right to unload at another country's airport and then to pick up passengers bound for other countries farther on. It was a right that Pan Am, for example, had freely enjoyed from the beginning on all its routes through Latin America.

In war, transport looms big in the national interest. Commercial aviation, shaped by secondary figures like Postmaster General Brown in peacetime, commanded the attention of the likes of President Roosevelt and Prime Minister Churchill at the height of World War II. Both sensed that it would play a big role after the war. Both saw its importance for the new alignments of power. Churchill's kind of empire, with its splashes of red across the map, was receding; Roosevelt's kind, which spread wherever airliners could deliver businessmen, was on the rise. They conducted their first decisive exchange at the Quebec conference in late 1943.

Roosevelt was astonishingly clear on what he wanted. He had listened to the voices in Congress, to Trippe of Pan Am, and to Slater of American Export. He had obtained lengthy opinions from the Pentagon and other departments, and the practically unanimous agreement within the administration was that America should go for all five freedoms and for U.S. airline competition overseas. After his talks with Churchill, he summoned civil air aides to a White House meeting in November 1943. Present were Undersecretary of State Edward R. Stettinius; Assistant Secretary of State Adolf Berle, his international air policy adviser; Assistant Secretary of War for Air Robert Lovett; and CAB Chairman Welch Pogue, his civil air policy expert.

First of all, FDR said, he did not want Germany, Italy, or Japan after the war to "fly anything larger than any of those toy planes that you wind up with an elastic."

As to air rights in general, he favored a "very free interchange." Each country — he cited Brazil as an example — should handle its own internal air traffic. He thought that all airliners should have the right of free transit and the right of technical stop everywhere, that planes of one country should be able to enter any other country for purposes of discharging traffic and accepting foreign bound traffic. He thought that each country should open several airports —"in the U.S. quite a large number" — to other countries' airlines. And he declared himself for the fifth freedom. His example was a Canadian airline that might be flying to Jamaica with stops at Buffalo and Miami. The Canadian planes should be able to discharge traffic of Canadian origin at Buffalo and take on traffic at Buffalo for Jamaica; but they should not be allowed to carry passengers from Buffalo to Miami. There would probably have to be a "United Nations organization to handle such matters as safety standards, communication, weather reports . . . and fares."

Midway in his discourse, Undersecretary of State Stettinius left the room to make arrangements for the arrival from Moscow that day of Secretary Cordell Hull. The President paused until the door closed behind Stettinius, who was Trippe's brother-in-law. Then he made his pronouncement about how American overseas aviation should be handled.

The President said that overseas aviation was too big in scope for any one U.S. company or pool. Certain companies, to speak frankly, Pan Am, wanted all of the business. He disagreed with Trippe. He was willing to agree on the record that Pan Am was entitled to the senior place, and perhaps the cream of the business; but he could not go along with the idea that they, or anybody, should have all of it. The plan by which various companies would have zones appealed to him — one company flying the North Atlantic, another the Mediterranean, and so forth. He saw no need for government participation in the companies, he said, except for subsidy on a route like that from the United States to South Africa, a route that would probably not be a paying proposition.

That session set the lines of U.S. policy, although bureaucratic bickering and airline infighting went right on. When Assistant Secretary of State Berle said it was time in late 1944 to call an international civil aviation conference at Chicago, Roosevelt gave it his blessing. And when his CAB Chairman Welch Pogue said he wanted to go ahead and invite domestic airlines to submit applications for postwar overseas routes, FDR told him, "You do what you want, and I'll take the heat."

While Berle skirmished with the British, Pogue invited applications for five specific areas — three Pacific and two Atlantic, the big one being the North Atlantic. Hearings began, and Trippe was on the defensive as never before in his life. Pan Am had long held the inside track with the executive branch as the sole holder of overseas mail contracts. Now, with that advantage slipping away, Pan Am turned to Congress.

All the airlines had friends in Congress. For most, these ties stemmed from roots back home. TWA, the Kansas City airline, had a particularly staunch friend in the senator from Missouri, Harry Truman. Northwest owed much to Warren Magnuson of Washington, always a power in the Commerce Committee. Pan Am, the airline that began at the water's edge, had relied on the widely acknowledged persuasiveness of Juan Trippe and on forceful lobbying. Though Pan Am did not fly to Maine, Trippe had gone fishing there with Senator Owen Brewster. Ever since, Brewster, ranking Republican on the Senate Commerce Committee, had been perhaps Pan Am's most outspoken congressional friend. But it was a Democrat, Senator Pat McCarran of Nevada, a leader in drafting the 1938 CAA act, who brought in the bill embodying Trippe's "community company" idea. Senator Josiah Bailey, Democrat from North Carolina and chairman of the Commerce Committee, gave the bill, known as the "All American Flag Line Bill," top priority.

President Pat Patterson of United, the only domestic line to side with Pan Am, was the strongest witness against the policy of extending competition overseas. He declared in 1944 that twenty-three airliners would be able to carry all the people flying across the North Atlantic in 1955. That was too much for former Senator Josh Lee, who had recently joined the CAB. Lee said, "It would have been just as logical to have determined how many people crossed the American desert by stage coach, projected the figure, and announced that this is the number of passengers who may be expected to cross the United States by rail." General Harold George, whose Air Transport Command had lifted three million members of the armed services around the world, pooh-poohed the conservative airline estimates. The size of transocean travel after the war would "stagger the imagination," he predicted.

While senators and airline presidents volleyed and thundered, Berle led the U.S. delegation to the international civil aviation assembly in Chicago in December 1944. Senators Bailey and Brewster were prominent among the American representatives. The Russians, in an early signal of their postwar stance, declined to appear. "The air has been used as an instrument of aggression," proclaimed Berle. "It is our opportunity to make it hereafter a servant of peoples."

For six weeks the meeting got nowhere. The British, sure their airlines would be ground under by the Americans if the open skies idea prevailed, remained tough. They were not giving away any freedoms. They wanted all transocean flights, just like transatlantic ship crossings, to terminate at the British "gateway." Travelers to or from continental Europe would make their way from London by some other plane or cross-Channel boat-train. But what the British really wanted was to prevent the United States from running away with all the North Atlantic business until they could crank up and build some ocean-flying airliners of their own — perhaps even jets. They therefore began talking about quotas on North Atlantic air traffic, and there wasn't much doubt that they had the power and the means to impose them: the prewar United States–British agreement authorized Pan Am to make three landings in Britain per week.

At the last minute the Dutch, as fearful for KLM's future as the British were for BOAC's, moved that the conference adopt only the first two freedoms — the right of transit and the right of technical stop. It was little, but it was enough, because Europeans and Americans alike saw the whole idea of a multilateral setup heading for the dustbin. Instead, all now expected to fall back on old-style bilateral agreements. Also, the British, Dutch, and others who feared American domination of international civil aviation saw another way to curb "cutthroat" American competition. In its windup session, the conference voted to establish a permanent international body to administer the presumably footling technical matters that Roosevelt had forecast for it.

An international agreement that gave airmen just two of their freedoms was very small beans indeed. Disappointed, Adolf Berle left Washington as ambassador to Brazil. Senators Bailey and Brewster, who would assuredly have blocked Senate ratification of a five-freedoms multinational treaty, grudgingly let Roosevelt pass off the piddling Two Freedoms Pact as an executive agreement not big enough to require assent by two-thirds of the Senate. Amidst these maneuvers, the "technical" organization to administer what was left of global air cooperation came into being. Two units were formed. The first was the International Civil Aviation Organization (ICAO). The second was the entity that came to be known as the International Air Transport Association, or IATA. Governments took part in the first, airlines in the second. Only Russia was not represented. Their headquarters were established in Montreal. Together the new bodies exercised "nonexecutive powers" over such technical matters as safety standards, navigational controls, air maps, and (this was claimed for IATA) the setting of international air fares. Disarmingly, IATA's rulings were required to be unanimous, and to be renewed each year.

In 1945 President Roosevelt died, the war ended, and a Labour government took office in Britain. In the midst of these great events, the policies that Roosevelt had laid down for civil aviation in the war were put through almost exactly as he had framed them. The All-American Flag Line Bill went down to defeat in the Senate. Trippe, seeing the way the wind blew, tried an uncharacteristically short-sighted stroke against the domestic airlines' applications for international routes. Pan Am entered a wholesale counterclaim for a whole network of routes crisscrossing the United States — the claim was duly rejected. In July the CAB announced its long-awaited decision on who would fly the North Atlantic.

Three companies received seven-year certificates to carry transatlantic passengers. The first, as Roosevelt had signaled, was Pan Am. Pan Am was confirmed in its pioneer route to London and authorized in addition to fly onward across southern Germany and the Balkans to Istanbul, Teheran, Karachi, and Calcutta. The second was a formidable newcomer, entering at the very last moment. When Pan Am won a suit compelling American Export as a steamship company to sell its controlling interest in American Export Airlines, American Airlines had stepped in and picked up the majority holding. On June 1, one month before the CAB decision, Amex was re-formed as American Overseas Airlines, the international division of American, the biggest domestic airline. American Overseas was authorized to fly to London, Scandinavia, northern Germany, and on, if the Russians ever allowed it, to Moscow. The third winner was the domestic company that got a jump on wartime transocean flying because it possessed the only four-engine landplanes. TWA won a certificate to fly to Paris, and onward to Rome, Athens, Cairo, and Bombay.

Thus Roosevelt's zone plan became the basis for U.S. overseas air policy. American Overseas would fly to northern Europe, Pan Am to central Europe, TWA to southern Europe. Competition among Americans on the North Atlantic was assured, but it would be "area competition." All three would fly the ocean, but to different sectors of Europe.

It remained for the U.S. government to obtain access for the three airlines to the places the CAB said they could fly to. The months had passed. The war was over. The U.S. bargaining position on the North Atlantic had improved. Under the Two Freedoms Pact, the United States had obtained from Newfoundland the right to use the field at Gander and from the Irish the right to land and take off at Shannon. In Britain the Churchill government had fallen. The new Labour government, taking office in a severe winter of shortages and dislocations, found it necessary to ask the United States for an enormous loan.

When the U.S. and British government representatives met in Bermuda in early 1946, they moved quickly to a compromise. The pending British loan speeded things up. All the five freedoms that Roosevelt asked and Berle sought in vain by multilateral arrangements were now gained by bilateral agreement between the two nations. The British agreed to let Pan Am carry passengers not only to London but also between London and certain points on the continent. That was the beginning of the end of the gateway idea. The Americans obtained thirteen international routes in trade for seven for the British, one of which opened Hawaii to outsiders for the first time and thereby made a Union Jack round-the-world line via Australia possible. Of great importance to Pan Am and American Overseas, the British conceded unlimited flight frequency on the North Atlantic, subject only to discretionary controls by the two governments.

But the British, even in their first dismal awakening to their postwar weakness, were by no means routed. They won formal U.S. assent that IATA would set fares. For Britain this was a big victory, but not only for Britain. It saved European transatlantic airlines from ruin and likely extinction. Without question the United States had all the airliners in 1945, and if the Europeans wanted to compete they could only take their turn placing orders with the manufacturers in California. Given half a chance, the U.S. airlines with a big edge over their rivals in prostrate Europe would have flung themselves into a no-holds-barred fight for business.

The power to set rates was pivotal. Under the firm direction of Sir William Hildred, former head of British civil aviation, IATA set fares high and kept them high. For all the requirements that its rulings be unanimous, and renewed from year to year, IATA grew into a pretty good facsimile of a cartel.

The Bermuda bilateral agreement was not just a compromise, it was a landmark compromise. It ratified the new balance of commercial air power for the Western world. When the French quickly followed the British example and in March 1946 signed a Bermuda-like pact with the United States that opened the door to Paris and beyond, the pattern was set. The kind of United Nations of the air that seers like Henry Wallace had conjured up during the war, which would have organized and even enforced freedom of flight everywhere, simply faded away. By innumerable bilateral pacts, the United States gained the five freedoms of the air anyway. And in its hour of victory and vaulting power, the United States agreed to live and let live. By conceding the Europeans effectual power to keep fares high on the North Atlantic and elsewhere, the Americans enabled non-U.S. airlines to survive.

Before long the Europeans were buying Douglas, Lockheed, and Boeing airliners under the Marshall Plan. After that it seemed to become a habit, and for the next thirty years they were as faithful customers of the West Coast airplane builders as any domestic U.S. airline. Over the years foreign (chiefly European) airlines bought $23 billion worth of commercial planes for their routes. Looking back, it can be seen that by enabling the non-U.S. airlines to win their breathing spell in 1945 the Americans unwittingly opened the most durable market for any U.S. export since World War II — the market for U.S.-built airliners.

Nowhere was Pan Am's monopolistic presence more acutely felt than in Latin America, but when America reshaped its international air routes after World War II, even that situation changed. Throughout the hemisphere Pan Am's mail contracts carried the maximum rate — many times higher than any paid to domestic lines. And in the actual business of flying down to Rio, the airline's operations, for all their professional competence, were not very enterprising: airmen still joked that Pan Am flew at night only when the moon was full over Miami's Biscayne Bay.

By 1945 the highest powers in the U.S. government were convinced that Pan Am needed competition south of the border. Its foot-dragging in the 1940 SCADTA takeover in Colombia had exasperated Washington and driven the State Department and defense establishment so far as to seek a change in the company's management. In other Latin American countries, where Pan Am had a stake in local airlines (notably Brazil), U.S. officials viewed its actions as anything but identical with the national interests of the United States. Welch Pogue, the astute chairman of the CAB, said later, "There was enormous executive branch and State Department interest — as strong as I've ever seen — in bringing in another carrier alongside Pan Am."

In the official view, this second airline could not be Panagra. Owned half by Pan Am and half by the Grace shipping interests, Panagra had flown up and down the west coast of South America as long as Pan Am had flown the east coast. But every time the Grace people pressed for extension of Panagra flights north from the Canal Zone to connect with domestic airlines at Miami, Pan Am, guarding its trans-Caribbean monopoly, had fenced them off. Had Pan Am relented a little and played along with the Grace demands, Panagra might have evolved into the second U.S. hemispheric airline. As it was, Pogue said, Panagra, although "technically very capable," was "a helpless eunuch."

So the CAB looked for some other carrier to enter the lists against Trippe's outfit. All thought the new contender would be Eastern. Under the iron-fisted leadership of Captain Eddie Rickenbacker, Eastern flew profitably to Florida and other southern states. In the early part of the war, Eastern had shown how to fly through tropical nights making contract trips for the Air Corps to Brazil. But on the first day of hearings, Eastern withdrew. Captain Eddie had made up his mind only the day before. His word was law at Eastern, and he never said why, although his staff people later explained it was a "marketing decision" — there didn't seem to be enough traffic for two airlines.

In Chairman Pogue's view Eastern's withdrawal left only Braniff to carry the flag of competition to South America. But Braniff had only asked for some routes in Mexico and the Caribbean, and the majority of the CAB decided to recommend no second carrier for South America. Thereupon the State Department and Pentagon, in Pogue's phrase, "put Truman's feet to the fire." Under the Civil Aeronautics Act of 1938, the CAB decides where airlines fly inside the United States. It also decides where they fly outside the United States — but the decisions must have the President's approval. Because awards for international routes affect the conduct of U.S. foreign affairs, the President must have the final say. Truman now ordered the CAB to put Braniff into South America.

Braniff had not asked for it. Oklahoma-based, Braniff was a regional carrier with a profitable route linking Chicago and Texas and an able chieftain in Tom Braniff, a former insurance man. Late in 1943 Braniff set up a subsidiary, Aerovias Braniff, and in April 1945 Braniff's DC-3s began flying between Mexico City and Nuevo Laredo, where they made border connections with its U.S. system. Then on July 1, 1945, Braniff extended its Mexican service from Mexico City through Veracruz to Mérida on the Yucatán peninsula.

To Pan Am this was sheer effrontery. Its wholly owned subsidiary, Compañía Mexicana de Aviación (CMA), had dominated scheduled flying in the country since 1929. Its own international line had operated south from Brownsville through Mexico and Central America ever since Lindbergh opened the way. Pan Am's men now defended their turf like spike-wielding nineteenth-century railroaders. When Braniff's planes landed at Veracruz they found the radio range turned off, the tower closed, and the gates to the terminal padlocked. Braniff finally got gas for refueling at Veracruz when the Mexican military showed up and smashed open the gates. At Mérida there were no troops to help, and, Tom Braniff said later, "our passengers had to climb through a barbed wire fence to get off at the airport." In Washington it was noted that both Veracruz and Mérida were among the forty Latin American airports Pan Am had developed with U.S. Treasury funds during the war.

Shortly after this came President Truman's order directing that Braniff fly not only to Mexico but also all the way to Buenos Aires and Rio. Technically, there were two certificates, one for the Mexican route, the other for the big line stretching from Houston by way of Havana to Balboa, Bogotá, Quito, Lima, La Paz, and Asunción to the two principal cities of South America. Thus in a single stroke, a regional airline was transformed into a full-fledged international competitor for Pan Am.

Precisely because Braniff was a small outfit with limited funds and no international experience, Pan Am fought on by all sorts of delays to discourage its advance. With connections in every one of the twenty republics, Trippe's company knew lots of tricks and used them resourcefully. When Braniff's men sought access to Havana's Ranchos Boyeros Airport, mysterious requirements rose up for landing fees, building permits, and insurance rules that were almost impossible to comply with. Pan Am also used its myriad ties to thwart efforts of the U.S. government to win permission of the sister republics to let Braniff in. After a rebuff in Mexico, Tom Braniff bellowed angry frustration. He said:

The negotiations between Mexico and the U.S. failed because Pan Am accomplished with the assistance of its Mexican satellite CMA what it had failed to accomplish in the U.S. — namely, perpetuation as a monopoly.

Though blocked in Mexico, Braniff did not turn back. After all the delays, one of the new postwar planes, a DC-6, made the first flight as far as Lima on June 4, 1948. A year later Braniff reached Rio; then, four years after President Truman gave Braniff the go-ahead, service started between the United States and Argentina in May 1950. Tom Braniff, the only personality in U.S. aviation whose planes still carried his name, was not aboard the inaugural flight. Because he was killed in a private plane crash in Oklahoma in 1954, he missed the sequel to his long effort. As distances shriveled and more and more Texans and Midwesterners discovered South America, Braniff went from strength to strength. In 1968, an upstart no longer, the company bought out Panagra. Thus it emerged at last as President Truman had intended — another major American airline flying down to Rio and Buenos Aires.

The decisions of 1946 cracked Pan Am's hemispheric monopoly. From the standpoint of governmental policy, authorization of Braniff to fly the intercontinental route was the most important decision. But in terms of the routes that Americans in large numbers pay cash to fly on, Washington's awards for routes nearer home threatened to pain Pan Am more. It was here, on these islands in the sun, brought close to the United States by airliners, that the greatest expansion in overseas flying was to take place. Juan Trippe, preoccupied with building and defending his transocean empire of the air, could be pardoned for underestimating the thrust of the domestics into his offshore preserves.

No fewer than six such airlines broke into the domain of surf and sun where Pan Am had always carried most of its passengers. Colonial Airlines, hitherto a minor operator flying between New York and Montreal, won the right to compete for the tourist traffic to Bermuda. Eastern, which had not even asked for it, gained permission to fly alongside Pan Am between Florida and Puerto Rico. Chicago and Southern, another small outfit, won routes linking Houston and New Orleans with Puerto Rico in one direction and with Venezuela in another. National, an even smaller domestic carrier, got the right to fly against Pan Am on the heavily traveled Miami–Havana crossing. Pan Am won one offsetting prize — a direct route between New York and San Juan that its DC-4s promptly and profitably turned into a bridge for Puerto Rican migration to the mainland.

As early as 1940 American Airlines broke through Pan Am's Mexican barricades by getting the Communications Ministry's permission to fly from Dallas and El Paso to Mexico City. The CAB held up approval, but after Pearl Harbor President Roosevelt stepped in. For the same war-emergency reasons that won TWA its breakthrough certificate to fly to Egypt, Roosevelt directed that American should start regular schedules to Mexico City in 1942.

It was a signal of what was to come. The war over, Washington authorized three airlines to start service across the border — Western between Los Angeles and Mexico City, Eastern between New Orleans and Mexico City, Braniff between Texas and Mexico City. On top of the existing Pan

Am and American routes, this was too much for the Mexican government, which balked at all three. Mexican fear of U.S. domination was expressed in the nineteenth century by the president who said, "Poor Mexico — so far from God, so near the United States." Six years went by before another Mexican president relented and let more yanqui airliners in. Even then, American Airlines had its border troubles. C. R. Smith, ever the great salesman, wanted to advertise New York–Mexico City flights but could not. Planes left New York with names like El Azteca and El Toltec and, after a stop at Dallas, carried their passengers on to Mexico City. But because the Mexican government had authorized service only to Dallas, these were technically not direct flights and American was prevented from proclaiming a direct New York–Mexico City service.

A month after Washington split Pan Am's Latin American bastion wide open, the CAB recast the routes of air commerce for the Pacific. Again the board followed the Rooseveltian dictum: give Pan Am good routes, perhaps the best routes, but domestic competitors must get their chance. In the Pacific, where Pan Am had sent out Eddie Musick in the 1930s in almost the same way Cape Canaveral sent out John Glenn in the 1960s, the airline was confirmed in the midocean route it had pioneered. Pan Am, using Clippers that landed on runways instead of lagoons, would fly through Hawaii to Hong Kong — and also to Shanghai and Tokyo for good measure. At the same time United, the one domestic airline that had been willing to let Pan Am keep all the international business, obtained the right to fly from California to Hawaii alongside Pan Am. United had flown the route during the war and saw commercial possibilities in what it insisted was only an extension of the domestic sphere of air travel.

The trans-Pacific competition that Washington imposed was galling for Pan Am to accept. As far back as 1930, when domestic airlines still flew open-cockpit planes, Juan Trippe had staked out the beeline northern route over Alaska. And if a lot of boys thought of Eddie Musick the way a later generation thought of the astronauts, what about the Lindberghs? The great survey flight they conducted from Anchorage to Tokyo for Trippe, memorably reported in Anne Lindbergh's 1931 best-seller *North to the Orient*, invited comparison with the expedition of Lewis and Clark. After these feats, only Soviet and Japanese refusal to grant permission to land had kept Pan Am from flying that way.

Now the CAB handed the northern line to the Orient to Northwest. Unlike Braniff, Northwest wanted it and lobbied for it. Indeed Northwest's prairie-bred stewardesses were said to have begun studying Chinese. Yet before the war Northwest had been a regional airline, headquartered in Minneapolis and specializing in cold-weather flying along the country's northern rim. Alaska was the one theater of war to which the short-legged DC-3s of the domestic airlines could fly, and the

Army had picked Northwest to open the overland air link to Fairbanks. Northwest was fast off the mark. Its aircrews racked up many hours on a supply line that soon led (with the delivery of the four-engine C-54s) far out along the Aleutian chain. The airline also had well-placed friends in Washington. Late in the war it won a big route extension eastward into New York City. The war over, it received the right to fly all the way from Seattle via Anchorage to the Orient. Suddenly Northwest, the frostbite outfit that painted the tails of its planes red to be able to spot them readily in Arctic snowbanks, was a big-time international airline.

The CAB didn't get around to South Pacific assignments until a year later. Then it confirmed Pan Am in the south sea islands route on which the trail-blazing Musick had lost his life. This track, by now familiar to thousands of wartime travelers, led from Hawaii to New Zealand and, by extension, to Australia. Meanwhile Pan Am had gained a prize that nobody much wanted: the route across the South Atlantic to Africa. It was an ironic sort of consolation prize. After Pan Am opened the wartime trans-Africa route for the United States, the Air Transport Command took over and tossed out Pan Am's local manager amid charges that the airline was trying too crassly to win itself a commercial toehold in the area. Then the huge flow of wartime traffic dried up, and commercial possibilities looked slim. But by then the war had thrust America into world power and responsibility: U.S. national interests required that a line to Africa be maintained. Pan Am, still the nearest thing to a chosen instrument the United States possessed, was called back and handed a hefty mail subsidy to show the American flag on the airways to South Africa.

23

North Atlantic Shakeout

In late December 1945, Pan Am's Boeing flying-boat Bermuda Clipper set off from Southampton on the last of 3,650 flying-boat crossings to New York. Since the twenty-five passengers and eleven crew wanted nothing more than to get home by Christmas, Captain Bill Masland streaked for home with all possible speed. Taking the obligatory track through Africa and Brazil, stopping only for fuel, the Bermuda Clipper made it home in a record three days and two and a half nights. It touched down on Bowery Bay off La Guardia Field at 2:00 A.M. the day before Christmas. Masland said later, "Only the night watchman met us — and that was the end of the proud Clippers on the Atlantic."

Hail and farewell? Not a bit of it. At that moment Pan Am officials were too busy revving up for the landplane contest on the Atlantic to pay any attention to vanishing flying boats. The American public had demanded competition, the government had demanded competition, and now Pan Am was getting it on all its old routes, but most of all in the North Atlantic. Already its prewar challenger American Export, taken over by American Airlines and restyled American Overseas, had started carrying paying passengers across the ocean in landplanes. Having picked up six Douglas C-54 Skymasters from the war surplus board, American Overseas on October 24, 1945, brought off the first scheduled commercial landplane crossing — and in less than half the time the ponderous Bermuda Clipper had needed for its last dash.

Pan Am had to contend with TWA as well. Of the seven domestic airlines that spread their wings overseas after V-J Day, this was the one that set out to take the whole world by storm. After an airman's war, TWA, an airman's airline, would carry on in peace. Favorite of the Air Corps generals who had grown so powerful in the war years was TWA President Jack Frye, a flying man who talked their language. Proclaiming his vision of a world-girdling airline, Frye already had his lieutenants starting to develop local airlines in Italy, Egypt, Turkey, Ethiopia, Central America, and the Philippines as feeder lines for TWA's intercontinental services.

TWA made plans to commence transatlantic schedules with Washington as its U.S. terminal because, its chairman said, "Washington is where the power is."

At the moment, TWA looked like a formidable rival. The company that had shown the way with the DC-1 and the Boeing 307 now prepared to introduce the only really new postwar commercial plane. This was the big and beautiful Lockheed Constellation, conceived just before war broke out. War had delayed its construction but in April 1944 Frye and Howard Hughes climbed into the first one built for the military and flew it across the country in the record-shattering time of six hours and fifty-eight minutes. That had been a proclamation of what the public could expect after the war from TWA — and now the war was over. The TWA pilots who had crossed the Atlantic ten thousand times in wartime C-54s could begin breaking records lifting passengers to Europe in powerful and high-flying Constellations.

But there would be a fight, because Pan Am had claims on the grand new Connie too. Before Pearl Harbor Hughes had reluctantly and secretly consented, under the force of Washington's program priorities, to let Lockheed contract with Pan Am for forty of these advanced commercial airliners. Hughes bowed to this change on the presumption that Pan Am, an overseas airline, would not be competing with TWA, which had ordered the plane to get the better of United and American on the transcontinental run. During the war a handful of these planes were built. Some of these the Air Corps assigned to Pan Am, whose contract fliers thus gained valuable experience in the new plane in the war's closing months. And Lockheed acknowledged its early commitment to Pan Am by making some 049s, as the first peacetime Connies were called, available to Pan Am. So when TWA won its Atlantic route and prepared to launch transocean service with the 049s, Pan Am threw its flying boats away and got ready to fight for supremacy with Connies too.

The contest began December 5, 1945. In Washington that day TWA loaded up its first 049 with a cabinful of government VIPs headed by Postmaster General Bob Hannegan and took off for Paris. Heralding the new era of air travel, the flight made headlines round the world. With a pressurized cabin, the 049 could fly at eighteen thousand feet and leap over the Atlantic's worst weather. Its four big Wright 3,350-horsepower engines enabled it to cruise the heights at 280 miles an hour. Captained by Hal Blackburn, a veteran of forty-four wartime crossings, TWA's plane broke every record along the way. It took just twelve hours and fifty-seven minutes to wing to Paris, just six hours, twenty-seven minutes to jump the ocean gap between Gander and Shannon.

If there was to be another contest like the ocean liners' race for the North Atlantic Blue Riband, Pan Am's Connies had to flash across just as

The Lockheed Constellation, designed secretly for TWA before the war, did not fly passengers until 1945.

fast. But to Pan Am's old pros, revenue flying was the payoff, and its airmen winged into pennant contention on January 20, 1946, when they carried the first paying passengers to Europe in the new wonder plane. That crossing — to Lisbon — was followed by another Pan Am first, the first Constellation flight to London on February 5. Thus, casting loose from their waterbound traditions, Juan Trippe's masters of ocean flying made a fast start in the new landplane race across the Atlantic. And to show that there was lots more to the operation than just flying, the old master himself proceeded to show the new boys some of the international airline tricks.

The CAB's solomonic decisions had assigned Pan Am and TWA flights to different parts of Europe; they were not supposed to be in direct competition. But Trippe was not averse to giving the TWA upstarts a first scare. With a fanfare of public relations bugles, he announced that Pan Am would carry passengers to Shannon, the one point served by TWA as well as Pan Am, for $275 — $100 less than TWA had already advertised. The impact of this announcement, coming right after the war when public attention turned for the first time to the very real possibilities of transatlantic flying, was lasting. For Trippe it asserted a reputation for advocating low fares for transocean travel that he repeatedly laid claim to thereafter.

Pan Am's wily president may or may not have intended to cut the fare — when the CAB asked his company to present figures showing it could make money at the lower rate it did not do so. By then Trippe knew he would not have to make good on his claim. On November 26, 1945, Britain's air attaché in Washington wrote Pan Am that if it went through with "its intention to operate at a reduced rate which has not been agreed, His Majesty's Government will be reluctantly obliged to restrict the number of [Pan Am] services to the United Kingdom to two a week." Since Pan Am would have lost money if limited to two flights a week, no more was heard of Trippe's low fare. Instead Pan Am was soon dickering with the British about raising fares — to perhaps $500 to $600 one way. John Slater, the chairman of American Overseas, who was visiting London, said his line could not be bound by such high tariffs. Thereupon Trippe rushed reinforcements to London to head off a possible Overseas deal with the British. In these shock tactics, TWA's freshman ambassadors were caught like sparrows over a badminton net. Out of the maneuvering came a cunning compromise engineered by the old hands of Pan Am and BOAC: IATA — the International Air Transport Association — became a powerful self-regulating body of international airlines which set uniform transatlantic fares that prevailed for the next thirty years.

With that the chance of a quick victory for any one airline evaporated. At first there was more business than anybody could handle. Pan Am, TWA, and Overseas planes flew, jammed with priority government passengers. When BOAC opened a hole-in-the-wall office in Manhattan, lines of customers anxious to fly over to look for relatives stretched into the street. "Three men stood at the counter just taking in money," manager Paul Bewshea said later. By hook and crook BOAC wangled four Constellations, Air France two. The Scandinavians, Belgians, and Dutch had to make do with DC-4 retreads. But they flew the Atlantic in 1946 too.

Pan Am led the pack. It was better organized. It called itself the "world's most experienced airline," and nobody doubted it. Its fifteen Connies carried 66,500 passengers across the Atlantic that year — more than half of all who went by air. TWA and American Overseas split most of the rest. As Trippe had predicted, both lost money even with sizable lifts from Washington in the form of mail pay.

And TWA the introducer of great new planes, TWA the darling of the Air Force, TWA the airline poised to girdle the globe with the help of a lot of lesser foreign-flag companies — TWA ran into nothing but trouble. In its high-flying 1945 report to stockholders, entitled "Horizons Unlimited," it talked of hiring 2,000 new employees — "We haven't the jobs for them yet, but we'll need them for our new opportunities."

But as Vice President Paul Richter said later, "About everything that can happen to a company happened to TWA in 1946." Inexperience cost the airline its seeming advantages. On domestic as well as international services, it never seemed to have enough planes. The engines of the first 049s, identical with those that powered wartime B-29 bombers, overheated when flown as long and hard as commercial schedules required. Operations chief John Collings growled, "Two things that engine does not like — heat and horsepower." On hot days, lugging heavy loads, as many as nine Constellations a day had to be pulled off for engine changes. At great 'expense the company installed firewalls around the engines to counter a problem the Air Force had shrugged off in the crisis of war.

With such troubles, flying these big planes nonstop across the country turned out to be a money-losing proposition. The first 049s had a compartment with berths for the crew on international flights. On the domestic flights eight passengers were shoehorned into this "bird cage," which was over the wings next to the exhaust stacks, with knee-bumping settees for seats. Although these planes had fifty-seven seats in all, cramming in passengers for eleven-hour transcontinental flights broken only by fuel stops at Kansas City lost passengers. And when the airline began stopping more places to pick up people, it forfeited the advantage of the Connie's speed.

In August came a bad blow. An 049 on a training flight caught fire and crashed at Reading, Pennsylvania. The fire was traced to a point in the fuselage wall where hydraulic and electrical conduits passed close together. In these first 049s, built under wartime conditions, aluminum had been used for the wiring instead of copper. When electrical connections in the cargo hold vibrated loose, sparks ignited fluid leaking from the hydraulic tube. Washington grounded all Connies six weeks until Lockheed could correct the design fault. Then a transatlantic 049, after discharging its passengers at New York, crashed and burned at Wilmington, Delaware. On top of all this the Air Line Pilots Association called its first strike — against TWA. There were lots of other strikes that postwar year, but an airline is peculiarly vulnerable to a work stoppage. It can't stockpile like other industries to cushion the impact: an airline seat unsold is lost forever, and it takes months afterward to regain customers who have gone elsewhere.

Golden dreams gone glimmering, TWA canceled orders for twenty-five more Connies, slashed flight schedules by 22 percent, and laid off the employees so optimistically hired twelve months before. "Our situation is as critical as it was back in 1934 when our airmail contracts were canceled," wailed Vice President Richter, "and finally the pilot strike coming when business itself was falling off."

In California Howard Hughes, TWA's biggest stockholder, had just recovered from an appalling plane crash. As he flew his twin-engine experimental X-F11 photo reconnaissance plane over Los Angeles that summer, the left propeller had run wild. Trying to bring the plane down on a golf course, Hughes smashed into a house, ripped off a wing. He was pulled out of the wreckage unconscious, bleeding, his skull fractured. Fresh scars of the accident marking his face, Hughes went east one cold bitter day in January 1947 to save his stricken airline.

Among other things he fired Jack Frye, and before he was done, he put up the whacking sum of $10 million to keep TWA from going broke. It was every bit as big a crisis for TWA as 1934, and bigger than the 1938 impasse that brought Hughes into ownership.

In the hour of TWA's adversity, Juan Trippe thought his chance had come. Of the two rivals that had muscled in on his North Atlantic domain, TWA was the airline whose routes he coveted most. If he could somehow gain TWA's routes, Pan Am would win access to Paris. Pan Am already had the London gateway, shared with American Overseas, which also flew to Germany. But there were two places above all where Americans wanted to go in Europe — London, where you spoke the language, and Paris, where you had the fun. London was important for government and business travel, but London after the war was a landscape of austerity. All too frequently, travelers checked in at the hotel, took one look at their thin breakfast next morning, and caught the first available plane for Paris. Beyond Paris were TWA's other big cities — Rome and Madrid, both centers of terrific tourist potential.

When TWA was in its throes, Trippe telephoned Hughes's financial man Noah Dietrich, and said he had an important matter to discuss. Twenty minutes later Trippe was in Dietrich's Manhattan hotel room proposing that Hughes consider a merger of TWA with Pan Am, or at least of its transatlantic operations with Pan Am.

At that moment Hughes was in a real bind. On the government front he was jockeying for a $10 million bail-out from the Reconstruction Finance Corporation, but CAB Chairman James Landis told him the price would be his withdrawal from direction of TWA. "I don't believe one man should own a public utility," Landis said. In the corridors of Wall Street he was hunting for a credit line for TWA, but to get it, it appeared he might have to put his Hughes Tool Company gold mine in hock. Both of these elements — the political maneuvering in Washington, the numbers game of Wall Street — were part of the picture as Trippe made his merger move. Of Trippe's overture Hughes said later, "Let's say I was considering giving in."

It happened that the Republicans had just captured control of the Senate and House for the first time since Herbert Hoover's presidency. Thirsting for blood, they boiled into Washington and opened fifty different inquiries into Democratic rule. It was a time of congressional hysteria. Senator Owen Brewster, Pan Am's great friend, took over the War Investigations Committee. He also cherished vice presidential ambitions and thought to use the committee to lift himself to the vice presidency just as Harry Truman had.

So, as Juan Trippe began to press Howard Hughes to make a deal with Pan Am, Brewster ordered his staff to look into the wartime doings of the Hughes Aircraft Company. Brewster was not the only Republican who thought there was pay dirt there. Ever since Elliott Roosevelt had married actress Faye Emerson at a 1945 party organized by TWA men on the rim of the Grand Canyon, members of Congress had been charging a Hughes payoff to the president's son. Ostensibly the purpose of the inquiry was to find out why the government got nothing for the $18 million it paid Hughes's company for the mammoth, eight-engine wooden flying boat built on a war contract, and only three aircraft for the $22 million it sank in Hughes's X-F11 photo reconnaissance plane. But the inquiry suddenly veered toward young Roosevelt's free rides in TWA planes, and a Hughes expense-account man named Johnny Meyer, who supplied telephone numbers — and girls.

In the summer of 1947 the committee put Hughes in the dock under the klieg lights of the Senate caucus room. By this time Hughes had found financing for TWA from the Mellon banking interests, and at a secret Palm Springs meeting with Juan Trippe he had broken off the merger dickers with the abrupt answer, "Hell, I thought you were going to sell Pan Am to me." Uncowed and defiant, he now picked Brewster as a personal antagonist and went on the attack. Coached by publicist Carl Byoir, he charged in headline stories beforehand that Brewster in his inquiry was only fronting for Pan Am, which wanted him out of the international airline business. The tactic succeeded. By the time Hughes took the stand, Brewster had relinquished his committee chairmanship to Senator Homer Ferguson so he could appear as a witness and defend himself against Hughes's charges.

Brewster didn't have a chance. Millionaire flier and man about Hollywood that he was, there was something very American about Howard Hughes. Lanky, dark-mustached, in his rumpled, ill-fitting suit, his scrawny neck sticking out of a misshapen collar, Hughes presented himself before the committee room cameras like Jimmy Stewart playing in *Mr. Smith Goes to Washington.* Sworn in, he scowled at the bald, paunchy Brewster lolling in senatorial ease at the corner of the committee table. In answer to Ferguson's first question he said:

Well, I charge specifically that during a lunch at the Mayflower Hotel with Senator Brewster in the week commencing February 10, 1947, the Senator in so many words told me that if I would agree to merge TWA with Pan Am and go along with his community airlines bill, there would be no further hearings in this matter.

"Inconceivable," roared Brewster. But Hughes, pressing the attack, demanded the right to cross-examine the senator. Ferguson declined, but he agreed to put Hughes's questions to Brewster. That night Hughes took counsel with his lawyer and Byoir. The next day Hughes began reading, "Senator Brewster's story is a pack of lies," and the audience, packed by Byoir with Hughes fans, cheered. Brewster, replying to forty-odd questions that Hughes handed Ferguson, said certainly he knew Juan Trippe ("a very able man") and Pan Am Vice President Sam Pryor ("a very close and gratifying friendship"). Yes, he had accepted a couple of Pan Am plane rides, had stayed at Pryor's "very modest, bungalow-type house" at Florida's Hobe Sound, had tasted Pan Am hospitality at the company's F Street mansion. Then Hughes took the stand. He said:

It stands to reason one of us is telling something which is not the truth. Brewster has been described to me as one of the greatest trick-shot artists in Washington.

I'm supposed to be capricious, a playboy, eccentric, but I don't believe I have the reputation of a liar. Brewster said he'd been a senator for 23 years; well, I've been in the public eye for 23 years. I believe in that respect I have the reputation which most Texans consider important. I believe, to use a corny phrase, most people consider my word to be my bond.

By this time the shrewd, truculent witness had almost wrecked the investigation. He certainly wrecked Senator Brewster, who departed abruptly for the Maine woods and never cut much of a figure in Washington again. At one point Ferguson asked Hughes, "Will you bring Johnny Meyer in?" Hughes replied, "I don't think I will," and laughed right in Ferguson's face. Ferguson said, "It's not something for laughter." Hughes said, "I didn't laugh; someone laughed back there." Then everybody laughed.

The show was over. It invited comparison with another hazing to which another airline tycoon had been subjected in the same caucus room fourteen years before. That time William Boeing had walked out and, vowing never again, sold out all his aviation interests. This time Howard Hughes walked out and, even more of a loner than Boeing, practically disappeared from public sight for the rest of his days. But Hughes, destroying Brewster in the warfare of Washington while fighting clear of the thickets of Wall Street, had clinched his hold on his airline. A less rigorous chairman succeeded Landis at the CAB, and the board allowed Hughes to exercise his option and turn the $10 million he had advanced to the airline into TWA shares. With that he emerged as incontestably the biggest airline owner, holding no less than 46 percent of all TWA stock.

After the TWA debacle Trippe's next try was more successful. This time the target was American Overseas, third of the three U.S. lines granted seven-year certificates to fly the North Atlantic after World War

II. American Overseas did not fly to glamorous places like Paris, but it certainly ate into Pan Am's share of the business on the Atlantic. It flew daily to London, and it flew also to Scandinavia and Germany. A great part of its traffic was military, but that was the name of the game in those years when the cold war had settled over Europe and the United States was treating everything west of the iron curtain as a bulwark that required lavish support.

Like TWA, American Airlines ran into trouble after expanding too fast across the Atlantic. But American's troubles surfaced a year or two later than TWA's. Its acquisition of American Overseas had occurred during the war. At the time President C. R. Smith was absent running the U.S. Air Transport Command, and his right-hand man Ralph Damon was temporarily away building fighters for the Air Force at Republic Aircraft. In 1945 Smith returned as chairman, Damon as president. Smith lived in Washington, handling policy, routes, and government relations; Damon bossed airline operations in New York. But from the start divergences developed. Damon wanted American to fly both domestic and international. Smith said:

I want to make American what it used to be — a good and profitable airline. And I don't want to divert attention from those goals. We'll have one set of people for the domestic, and one for overseas.

Thereupon Smith picked General Harold Harris, the old Panagra hand who had worked for him in the ATC, to run American Overseas.

More than any other airline president, C. R. Smith sold flying as a passenger-carrying enterprise. In making American the most traveled airline before the war, Smith stressed that his company could operate profitably without any airmail subsidy. At a time when passenger fares were down to four cents a mile, he declared, "What this country needs is a good three-cent airline."

But the DC-3 that had carried Smith's airline to the top of the heap in the 1930s was now outdated. The cost of shifting to four-engine planes (and converting forty wartime C-54s to take care of the customers while better planes were being built) used up a lot of the $40 million working capital American raised by selling stock after the war. And when there was nothing but delay and discomfort, and nothing but noisy old planes to ride in, the pent-up demand for air travel flagged. Then, as the queues at the ticket counters began to fade away, the costs of flogging harder for traffic ate into American's income. Mail pay was no help because the government, noting airline wartime profits, had slashed the rate. Wages were up, payrolls were up, costs were up across the board.

Then what happened to TWA happened to American. Just when the company was extended too far, the costly new DC-6s, the first pressurized airliners ordered from Douglas, were grounded. At this point the Chase

Bank informed American that it could extend no further credit. American's senior board member, Edgar Queeny, president of Monsanto Chemical, had seen it coming and told Smith, "We have to have someone on the board with a financial background." He nominated Charles Cheston, a Philadelphia investment banker and fellow director of J. P. Morgan's. Cheston brought in Bill Hogan from the H. J. Heinz Company. Smith informed Hogan, "We can't meet our goddam payroll tomorrow." Acting fast, Hogan lined up a $12.5 million credit. Hogan told Smith:

Your ideas are no damn good. You're trying to run an airline with planes half full. You're losing millions selling seats for about two-thirds what it costs to produce them. And yet you want to get more business, more routes, more planes. You'll only lose more millions at this rate.

Smith moved to New York, and sent Ralph Damon "to the field" — La Guardia Field — to run the airline while he tried to pull things together. Dropping his talk about a "three-cent airline," he joined other companies in an appeal to the CAB, which quickly granted a fare increase to five cents a mile. He also swallowed his strong feelings against accepting mail subsidy, and along with United and TWA accepted a quick $2.5 million retroactive mail pay made available by the CAB with the proviso that it might have to be repaid later. But the deciding factor in American's eventual turnabout from the precipice proved to be the performance of the new DC-6. In fact it took some time, after the faults that led to the plane's grounding had been corrected, to find out what a whiz of an airplane the DC-6 was. The version known as the DC-6B, flown in large numbers by both American and United, turned out to be an even more economical airliner than the fabulously profitable DC-3 — probably the most economical piston plane ever operated commercially.

Before all this became clear, however, Smith moved to unload the international part of his company. Critics have said he lost his nerve. He himself said later, "I made a mistake." But he had had enough of overseas aviation in the war. American Overseas seemed a long way from profitability, and the domestic airline urgently demanded attention. As he said later:

Every sonofabitch in the domestic operation found some company problem in Copenhagen or London. Every time I was looking for them, they were visiting some overseas place.

I finally made up my mind. I said to Charles Cheston, "I'm going to recommend to the directors that we peddle this goddam airline."

Pan Am was the only possible buyer. Cheston spoke to his good friend John Hanes, financial consultant to Pan Am. A few days later C. R. Smith got up from his lunch at Manhattan's Sky Club and crossed over to where Juan Trippe sat. They pulled chairs together, and soon the merger was in the works.

By the original deal Pan Am was to take over Overseas for $12 million. Export however did not want to sell out. John Slater, the old Amex chief who had stayed on as American Overseas chairman, said later, "We scared Trippe and made him put up the price to $17 million. They would have paid *anything* for that airline." The political situation was also vexing and the merger process dragged out. The time had passed when Pan Am's clout could accomplish almost anything in Washington. The company's prospects were hardly improved when Sam Pryor, Pan Am's leading wire-puller and a Dewey campaign leader in 1948, saw the President trailing in election night returns and asked in a midnight TV appearance, "Why doesn't Truman concede?" Not until 1949 did the CAB examiner bring in a recommendation for approval. And when the board finally got round to forwarding its decision to the President in May 1950, it was a three-to-two verdict against allowing the Pan Am–American Overseas merger.

What followed was one of the most amazing passages in American airline history. For nearly a month the North Atlantic Merger Case remained at the White House. The Bureau of the Budget quickly rounded up opinions from seven agencies concerned with foreign relations: all but the Commerce Department concurred in the CAB decision. The matter rested on Truman's desk while he dealt with the Korean invasion that erupted after June 20. Then, in the midst of the emergency meetings on Korea, the White House sent the North Atlantic Case papers back to CAB Chairman Joe O'Connell with the President's signature of approval.

That afternoon Stan Gewirtz, a former CAB staffer, dropped by O'Connell's office, and found the chairman having a celebratory drink. "Stan," O'Connell said, "here's the decision, signed by the president. We've won." Truman's covering letter indeed approved the board's finding. But it also said:

I do feel strongly that, if only two U.S. carriers operate in this area, their respective routes should be sufficiently balanced so that they can maintain effective competition on a sound economic basis. I am not satisfied that the merger as now proposed would result in such a system.

Gewirtz said, "Aren't you going to release it?" O'Connell's aide Woody Donnelly said, "We don't have enough copies." Gewirtz said, "You can't tell about the White House, they might change their minds. Why don't you issue a press release?" O'Connell took another slug of whiskey. Donnelly said, "We don't have enough time."

Between sunset and dawn the North Atlantic Route Case was called back to the White House. On June 30 the presidential signature of approval was erased from the document at Truman's order. At a phone call from Presidential Secretary Matt Connelly, Oswald Ryan, one of the two dissenting CAB members, went to the White House and talked with the

President. O'Connell heard about this first when Ryan told him the President had changed his mind. Truman sent the case back with orders to approve the Pan Am–AOA merger — and to authorize both Pan Am and TWA, the "remaining carriers," to fly to London, Paris, Rome, and Frankfurt in face-to-face competition.

In the climactic hour of the postwar North Atlantic airline shakeout, the three rivals had scrambled for their lives. Later Pan Am's Henry Friendly said mildly:

The way the law is written, presidential approval was necessary. But there was no procedure, no way for each side to have its say. You didn't know who was handling the matter. You didn't know who was saying what to whom. It was just rough and tumble.

Stan Gewirtz, who had left the CAB for Landis's law firm, said later:

That felon, that charming felon, Matt Connelly — he did it. He just took ink eradicator to remove the president's signature in two places. I saw it later. We all saw it when Jim Landis appealed the decision and the papers were brought into court in Connecticut.

I always heard that Connelly was standing over the president at his desk. Truman couldn't have cared less. Connelly says, "Boss, you didn't sign this paper." And he pushes these new documents before him.

That put the case back to the CAB. Oswald Ryan at the CAB worked with Connelly. And Carlene Roberts worked things out with Matt Connelly.

Ironically, it was American and not Pan Am that engineered the reversal that achieved Trippe's goal. Carlene Roberts was American Airlines's representative in Washington. C. R. Smith, in the showdown, was bent on getting rid of his overseas operation. Smith himself had been living in Washington, actively politicking for his airline. He was a Texan, and regional ties counted heavily for domestic airlines in the Washington infighting. In this case Smith turned to his fellow Texan, Speaker Sam Rayburn, to bring pressure at the White House.

But nobody at American was as close to Matt Connelly as Carlene Roberts. A skillful member of American's customer service department in Chicago, Carlene Roberts had moved to Washington at the end of the war and built up a terrific name as a lobbyist. As intelligent as she was beautiful, she seemed to have everybody's number. She was on good terms with Presidential Aide John Steelman, who was in a position to put in a word on airline route awards. And she was a great favorite and dinner companion of Matt Connelly, who later went to jail for accepting payoffs for fixing income taxes. "She was the best," said Wayne Parrish, most knowledgeable of aviation editors. "The most effective operator in Washington," was the admiring judgment of Ralph Damon, who was in a position to know.

TWA also had roots back home, in Kansas City, the President's home town. When the North Atlantic Route Case got snagged at the White House, Howard Hughes hired Missourian Clark Clifford, Truman's ex-counsel, to make sure his airline's voice was heard in the Oval Office. It was one of the first assignments that Clifford took in establishing himself as the most super of Washington's "superlawyers." TWA's voice *was* heard — and the decision was tailored to put TWA in London.

Having been bypassed so ruthlessly, Joe O'Connell had no choice but to resign. Then the board, with Oswald Ryan as acting chairman, not only upheld the merger but also, as the President directed, gave TWA the right to fly to London while Pan Am gained access to Paris and Rome. On behalf of protesting American Overseas pilots, former CAB Chairman Landis took the decision to court. His plea was turned down.

Thus Trippe finally gained a victory in his long, dogged campaign to reduce the number of U.S. rivals flying the Atlantic. And by eliminating American Overseas, he put Pan Am in position to consolidate its pioneer place as the foremost carrier on the Atlantic run. Taking all its world routes together — European, Latin American, Pacific, Alaskan, African — Pan Am stood out more than ever after the 1950 decision as the preeminent international airline.

But the scene was changing, and not necessarily to Pan Am's advantage. The seven-year certificates issued for the North Atlantic routes at the outset of the postwar race were now made permanent. By Truman's intervention the number of contending airlines was reduced from three to two. This might have been construed as a step toward Trippe's goal of one chosen instrument flying U.S. passengers across the ocean. But the effect of Truman's action was to substitute the most direct and intensive face-to-face competition for the discarded system of area competition.

In short, at the very time that transatlantic air travel looked like it was becoming big business, Harry Truman lifted his hometown airline into frontal contention with Pan Am. Transcontinental and Western Air, the airline born of a shotgun marriage and reared in Kansas City, now won access to London, the principal terminal for overseas travel. Transworld Airlines, as Hughes had renamed it in May 1950, had survived the Atlantic shakeout despite all its vicissitudes and emerged as a feisty rival for Pan Am.

In the big postwar Atlantic shakeout, Trippe had won a victory that appeared to assure Pan Am of nearly two-thirds of U.S. transocean traffic. But TWA had survived and was now the only airline that could offer both domestic and transocean passenger service. Even before the revised decision took effect, TWA was advertising, as the President noted with interest, "direct service from American cities to London and Frankfurt." That, in the long run, was the greater gain.

24

The Postwar Airliners

The Constellation was a lovely sight. Lockheed had a tradition of building breakaway planes. Wiley Post and Howard Hughes had broken records circling the world in Lockheeds, Post in a single-engine Vega, Hughes in the twin-engine Electra. The Connie, designed by Kelly Johnson, the best designer of his generation, had that handsome, racy Lockheed look. Its wing grew out of Johnson's design for the P-38 Lightning, a plane so fast that trying to make it faster in World War II brought Johnson right up against the mysterious problem of compressibility. The downward curve of the fuselage behind the wing, so pleasing to the eye, responded to the force of a mathematical imperative: it reduced drag by some 2 or 3 percent. The shape also ensured that the tail surfaces rode high above the plane's slipstream and thus made it easier to control in flight. The Connie's nose dipped downward for a structural reason too — it shortened the nose wheel, whose gear would otherwise have been too long.

But the Connie's contoured profile, Johnson said later, turned out to be an economic mistake. A straight-line fuselage like that of Douglas planes was cheaper to repair, maintain, and fly. It cost more to store odd-shaped parts, and such details mattered in the competitive race. At the time there were strong reasons for the Connie's three tails, though they tripled the number of spare parts an airline had to keep around. Johnson said later:

We had so damn much power you couldn't control it with just one tail surface. Besides, a single tail would have been too tall to get into a hangar — one of TWA's requirements for the plane.

Every airplane is designed around its power plant. The Connie packed all that power because it was built to carry four Wright 3350 engines, the biggest engines of World War II. But at first the huge engines created nothing but problems for TWA. In Kansas City, Gordon Parkinson said later, "in one day we had ten engine failures." Luckily, the Connie could keep going on three or even two engines.

The first 049 model Constellations could climb to 25,000 feet, cruise at 280 miles an hour, and fly substantially farther than any previous airliner. That its great range was not put to better use rankled at Lockheed. Hall Hibbard, the company's chief engineer, always held it against TWA that the airline did not persist in operating the plane nonstop across the United States. He said:

The 049 with the Wright 3350 engine was a leadpipe cinch to fly to New York nonstop. It was designed to do it, and they wouldn't do it. No other airliner had this capability for years. American and United had only the DC-4 — and the DC-6, when it arrived in 1947, lacked the range.

But Lockheed's President Bob Gross himself argued at the time that a ten-hour transcontinental ride without a stop at Chicago or Kansas City was out of the question. He said, "Goddam it, you can't fly to New York without getting out and taking a little walk."

Kelly Johnson identified the problem in different terms:

The 049 was not fast enough to fly across the country in less than 9 hours, certainly not against a 100-mph headwind. And the pilots union put in a clause in their contract that an aircrew could not be kept aloft more than nine hours. That would have required the airline to double-crew, and they wouldn't do it.

In crossing the ocean, the same difficulty got in the way. On transatlantic flights three pilots took turns in the cockpit of the 049s, which stopped at Gander and Shannon; there were also two navigators and two flight engineers spelling each other. To have flown across nonstop, TWA and Pan Am would have had to double-crew, and the Lockheed engineers could never talk them into it. But Johnson was sure the 049 could have done it. He said later, "We did 6 or 8 wind-tunnel tests and found the plane could go 3,000 miles."

TWA Vice President Floyd Hall thought his company should have exploited the Connie's range and the airline's route patterns so that the traveler could have boarded the O49 in Los Angeles or Chicago and flown the same plane onward from New York to Paris. He was ahead of his time. TWA's international division was practically autonomous, and the airline had troubles enough at the time without shuttling planes back and forth between domestic and international. Anyway, the sales department argued that not enough customers bought through tickets to justify the shift. But as Hall pointed out later, "When through service to Europe *was* started in 1955, you couldn't get enough planes to meet the demand."

In other words, public expectations grew, and the capabilities of airliners grew with them. In fact, Johnson said later, "the question with commercial airliners is how you *grow* them." The airlines kept asking for more, and Lockheed gave it to them. Lockheed started with a relatively low wing-loading on the 049, and in five successive model changes had it

Lockheed designer Clarence "Kelly" Johnson in 1965.

doubled. Still using the same original wing-platform, Lockheed boosted the weight almost 100 percent in fifteen years, lifting the engine power, stretching the fuselage by nineteen feet, and increasing seating capacity from fifty-four to ninety-nine. It might look like the same triple-tailed Connie, but by the 1950s it was a much heavier, stronger, faster, and rangier airplane. Quite an evolution from one basic airliner design.

At Douglas it was just the same. The granddaddy of Douglas's postwar family of airliners was the DC-4. Completed in time to be harnessed as the overseas workhorse of World War II, this sturdy plane solved the problems of four-engine design so well that later Douglas piston aircraft — the successive models of DC-6 and DC-7 — were all "stretched" versions of this basic airplane.

The DC-4 introduced the nose wheel and tricycle landing gear to airline travel, and that was a big change. The old DC-3 landed on two wheels, then dropped its tail with a thump that could disconcert the inexperienced passenger. At rest the old DC-3 sat back on its haunches, so that upon entering the cabin you had to climb like a goat to get to one of the forward seats. With the nose wheel, three-point landings became possible. Also, because the nose wheel was steerable, planes became more manageable on the ground.

The DC-4 was also the first airliner big and complicated enough to require power-boosted controls. It was the first U.S. airliner with flush-riveted skin — up to this time mushroom-headed rivets had been used. With its big, strong Pratt & Whitney R-2000 Twin Wasp engines, it set the standard of performance that became the basis for all airline flying after the war. That is, for the first time with four engines, you had a plane that could fly even if as many as two engines conked out.

Like the Connie, the DC-4 was born with wings that could spread. They were built up by the same multicellular method of construction that had proved so successful in the DC-3. Tail fin and rudder were a single unit of simple, clean design, and the fuselage, slightly egg-shaped in cross section, was laid out in a long, tubular form that was so practical it soon became standard for American airliners. In a cabin not quite ten feet across and twice the length of a DC-3's there was space on each side of a central aisle for forty-two seats, compared with the twenty-one of a DC-3.

Ancestor though it was of a noble line, the DC-4 was a rough-and-ready beast, bred in wartime conditions. It was slow (200 miles per hour compared to the Lockheed 049's 280-mile-per-hour cruising speed), it was noisy, and it was unpressurized. Its peskiest fault, leaky gas tanks, had been licked by the contract carriers that flew it in the war. Though DC-3s had also had gas tanks integral to the plane's wings, TWA and American soon found out that the twisting point in the DC-4 wing was different, and not even repeated spraying with sealant would plug the leaks that kept breaking out in unexpected places. Abe Hoyt, American maintenance troubleshooter, finally found out that the trick was to lay a sufficiently thin first coat over the cracks. With a first layer no thicker than one-sixteenth of an inch, and any number of layers atop that, a DC-4 could toss and turn through an Atlantic storm and still get home with tanks intact.

When peace broke out, American and United bought up war-surplus C-54s in lots of fifty, put in seats, toilets, and galleys for passengers, and hoped that the customers would keep their seat belts fastened through the usual low-altitude buffeting until they could put pressurized planes like TWA's Connies on their runs. It was a struggle, but it had been so long since businessmen could get a seat on a plane that nobody had to look for customers, and then TWA ran into such trouble digesting its new worldwide routes that its rivals more than held their own. Then in early 1947 United took delivery of the first DC-6, and at once United and American experienced *their* transition problems.

On April 27, 1947, United put the first DC-6 into coast-to coast service, stopping only at Lincoln, Nebraska, for fuel. Stretched seven feet longer than the parental DC-4, the new plane had Pratt & Whitney R-2800 engines that turned up 2,100 horsepower on takeoff and was thus a shade faster than the 049 Constellation. In the daytime version there were fifty-two seats, and when operated as a sleeper plane it had berths for twenty-four. American followed soon after with both types of DC-6; for a time the new DC-6 was advertised to fly from east to west in a flat ten hours. TWA's best time with the 049 Constellation was ten hours, ten minutes. The advantage seemed to have shifted to United and American.

Then, like the Constellation before it, the glorious new DC-6 met with disaster. With fifty-two persons aboard, a United transcontinental plunged into Bryce Canyon, Utah. Fire aloft, the most feared of all flying perils, was suspected. Then an American Airlines DC-6 bound from San Francisco to Dallas caught fire in the air. This time all aboard escaped because the plane happened to be passing over Gallup, New Mexico, and quick action by Captain Evan Chatfield got the DC-6 down on the ground in time. A big hole showed where the fire had started — down in the belly where the plane's gasoline-fueled cabin heater was, just where the United men suspected their fire had occurred.

The investigation focused on the heater. But it was known that the crew had been transferring fuel between tanks in the wings before the accident, and Abe Hoyt remarked that perhaps they spilled some. Could there be a link between a gasoline spill at the front of the big plane, and a fire far back in its belly? American's Captain Glenn Brink took a DC-6 up over Los Angeles and transferred fuel until he was sure it was overflowing. Making the same emergency descent that pilot Chatfield did, Brink and his colleagues found that gasoline from the overflow vent near the nose had streamed back into the plane's big air scoop behind the wings and along the ducts to the heater. Had the heater been working, the gasoline would have ignited. The CAB investigators repeated the test with red-dyed water, got the same results. The fatal flaw had been found. The DC-6 remained grounded three months while Douglas moved the fuel vent to a safe distance from the air scoop.

↳The DC-6 returned to the skies, and it soon became apparent that Douglas had brought in another winner. The fuselage was a vital nine inches narrower than that of the Constellation, hence not so well suited to five-abreast seating, but a few years later the airlines persuaded Donald Douglas to stretch his plane another six feet and install a slightly more powerful engine. By that time, between the pitiless scrutiny of their mortgage holders, the exhaustive research of their regulators in Washington, and the day-to-day analysis of their own experts, all the airlines knew to the last decimal point just how many passengers each plane had to carry on each leg of each trip in order to make a profit. Every headquarters had file cases full of data on which type of plane — taking into consideration the purchase price, flight-crew pay, fuel and food cost, and the rest — flew each trip most economically. The statisticians liked to divide the number of revenue passengers by the sum of these charges. The final result was the average per-mile performance of the airliners, and by this measure the DC-6B was a more economical performer than the Lockheed Constellation, more economical indeed than any piston plane ever built, including the legendarily profitable DC-3.

Nobody ever made such claims for the third entry in the postwar sweepstakes. Boeing's 377 Stratocruiser, substantially bigger than either the DC-6 or the Constellation, was developed for Pan Am out of the B-29 Superfortress bomber at the end of the war. Having set a standard of luxury design with Pan Am's prewar flying boats, Boeing's Wellwood Beall undertook to do the same for landplanes. In those days all airline passengers were well to do, and expected first-class attention. So Beall designed a deluxe cabin that could easily seat sixty and included spacious sleeping berths, as well as a honeymoon suite in the tail. He spent a quarter of a million dollars, an unheard-of sum, working up a really satisfactory chair for the passengers. And — the one thing everybody remembers about this plane — down in the belly he put a lounge to which passengers, tired of sitting so long in one place, could descend by a circular ladder. "The conversation pit," he called it.

The airplane has been defined as a machine that almost doesn't fly. If there ever was an airliner that waddled down the runway and still managed to take off, that plane was the Stratocruiser. Lantern-jawed, double-chinned, even dewlapped in profile, it looked like a stranded whale on the ground. Rising from the runway, it seemed to ram rather than knife its way through the air. And, in fact, Pan Am Captain Francis Wallace said later, a Stratocruiser takeoff at a tropical capital like Port-au-Prince was

The DC-6, Douglas's postwar airliner.

The luxurious Boeing Stratocruiser featured a lower deck cocktail lounge.

Wellwood E. Beall (left) and Edward C. Wells in 1948.

"something to experience — marginal." In the hot temperatures prevailing "on the way to Rio we could never climb above 12,000 feet." In cooler air with lighter loads, say toward the end of a journey to London or Paris, "the Stratocruiser would go up to 25,000 feet very nicely and really move. That was about the only time we could beat a DC-6."

What with miscalculations of output capabilities, more lucrative military orders, and a prolonged strike, Boeing did not deliver its first Stratocruisers until more than two years after the promised date. But from the first the plane was so expensive ($1.5 million apiece) that only a few airlines ordered it; Pan Am took twenty, American Overseas ordered seven, and United seven. By then Northwest, strapped for funds, was under some pressure to cancel its order for ten at $1.75 million apiece. In those days the political contributions of such giants as Lockheed, Boeing, and Northrop were not policed as sternly as later, so after Boeing made a timely contribution to President Truman's 1948 reelection fund, the CAB and Reconstruction Finance Corporation "looked into" the financial plight of the airlines. The CAB gave Northwest a route with mail pay between Portland, Seattle, and Hawaii, and the RFC advanced the airline $12 million to pay for the ten Boeings.

When the planes were delivered, the airlines also found them costly to operate. The first reason was that Boeing had equipped them with a huge and complicated engine too new to have been tried out in the war. This was the Pratt & Whitney Wasp Major. Designated the R-4360 because its twenty-eight cylinders contained a total of 4,360 cubic inches of "swept volume," the engine yielded no less than 3,500 horsepower on takeoff, or nine times that of the original Wasp engine that powered America's first commercial airliners. Obviously, you couldn't build piston engines much bigger.

Breaking in these huge power plants was a costly experience. Northwest Captain Walter Bullock made three successive starts one day from Seattle, returning each time because of engine trouble. It was bad enough to keep changing 112 spark plugs, but Pan Am was changing cylinders every 150 hours. Even so, cylinders were known to fly off, sometimes through the fuselage. Standard Oil of New Jersey had to invent a super high-octane fuel to help the airlines fight fouled spark plugs and exhaust-valve failures. "Best three-engine plane flying," the wags said. In the first rough shakedown year, Northwest's President Croil Hunter remarked, "You're safer flying than on the ground around that plane — you're so liable to be hit by a falling engine."

A second reason Stratocruisers cost so much to keep in the air was their propellers. This whirling toothpick produced the thrust that got the Wrights off the ground. In Lindbergh's day it was still a toothpick. Even so, it was something to watch out for; a lot of airmen, and a good many early passengers, got killed or maimed when they stepped in the path of a blade. In the air, props could also be lethal. Once in Army training days in the 1930s Pan Am Captain Horace Brock's best friend, flying in formation with other pilots, was decapitated when the next plane's propeller knifed through the cockpit.

With the advance in engines, propellers were made of metal, regulated to different pitch for different jobs, and refined to amazing aerodynamic efficiency — something of the order of 92 percent. And as the planes got bigger, designers took advantage of cabin size to reduce the chance of props flying off and hitting the passengers.

Boeing's engineering chief Ed Wells said later:

We arranged the airplane so that we didn't seat passengers in the plane of the propellers. Exposure to the blades of the inboard engines was of course much greater — an inboard engine with a propeller failure in a range of 90 to 100 degrees might damage the cabin, whereas with an outboard engine it might be in the range of at most 35 to 40 degrees. We placed galleys and clothes closets in the propeller planes in our airplanes, and we always calculated that the passengers were shielded by the inboard engines.

After the war pressurization increased the risks of malfunctioning props. And propellers grew bigger, with a sweep sixteen feet across; some came with three blades and even four. The manufacturers turned from aluminum to steel, and then the blades became vulnerable to metal fatigue. The props also got so heavy that two men could not lift them. Weight being anathema in airplane design, the companies made their props hollow — with a mastic core filler.

Even before the Stratocruisers ran into their troubles, there were some spectacular prop failures. On the night of August 22, 1950, American Captain Bob Baker was flying a DC-6 eastward across the Continental Divide when he heard a loud racket behind him and his cheeks began to flap. Baker knew at once his plane had experienced explosive decompression, the phenomenon that occurs when the pressurized seal of a high-flying plane is suddenly broken. The air inside rushes out the hole in the fuselage, and the air in your lungs is sucked out too. Back in the cabin, the passengers' cheeks fluttered like butterfly wings.

A piece of a blade on the No. 3 engine had broken off and ripped through the fuselage. Since this engine was barely six feet away from the left cabin wall, the severed part might have hit the plane anywhere in the ninety- to one hundred-degree arc. As luck would have it, it tore through the fuselage from down to up. It passed through one of the lavatories located just behind the first-class seats and smashed a twelve-foot hole in the roof. Out the hole went pillows, insulation, blankets — but no passengers. The lavatory was unoccupied. One passenger died of a heart attack.

The tremendous vibration induced by losing part of the propeller tore the DC-6's No. 3 engine off almost instantly. The flying hunk of metal also broke some control cables. But Captain Baker, a cool-headed man, brought his DC-6 down safely at Denver.

The missing engine was found on an eleven-thousand-foot shoulder of the Rockies, and it was established that a fifty-inch segment of the hollow-steel prop blade had snapped off. American Airlines offered a $5,000 reward for the missing piece. Although the chances that anyone would ever find that small hunk of metal in the rugged, wooded mountains of Colorado were slim, two kids out fishing found it, heard about American's offer, and got the money.

The investigators found a very clean break. The hollow blade was formed by welding two long steel clam shells to a mandrel that extended from hub to blade tip. Government regulation required the propeller manufacturer to take X-ray photographs of his work. The investigators therefore went to Curtiss's Electric Propeller division and dug out the picture. They found that the tip of the mandrel when rotated in the process of welding had scratched the inner surface on each side of the hollow blade. Under the stress of constant flexing as the variable-pitch mechanism kept altering the blade angle in flight, the blade finally broke — at the point it was scratched.

The Stratocruiser also started out with Curtiss electric props. "I had one of the first Boeing propeller emergencies," Pan Am's Captain Hack Gulbransen said later:

I was 800 miles out of London in the middle of the night when the engineer noticed a sudden drop in oil pressure in the No. 1 engine. He leaned forward to jab the feathering button. But before he could act the engine seized. I tried to feather the prop, but the gearing in the hub fouled up and the prop became a pinwheel. The heat of the friction began to melt the magnesium housing. Then the propeller began to wobble, and the propeller shaft ground against the housing. Oil in the casing caught fire and fed the flames. Shaft and all, the propeller spun off into the darkness.

Flames still shot out and spread over the wing. I thought, "Well, here goes two million bucks." I alerted the entire crew to the emergency. But because of lack of continuing friction after the prop flew off, the fire went out — and I flew back to Shannon.

What had happened was that the oil pump feeding the engine had suddenly failed, while the scavenger pump draining oil out of the engine had kept going. Suddenly sucked dry of lubricant, the engine seized, and the propeller ran wild.

A couple of years later Gulbransen was over the Atlantic when he heard a BOAC pilot report a runaway prop on his No. 4 engine. Gulbransen picked up his mike and told the British captain exactly what would happen. The propeller housing, made of magnesium, caught fire just as it had on the Pan Am Boeing, and burned with a brilliant white light. The four-blade propeller blew off in front like a giant catherine wheel before dropping into the Atlantic. The Stratocruiser made a successful emergency landing at Sydney, Nova Scotia.

Though the airlines eventually tamed the Stratocruiser's unruly engines, the props remained a worry. Pan Am's maintenance men used to go out before and after every flight to tap each hollow blade with a silver half-dollar to make sure the mastic was in place inside. The airline tried four or five different kinds of prop without ever getting a satisfactory match with the engine. One Pan Am plane was lost over the Brazilian jungle with fifty people aboard when the No. 2 engine and prop flew off in midflight. Another suffered blade failure and loss of the No. 4 engine before crash-landing at Johnston Island in the Pacific. On another, the blade of the No. 3 propeller razored through the cabin, killing one passenger and injuring another during an approach at Manila. When Pan Am finally retired its Stratocruisers, operations boss John Shannon said he was glad to see them go for one reason. "Every hour that plane flew it cost me $10 just for propellers," he said.

Despite such headaches, the airlines were slow to part with their Stratocruisers. Northwest hung onto its 377s for years, even when they lost money. The reason was that from the standpoint of passenger comfort, they couldn't be beat. Riding in a Stratocruiser was real first-class travel. A flight to London or Paris took about twelve hours, which gave the passenger leaving New York in late afternoon time for a leisurely drink, dinner, a visit to the lounge, a reasonable night's sleep, and breakfast. On arrival in the morning he was ready to go. Passengers flying Northwest from Seattle to New York reported much the same thing. They loved that big cabin, those comfortable seats, and above all that lounge below. Pilot Lee Smith told Croil Hunter that their airline could well charge off the five-cents-a-passenger-mile deficit on Stratocruiser operations to advertising. "That plane *made* Northwest," Smith said later.

United looked at the barrel-chested Stratocruiser with a harder eye. It placed seven on its Honolulu run, complete with the honeymoon suite and other comforts. Some travelers afterward recalled this as air travel at its finest — a ten-hour cruise during which stateside stewardesses and Hawaiian stewards plied them with champagne and five-course meals, the bar throbbed enticingly below, and everybody strolled the spacious aisles festooned with flowers. But the airline noted in its 1952 report that it lost two cents a mile even when operating these flights with fifty-five passengers. Shortly afterward United sold off its Stratocruisers.

In all Boeing built only fifty-five of these luxury liners. Unlike Douglas and Lockheed, it was never asked to "stretch" its design. No doubt the firm recouped its losses by selling nearly a thousand C-97 cargo carriers — the 377's sister plane — to the Air Force. Comfortable the Stratocruiser certainly was, and possibly economic by the measure of Lee Smith's kind of invisible-entry bookkeeping. But the plane was less than half as safe to fly as either the DC-6 or Constellation. In twelve years Pan Am lost five of them. One was when Captain Richard Ogg, in the airline's first landplane ditching since Key West days, eased his Stratocruiser into the ocean alongside a Pacific weather ship; all 31 aboard were rescued. But there were six fatal Stratocruiser crashes. In these 108 passengers and 24 crew members lost their lives.

American aviation's long love affair with the piston planes continued; a successor had to be found to the DC-3 for the shorter runs. First to appear was the Martin 202, and it spelled still more trouble for Northwest Airlines, which had ordered some to complement its big Stratocruisers. "Martin, after building military planes, was trying to get into the passenger field — and they goofed up," Lee Smith said later. "The pilots wouldn't have it."

The Martin 202 made its first flight November 22, 1946. It was fast — fully a hundred miles per hour faster than the DC-3 it replaced. But it was unpressurized, an interim plane built under wartime conditions. For Northwest, first to get them in 1947, the Martins were a disaster. The airline lost four in a year. Even before the CAA grounded them for a structural defect in the wing, the pilots had presented a list of 140 shortcomings that needed attention. Captain Vince Doyle said later, "There was one place below the cockpit where hydraulic, gas, electrical, and oxygen lines all came together, and not even a hole to drain the hydraulic fluid that you knew sooner or later would leak there." When passengers left the airport at Minneapolis rather than fly on the Martins wheeled up to the departure gate, the airline abandoned the planes altogether.

Convair had better luck with its twin-engine postwar entry. American Airlines put the CV-240 into service in June 1948, to complement its DC-6s. The airplane carried forty passengers, nearly double the capacity of the old DC-3, and was almost as fast as the Martin. The Convair, pressurized, was so successful that a few years later it was "stretched" to make room for more seats, and then United used it too. Just about everybody that flew much in the prejet years climbed the foldout stairs of the Convair, the Yellow Cab of the airways, a rugged, compact plane that charged up and down the runways of Midway and O'Hare like a linebacker for the Chicago Bears.

25

"The Gush of Dollars Is Over"

At the end of World War II, the airlines thought their future had arrived. As late as the end of 1945, their 397 airliners still carried only an eighth of the traffic that went by Pullman. But in the years after Pearl Harbor, 1.5 million Americans had learned to fly, another million had flown the oceans and continents as military passengers, and uncounted millions at home, kept from the air by priority restrictions, had witnessed the spectacular advance of aviation round the globe. All that seemed necessary now was to wheel the planes up to the gates and stand back for the flood of people crowding to travel the fast, modern way. So the airlines placed huge orders for new planes. They hired extra help. And they opened their doors for the big postwar ticket-buying rush.

A year later the *New York Times* reported: "The gush of dollars is over. The problem is one of drastic reconversion — back to customer appeal." With one or two exceptions, all airlines were in the red, and some nearly bankrupt. Although the number of airliners had nearly doubled in 1946, many of them bigger and faster four-engined Douglases and Lockheeds, the airlines had miscalculated their appeal. They had still not induced a profitable number of the American public to travel in their planes. The influx of passengers had failed to keep pace with their inflated payrolls and the cost of their expensive new equipment. Neither in comfort nor in punctual performance did the airlines satisfy the customers, who turned away in droves to rail and steamship lines.

For the troubled airlines, the year of postwar transition stretched out for the rest of the decade. American canceled multimillion dollar orders for Republic Aircraft's glowingly advertised, 440-miles-per-hour Rainbow airliner, thereby knocking out half of Long Island's aviation industry. TWA slashed schedules, Western surrendered routes. Northwest scrapped plans for a new global capital at Seattle and crept back into a modest shed in Minneapolis. United cut back by centralizing operations in Chicago and Denver. Tough Eddie Rickenbacker, whose Eastern Airlines alone made money all these years, shaved costs by shortcutting across water between New York and Miami while putting in for mail pay to all shore points along the way.

Though faster planes were coming on line, the DC-4 remained the mainstay for most airlines in these awkward years. If the public would not fly in a plane whose reliability had been proved in three years and 350 million miles of wartime operations, what could the industry hope for? In the midst of other setbacks, this question was brutally posed in the early summer of 1947. Suddenly, in the space of fourteen days, three DC-4s plunged to earth in crashes that killed 150 persons. Ground the DC-4? The usual CAB investigation took months, and that would wreck the airlines. The President and Congress opened inquiries. Passengers were refusing to board DC-4s. Concern had already been growing over fatal accidents — eight crashes in 1945, nine more in 1946. Reassurance was needed — fast.

Investigators already knew that the plane was not to blame for the first crash. United's DC-4 had crashed after takeoff from LaGuardia Field because Captain Ben Baldwin left the gust lock on. All airliners had gust locks clapped on their tail surfaces to keep them from being blown around while on the ground. On the old DC-3 it was the copilot's job to go back and pull the blocks off the stabilizer just before climbing aboard. On the DC-4 a mechanism that the pilots could operate just behind their seats took the place of this primitive arrangement. Baldwin, unable to lift the plane's nose because he had forgotten to disengage the mechanism, bounced across Grand Central Parkway and crashed in a field eight hundred feet beyond the runway. One of the five survivors (out of forty-eight aboard), Captain Baldwin denied the error, but the evidence was conclusive and he was dismissed. The second accident, less than twenty-four hours later, was far less easy to explain. But by a remarkable coincidence, government accident investigators, flying some three miles behind the Eastern DC-4 over Bainbridge, Maryland, had witnessed the crash. It occurred in broad daylight. Eastern's plane, flying straight and level at four thousand feet toward Baltimore on its way to Miami had suddenly nosed over and dived into the ground. All fifty-three aboard had been instantly killed.

The investigators noted that in its fall the Eastern DC-4 had not just gone into a vertical dive, it had "tucked under." That is, it had started to go over on its back, as if beginning an outside loop, as it hurtled to earth. Something about that uncontrollable plunge made them suspect, especially when investigation showed everything else in order, that the Eastern plane's gust lock had been engaged too. But why in midair? They speculated as follows: The pilots had been talking about the preceding day's crash. Captain Bill Coney, an old Navy test pilot and keen flier, knew it was perfectly possible to fly a plane with gust lock on. The pilot simply made small adjustments in the trim tabs to hold the plane level. And Coney had been demonstrating precisely this to First Officer K. V. Willingham when the accident occurred. Since passenger-laden planes are normally a bit tail-heavy, he would have given the trim tab regulating the elevators a few turns with his thumb and forefinger to hold the nose down as he talked. Then, demonstration over, he would have disengaged the gust lock — and suddenly the plane pitched right over. Pull back on the yoke as hard as the pilots might, the rush of air across the tail surfaces in the dive was too great for them, and the plane crashed.

To test their theory, the investigators took a DC-4 up to a safe twelve thousand feet over Chesapeake Bay and went through their scenario. Flying level with gust lock on, they took only one and a half turns on the trim-tab screw, then released the gust-lock mechanism. Over the plane went, and only fast action enabled them to right it. Had they found the answer? It was not conclusive, but nothing they unearthed at the crash site hinted at any structural failure in the plane. Meanwhile, the third of the fatal accidents, a crash of a Capital Airlines DC-4 into a shoulder of the Blue Ridge during a night approach from Pittsburgh to Washington, had been identified as an "altitude problem." That is, it was a question of how high the pilot thought he was flying, not of a defect in his plane. So the President's Safety Committee report was rushed to the White House, the public was assured the DC-4 was all right, and the airlines carried on.

When the CAB duly issued its official finding on the Bainbridge crash, engagement of the DC-4's gust lock was listed as only one of ten possible causes. That fall, American had an incident with a DC-4 that greatly embarrassed the airline, but dissolved the investigators' last doubts about the DC-4 and enabled the industry to breathe easier at last. On October 9 Captain Jack Beck joined Captain Chuck Sisto's DC-4 in Dallas to familiarize himself with the aircraft during the flight to Los Angeles. As they passed over El Paso at eight thousand feet, Beck was flying the plane and Sisto, riding in the jump seat behind the pilots, thought to play a trick on his friend. He quietly engaged the gust lock. With forty-nine passengers in the rear, the nose rose a bit. Beck corrected the slight tilt with a turn of the trim tab. The nose lifted again, and Beck, muttering to himself, had to make a further correction. Before long he had to repeat the move, and said to copilot Mel Logan, also a captain, "What's the matter? This plane wants to climb today."

At that, Sisto testified later, "I decided the joke had gone far enough. I disengaged the gust lock — and we went over like a shot out of a barrel, BANG." By luck Beck's seatbelt was loose. Flinging his hand up to avoid hitting the roof, he accidentally pushed the levers that feathered three of the plane's four engines. That averted a power-on dive. Although copilot Logan braced his feet against the instrument panel and pulled hard on the yoke, he was powerless to move it. The plane shot vertically to earth.

Then it began to tuck under. Finally it went over on its back in the start of
an outside loop. At that moment, Mel Logan's quick thinking saved fifty-
two lives. Though he could not move the elevators, the ailerons were
working perfectly. So he rolled the big plane over — and came out flying
straight and level at four hundred feet. People said that the passengers
looked "shaken" when the plane landed at El Paso Airport. A man from
New York said, "I have been through eternity." A young French student
staggered out of the lavatory plastered with excrement. Amazingly, although thirty-five aboard received minor injuries, no one was badly
enough hurt to require hospital attention. The airline accepted Captain
Sisto's resignation, and the DC-4, surviving man's frailties, flew on.

For resolution of some of their problems, the airlines turned to the new
International Air Transport Association, founded at the end of the war to
"prevent chaos" in setting fares. It contrived so well to put a floor under
Atlantic and European air fares that for years many thought of IATA as
an OPEC of the skies.

The airlines had less success when they tried to deal with domestic difficulties through IATA. One was "no shows." From the earliest days, the
prospective traveler had only to pick up the phone and reserve a seat. The
airlines wanted business so badly they were loath to add that the traveler
should also telephone, if he changed his plans, to let them know so that
they could resell the space. Until World War II the number of air travelers was small, and the airlines grandly forgave their no shows; you
could always get your money back for your unused ticket even if you
failed to cancel your reservation.

After the war the airlines were swamped with reservations. At first, as-
suming some would be no shows, they resorted to overbooking. Airline
ticket sellers had grown used to turning away civilian patrons under wartime priorities with the words, "Sorry, the plane is filled." But that did not
quiet the businessman who bustled up to the counter in peacetime with a
ticket for a flight only to be denied a seat. Still persuaded theirs was a
seller's market, the airlines decided it was time to make those who didn't
show up pay. In September 1946 IATA announced that 25 percent of the
sum paid for a ticket would be kept by the airline when passengers failed
to notify of cancellation in time for the airline to resell the space. This had
long been Pullman practice, and European airlines insisted on it. But that
fall U.S. air travel dropped off. American airlines, painfully rediscovering
that air travel had to be sold to their public, abandoned all efforts to en-
force the penalty.

The airlines' postwar transition to bigger planes brought a showdown
with their pilots and strained the whole system that had been built for
smaller and slower airliners.

Many find it hard to think of the Air Line Pilots Association as a union. Measured by members' incomes, it's more like the American Medical Association. But when ALPA rose in the Depression, it was a rather rambunctiously militant affiliate of the American Federation of Labor. It first made headlines when United Airlines pilot David Behncke popped into company headquarters armed with the signed resignations of fifty pilots and demanded to bargain with the management.

Up to that time airlines paid their pilots as the Post Office Department had paid them — a monthly base rate ($250), with more for miles flown, and still more for hazards like mountain and night flying. Average pay was $600 a month. In 1931 the Wall Street speculator E. L. Cord startled the Century Airlines with the declaration, "Any normal person can safely and easily handle a plane." Cord slashed pay to $150 a month. The pilots claimed a lockout and struck. In a new kind of picketing, they rented a plane, painted CENTURY IS UNFAIR TO PILOTS on its side, and flew it alongside every Century plane arriving in Chicago.

Cord's company went out of business, but another threat soon arose — the airlines began introducing the first modern airliners. TWA had just switched over to an hourly rate of payment for pilots and the other airlines followed suit: the airlines would harvest for themselves all the productivity gains obtained with faster planes. That was when the pilots got the AFL behind them. The pilots found another friend in Senator Robert Wagner, father of the New Deal labor legislation. With the help of Bob Wagner and the AFL, the pilots won an 85-hour limit on monthly flying time — instead of the 140 hours proposed by the airlines — and a formula that combined a basic hourly pay rate with added pay for mileage. Thus the pilots helped themselves to a big share of the productivity gains associated with faster planes and assured themselves the handsome salaries they have enjoyed ever since.

Next the union told United it wanted a better way of scheduling pilots' working time. President Pat Patterson said later:

I met these pilots at a pilot's home [in Cleveland]. They said their time off was interrupted too often. Often they were asked to fly with less than an hour's notice. And I swear, while they were meeting, five or six men got phone calls that they might by flying at 5 that afternoon.

Now that's a helluva way to spend your day off. I said, "Tell me what we should do." They came up with a draft. I brought it home, and our people said it couldn't be done. We worked it out, and it didn't cost us a nickel. We put men on standby, on reserve.

Soon after, the pilots produced new demands. Again they were bucked to Pat Patterson. Patterson said later, "You could see in a minute the way we had been doing it was all wrong, and we agreed." So the seniority system that prevails in all airlines was established. Pilots were numbered according to their date of first employment and their first promotion to captaincy. On the first day of the month the airline posted its schedule of the month's expected flights and the captains and copilots "bid" for the trips they preferred up to the maximum of eighty-five hours. The most senior, of course, won whatever trips they bid for, and those less senior quickly learned which trips they could put in for and expect to win. The system, proposed by the pilots and accepted by Patterson in 1940, prevails to this day, though nobody on the major airlines now flies more than seventy-five hours a month. All air crew members — flight engineers, navigators, stewardesses — sought and won the same "bidding" arrangement.

By the end of World War II, the Air Line Pilots Association had contracts with all major airlines. In the move to four-engine planes, they struck for more pay and totally grounded TWA for six weeks. There were more labor stoppages in 1946 than in any postwar year, and the public did not take kindly to a strike by men whose income averaged over $10,000. CIO President Phil Murray snorted, "Labor dispute, hell. That's a row between capitalists." The strikers did not gain their objectives, but the union continued to exert strong influence in Washington, where congressmen flying to and from their constituencies had personal reasons to listen to the lobbying for better working conditions and flight safety.

After the war the airlines hired a lot of new pilots. In this time of transition, flying men ceased to run the airlines. The most famous of these, Jack Frye, stepped down at TWA in early 1947. When he did, some of his best fliers went off to help start other airlines overseas: Otis Bryan ran the Philippine Airlines and Swede Golien helped organize the Ethiopian Airlines for Emperor Haile Selassie. One of the greatest old pilots, United's Ham Lee, flew only one or two trips in United's four-engine postwar airliners before he decided to retire to California — he had already logged thirty thousand hours, more than any other pilot, and his son Bob was flying for United. (The day came in 1973 when son Bob reached retirement age, and Ham Lee, erect and unspectacled at eighty, flew copilot on his son's last trip to Portland.)

Of those newly hired, Eastern's Floyd Hall said later:

We didn't care what background, whether fighter or bomber pilot, if they had a sizable number of flying hours, a good educational background and a love of flying. You have to bear in mind that a lot of the casualties of World War II were not exactly the walking wounded. We tried to stay away from such people, but we couldn't succeed wholly. Some we hired took a drink.

Not all the World War II airmen joined the airlines. A legion of ex-pilots founded their own companies — 2,730 "airlines" in 1945 alone, by CAB estimate, with 5,500 planes. The "airline" — sometimes the president was pilot, the vice president, copilot — could pick up a Douglas C-47 for $25,000 and pay for it at the rate of $4,000 a year. Most of these ventures were doomed from the start. Good military pilots don't necessarily adapt to straight-and-level cross-country flying. Many of these nonscheduled promoters could not fit their free-wheeling style to the exacting requirements of everyday navigation, safety, maintenance.

"Nonsked" was a term just entering the language when President Truman, on July 1, 1946, named James M. Landis chairman of the Civil Aeronautics Board. Landis at the time knew little about either scheduled or nonscheduled airlines. But on the face of it, he was the strongest figure ever to take up the task of airline regulation. With his broad brow and piercing eyes, Landis, one of the architects of the Securities and Exchange Act, was a master of the kind of administrative law that flowed from agencies such as the CAB. Dean of Harvard Law School since 1941, he had been called back by Roosevelt to head the nation's civil defense program and serve as chairman of the multinational Middle East Supply Committee; he was accredited as U.S. minister to the seventeen countries in that area. Landis liked public service and returned to Washington when Truman, just finding his feet as President, needed some strong New Dealers around him.

Landis, however, was a crusader who came to office when the crusade had faded. Truman's own authority had sunk to ebb tide. In the brutal warfare between the three power centers of postwar Washington — the Congress, the lobbyists, the bureaucracy — Landis lasted just eighteen months. The fall was marked by a personal collapse. Almost at the time he learned Truman was not going to reappoint him, his marriage went on the rocks. He drank heavily and had to resign his Harvard deanship.

The personal tragedy of a man an adversary later called "one of the best legal minds of the century" was not without consequences for airline history. No figure of comparable attainments ever held the office again. And into oblivion with Landis went important policy initiatives that only resurfaced in congressional measures for the industry's overhaul thirty years later.

In his brief tenure, Landis set out to regulate an industry that had pretty well regulated itself. On the one side he called on six of the sixteen domestic lines — Colonial, Northeast, Continental, Western, Capital, and National — to show why, since all needed mail-pay subsidies to survive, they should not be forced to seek merger partners. This was music to the ears of the big airlines, especially United, Eastern, and American, which had been actively trying to buy out some of these lines in the postwar shakeout.

But even while prodding the weaker sisters in this fashion, Landis harbored other notions that frightened the big operators. He had the idea that the air transport industry should be opened up to fresh competition. Within the CAB he talked about perhaps awarding certificates to new airlines. He took the upstart nonskeds seriously, listened to their pitch, and decided they deserved a chance. Here he went too far. It was one thing to let down the bars to the little fellow who ran what amounted to an air-taxi service. (In the name of "public convenience and necessity" the board had even authorized outfits operating "large" planes to carry passengers between any two points on an "irregular" basis.) But now Landis granted exemptions that allowed the first of these nonscheduled operators to fly passengers across the country on an experimental basis. A fight started: the big airlines took off the gloves to defend their aerial turf. They found a sympathetic reaction among Landis's fellow board members who resented the fancy Harvard logic. Some of them had also served in the airlines. They got the CAB order worded so that no one "irregular" company could fly more than one trip a week.

The nonskeds got around that roadblock by a device worthy of Landis's legal wiles. They organized a combined outfit in which one company flew Mondays at 9:00, another Tuesdays at 10:00, and so on through the week. This was a daily service. The only thing unscheduled about their scheme was the hour of takeoff. Soon nonskeds with names like North American, Standard, and Martin Air Transport had hired people, acquired goodwill, and worked up a fair business — mostly, they claimed, among people who had never traveled by air before. They flew from the dustier hangars, picked up their passengers at the shabbier hotels, and tended to hold off departures until their planes were chock-full. But they were delivering daily cross-country DC-3 and DC-4 service at the bargain-counter price of ninety-nine dollars one-way, evidently at a profit. "The public always likes to get things cheaper," an airline lawyer complained. What the country was witnessing, of course, was the arrival of what the airlines later adopted as "aircoach" travel.

But in pressing this innovation Landis lost his majority on the board. American's C. R. Smith, still headquartered in Washington, led the drive. Carlene Roberts, his able aide, won her lobbying spurs helping turn Landis's colleagues against this disquieting initiative. At the same time, Landis got into a damaging hassle with Pan Am's Trippe over the United States–Brazilian air agreement. Worse, Landis became expendable to Truman when the administration lost the midterm election to the Republicans and the President, battling to survive for the 1948 reelection match, swung right. With Landis went the last chance for the public to see the airways opened up to new entrants. Meanwhile, C. R. Smith saw that the way to handle the nonsked threat was to put down the fare. He slashed American's New York–Los Angeles rate to ninety-nine dollars — and the nonskeds were banished to insignificant, fly-by-night existence.

With faster planes, everything on the ground, from reservations to engine changes, had to be handled faster. Aloft, the postwar increase in airliner speeds meant communications, weather forecasts, and above all, landing and takeoff procedures had to be handled faster. You would have thought, after the tremendous wartime advances in electronics, that radar would have been the instrument grasped to speed planes through weather at airports.

Not so. That was the military's solution. The Air Force and Navy had perfected Ground Controlled Approach. But the airlines and their pilots shied away from the very idea of anybody guiding their planes from the ground. GCA involved the use of two radars — a long-range unit to locate the plane and guide it to the field and a short-range, precision radar used by the ground controller to "talk" the plane down the last ten miles to the runway. Compared with GCA, the CAA's Instrument Landing System (ILS), which left responsibility for the landing in the pilot's hands, was far less developed at this time. In 1946, when transatlantic flights were having trouble getting through the fog at their Gander stop, Pan Am obtained permission to use GCA there. It hired Air Force controllers, and for the next five or six years these men on the ground "talked down" Pan Am and other airline planes at Gander when ceilings were as low as one hundred feet and visibility as little as an eighth of a mile — conditions in which you would hardly want to drive a car. Captain Chilie Vaughn said later, "We never had one incident that would have damaged anyone's confidence."

ILS, by contrast, was a radio system. Two ground transmitters provided an approach path for the arriving plane's exact alignment and final descent to the runway. One, called the "localizer," gave the pilot course guidance that enabled him to line up his plane with the runway. The other, located about a thousand feet short of the runway, was the so-called glide-slope transmitter. This unit sent a narrow radio beam directly at the incoming plane at an angle of about three degrees off the ground. Aboard the plane a pair of intersecting, phosphorescently glowing needles presented these guiding signals on the pilot's instrument panel. When the vertical needle for course and the horizontal needle for descent crossed at right angles, the pilot figured his plane was lined up just about right, and all he had to do was to keep the needles that way to come down to a perfect landing.

Under development since the 1930s, ILS was not ready when postwar airliners and traffic demanded it. When the CAA started installing it at major airports in the summer of 1947, it was still pretty rudimentary. Notwithstanding their costly traffic delays, the airlines did not at once make use of it. ILS was subject to the imperfections of an all-radio setup; a change of weather, especially rainfall or snow, could knock it off line. The most careful maintenance was required. Congress, listening to the impassioned arguments from both sides, took a dim view of the idea of funding two different landing systems. When the CAA persuaded Braniff to try out the ILS needles in its planes, Carl Hinshaw, the House aviation expert, contrived to hold back funds. He demanded a "common system" for military and civilian flying. Meanwhile, glowing praise for GCA came back from Germany, where ground controllers kept the Berlin Airlift going like clockwork with landings and takeoffs at three-minute intervals through the thickest weather. Under such pressures a special committee fetched up a compromise: let the precision-beam radar of GCA — "the most dependable landing information" — be used along with ILS. Under heavy prodding the CAA put GCA gear in most airport towers.

In this dispute air transport progress was held up for several years and thousands of passengers suffered unnecessary inconvenience and delays. Finally jobs, not technology, were decisive. Captain Sam Saint, American Airlines' leading proponent of better procedures, spoke for his colleagues when he said, "Let the motorman do the approach and landing." The pilots took a labor union view of the issue. To them, ground control threatened their high pay and command status. Many, especially copilots, were "weekend warriors" in these cold war years, and they submitted readily and even gratefully to GCA ground controllers when flying for the Air Force Reserve. But on the job the next day they closed ranks in hostile opposition to direction from below. They said, "We don't want to put ourselves in the hands of some clerk sitting under a waving antenna." So much for all-seeing radar. Time and again CAA inspectors riding in jump seats saw how they worked. Commencing an instrument approach, the airline pilots would ask for GCA support. Then they would tune the radar operator's voice down to inaudibility and take the plane down, as they said, "by the needles."

The airlines, who had had more than enough of pilots' strikes, went along, and the ILS system gradually gained in precision. GCA soon vanished from the airports, although the search-radar component, redubbed "surveillance" radar to placate the pilots, remained a valuable adjunct to ILS. For two decades, GCA continued to be the only precision landing aid in use at military bases.

26

Air Travel in the 1940s

A *month* after V-J Day the War Department abolished its priority travel rules, and a horde of civilian passengers descended on the airlines. It seemed as if all those who had been denied the chance to travel by air suddenly flocked to fly — and the airlines were not ready for them.

Often it required eight to ten phone calls to reach reservations, and then a weary voice explained that the first flight on which seats were available was three weeks hence. The terminal waiting room was a madhouse of people seeking information, clerks who didn't know the answers, lost passengers looking for their luggage, and families meeting arrivals.

Since it was just as hard to get a line into reservations to cancel a seat as it was to reserve one, there were a lot of no shows. If you simply had to fly, you could go out to the field and be a "go show" — a hoper who hung round the gate on the chance of an unclaimed seat. A passenger heading east from San Francisco at this time simply went out to the airport and waited his chance. During the war the airlines had stretched their slim fleets — United, one of the biggest, operated a total of thirty DC-3s — by flying all night as well as all day. It was the midnight flight to Chicago that offered the determined passenger the prize he waited for: a seat he might conceivably keep all the way. Sagging into his chair, he thanked heaven it was not the cold metal bucket seat of wartime. He asked for no tray of hot food in his lap, no stewardess tucking little white pillows behind his head. He covered himself with a blanket yanked from the overhead rack. Somewhere over the Sierra Nevada he fell asleep, only to jerk back to consciousness with ears ringing as the plane jammed steeply down to the Reno runway. Again he dozed, only to be shaken to wakefulness again — at Salt Lake City, at Cheyenne. Then at Omaha, where dawn broke cold over the Aksarben Airport, he was off-loaded to make room for another passenger with a prior booking.

This was first-class travel. It was the only class of travel the airline offered and the passenger, grateful that he had been lifted so far toward his destination, did not at first complain. He rounded up his luggage, and joined the line at the ticket counter. Five hours and six cups of gray airport coffee later, he caught another DC-3 that delivered him at nightfall to his destination in Chicago.

By then he was fed up. The war was over, he said to himself. And the sight at Chicago, as *Fortune* reported that year, was enough to turn the traveler from the very thought of flying again.

Chicago is the worst; its airport is a slum. Chewing gum, orange peels, paper and cigar butts strew the floor around the stacks of luggage. Porters can't keep the floors clean if people are standing on them day and night. At almost all hours every telephone booth is filled, with people lining up outside; the dingy airport cafe is filled with standees. To rest the thousands, there are exactly 28 broken-down leather seats. One must line up even for the rest rooms. The weary travelers sit or even lie on the floor. The drooping grandmothers, the crying babies, the continuous, constant push and hustle of the new arrivals and new baggage tangling inextricably with their predecessors, make bus terminals look like luxury.

In such an atmosphere the beat-up traveler, interminably waiting for some unexplained reason, has no reason but to ponder bitterly on the brilliant ad that lured him with "Travel With the Easy Swiftness of the Homeward Winging Birds."

That was Chicago's Midway — 1.3 million passengers in 1946, twenty times more than before the war. New York's La Guardia — 2.1 million passengers — wasn't much better, and New York's International Airport was still abuilding in Jamaica Bay. Airports built to accommodate hundreds were jammed with thousands everywhere. Planes were late departing, much later arriving. When Leonard Raymond, who had flown American Airlines since 1926, tried to take his dispatch case into the cabin to do some paperwork on the New York–Boston trip, the DC-3 and its twenty-eight passengers waited until a supervisor returned with a state trooper to remove the dispatch case. "We were outnumbered, outgunned, we did not take care of our steady business," United Vice President Bob Johnson said later.

Aviation technology had leaped ahead in the war, much faster than the airlines could deliver it. All the companies were taking delivery of the four-engine airliners that had played such a big part in speeding men and equipment to the warfronts. All were rushing to install the new very-high-frequency radio communication systems that the military had developed to keep pace with the bigger and faster planes. But it took time to convert the big planes released by the Air Force, and it took much longer before the airlines and federal authorities were ready to use the tools of VHF and radar that alone could handle the rush of traffic. TWA had optimistically believed that the race would be won by the swift, and indeed a handful of Constellation greyhounds, ready by early 1946 to carry you at eighteen thousand feet in eleven hours from Los Angeles to New York, could probably have been filled four times over day after day. But the airline was unprepared for the problems that had to be overcome to operate at such speeds — problems of booking seats, of pulling out engines for repairs, of passing weather warnings in time, of landing the planes at congested airports. Although TWA was ready with the only big new postwar airliner, it seemed ready with nothing else, and most air travelers continued to cross the country in the slower planes of American and United.

Before the war the airlines had coddled their customers, making it easy to reserve flights by phone, offering free handling of baggage, serving free meals and snacks on board. The old customer soon discovered after the war that his Air Travel Card, while still entitling him to buy his ticket on credit, no longer gave him a 15 percent discount on the price of his trip. In Washington and a few other cities, the airport was so close to town that he went to his plane by taxi. In most cities he boarded an airport limousine, an ancient and swaybacked vehicle with suitcases lashed on top under a flapping tarpaulin.

If the terminal building at which he arrived was the same little old wartime structure, the airport itself was altered out of all recognition. Gone was the greensward with its line of small biplanes tethered along the fringes. Concrete runways built to support the twenty-ton monsters of 1946 now covered everything, and stretched almost as far as the eye could see. The airliners on the tarmac out front also looked different. There was the familiar DC-3, of course, sitting back on its haunches. But the silvery new DC-4s and Constellations stood high and level on tricycle landing gear. The passenger had to climb tall stairways to enter the main door at the rear.

Up front the captain and copilot were already to be seen at their stations, earphones clapped to their heads. At the top of the stair stood the stewardess, unchanging symbol of airline flying. An attractive figure she was, seldom far off the classic dimensions of 36–24–36. By this time the airlines had relaxed their requirement that only registered nurses be employed, and young women with a good education were now eligible for the job. This was the era when being a stewardess or hostess on an airline was the most glamorous and sought-after of young women's jobs, with a far-above-average chance of making a financially successful marriage.

It was a demanding job — making up berths, rushing back and forth with erp cups, holding the fort when the unexpected happened. One postwar American DC-4 flying east from Buffalo ran into a violent thunderstorm. Lightning flashed all around. St. Elmo's fire lit the propellers like blue pinwheels. Big sparks like rapiers darted from the wings. A woman began to scream at each lightning flash and almost caused a panic. Stewardess May Smith was looking for a way to calm her when Metropolitan Opera tenor Lauritz Melchior, sitting in a rear seat, began to sing. His great voice drowned out the sound of the propellers and seemed to fill the cabin until the fuselage bulged. The woman turned to look. Another woman joined in the song, and then others took it up. Soon the woman was not screaming any more. She was singing.

At such moments flying was still adventure, and stewardesses needed all the help they could get. It was a rare young woman who was not hurled against the roof on at least one bumpy trip during the two years she usually worked before she left to get married. Many also practiced their skills, in the air and sometimes on the ground, at handling drunks and wolves. But the stewardess's main task, on the new four-engine planes as much as on the old DC-3s, was serving meals. Had the DC-4 advanced at anything but deliberate speed, a single stewardess would never have had time to attend to everybody's wants in a flight between Chicago and New York. It was uptight service, and so was the eating. Adapting quickly, some passengers found an unintended use for the cellophane wrapper that encased the silverware. Fitting almost perfectly over the narrow necktie of the 1940s, it shielded the tie from spills incurred as the passenger shoveled in food while keeping elbows close to the body in such cramped quarters.

With the DC-6, United's service expert Don Magarrell succeeded in getting the entrance door moved from the rear to amidships just to facilitate on-board service to passengers. This divided up the big cabin, and thereby made everybody more comfortable. No longer did the stewardess have to stand at the head of that long metal tube, with forty-four pairs of eyes staring at her. Now there were two stewardesses, one to prepare meals in the enlarged galley, the other to serve them. But the airlines still had to rush hot food to their planes and serve it as quickly as possible; efforts to prepare deep-frozen food which could be thawed before serving had been frustrated because of the weight of the stove required. As it was, a typical DC-6 or Constellation flight carried 400 pounds of food and trays alone, and after liquor began to be served, bar stocks added a further 104 pounds. A total of 504 pounds for alimentation, to which the weight of those serving might have been added, mounted up to a serious factor in air transport economics, especially since every pound saved aloft was worth fifty dollars a trip to an airline in 1950.

Even when the new planes were pressurized and rode high over the weather, they had to come down to land. Then the promised comfort and ease of air travel were rudely ended. Delays in arrival at major airports were particularly trying. About half of all flights landed more than thirty minutes after the announced time. American Airlines Captain Hi Sheridan acknowledged in 1947 that airline schedules were a joke — "a fond and faint hope of the management." It was years before the CAA, the Air Force, and the airlines could settle their dispute over how much radar should be used, and meanwhile the air traveler was at the mercy of outmoded radio aids that worked slowly in bad weather, if at all.

The airways, just ten miles wide, converged on major airports like the ribs of a spiderweb. Airliners eased down through the overcast by following the beam to the cone of silence; there they made their procedural turns and then felt their way down to where they hoped the runway was. The maneuvers to make this final approach alone required from six to twenty minutes. At best only ten planes an hour could land, and sometimes only three or four reached the runway in an hour's time.

The rest were overhead awaiting their turn, slowly flying figure eights over some fixed point on the ground established by the CAA's traffic control men. The planes, mushing back and forth in their soupy limbo, were stacked every thousand feet up to a height of 8,000 feet. Whenever the tower passed the word, the pilot dropped his plane 1,000 feet. At the bottom of the stack he finally got the green light to head for the runway. At places like Washington there were sometimes two stacks, in which case the incoming plane had to work its way down twice before landing.

Passengers caught in this pickle could hardly do what Admiral Chester Nimitz did. Flying in a Navy plane from South Carolina, the victor of Midway was informed there was a four-hour stackup at Washington's National Airport. Instead of waiting he landed at Patuxent Naval Base in eastern Maryland, took lunch, then reclaimed his place in the stack and landed after only thirty minutes. Sweating it out in these lurching, groping orbits awakened all the old fears in nervous passengers. There were times when planes trying to get into La Guardia Airport reeled around in as many as four stacks, and other times when they were sent to such alternate fields as Albany, Philadelphia, or even Boston. One pilot counted twenty such diverted airliners on the field when he landed one day at Boston. Passengers with appointments in Manhattan found themselves as far away as when they began their journeys. When they caught trains to their destinations, they vowed they would stay on the rails. In one day 400 United passengers were dumped at alternate airports, at a loss to the airline of $6,000. But still the airlines, and even more the pilots, adamantly opposed adopting the Air Force's all-weather GCA landing system.

At 2:20 P.M. on December 29, 1948, the President of the United States was successfully landed by the ground control method with almost no delay. On that afternoon, when commercial airliners were milling around in the clouds, President Truman's DC-6 *Independence* arrived over Washington from Kansas City. The pilot was Lieutenant Colonel Francis W. Williams, and Truman gave him permission for a GCA landing and then went into the cockpit to see how it was done. From an altitude of ten thousand feet the *Independence* went into fourth position in the landing pattern. On the ground controller Thomas Breznay "talked" the presidential plane to a perfect touchdown although, the *New York Times* reported, "even the seagulls were grounded." Visibility was three-eighths of a mile, with fog and haze so dense that cabinet members waiting at the terminal only recognized the plane when pilot Williams reversed his props with a roar to slow the plane as it rolled past on the runway. "Were you worried?" a reporter asked. "No, oh my, no," Truman replied. But even after the President of the United States was brought down successfully by radar, ordinary passengers had to wait while the airlines gradually worked out the bugs in their own, pilot-managed all-weather landing system.

Overseas travel offered the greatest contrasts in this time of change. Only a handful of passengers had crossed the Atlantic before the war, almost every one of them a VIP traveling on official business of one kind or another. Now the spacious and luxurious flying boats of the 1930s had been left on their ramps at La Guardia's Marine Terminal, and the peacetime passenger looked for his chance to fly on the landplanes whose superiority had been demonstrated in the war.

Even before commercial service was launched, TWA — in the first of the gaudy junkets organized by the airline industry to celebrate all of its new beginnings after V-J Day — flew a load of Washington big shots across to Paris in late 1945. Aboard this first Constellation crossing were Postmaster General Robert Hannegan, members of Congress, government officials, corporation directors. The plane set records on both crossings, halving Lindbergh's flying time to Paris on the eastbound flight. But it was a Constellation of the early 049 type that Jack Frye and Howard Hughes had ordered from Lockheed back in 1939 when they were thinking only of TWA's transcontinental service. It had to stop at Gander both ways. On the westbound trip Captain Hal Blackburn informed his passengers that they would be spending the night in Ireland. Their stopping place was Shannon, and they put up at the St. George's Hotel across the river in Foynes. Since this was square in the middle of County Limerick and supplies were ample of what Blackburn called "that famous Irish blood plasma that sells for two or three dollars a fifth," he organized a limerick contest. The prize went to poet Archibald MacLeish for:

> There was a young lady from Foynes
> Who was fond of collecting coins.
> The thruppenny bits
> She made with her wits,
> The shillings and pounds with her loins.

That was a jolly beginning, and that fall a number of Americans hurried to the other side to see how relatives had made out in the war. But then the airlines ran into the North Atlantic winter. Many flights were canceled, more delayed. On top of this an American Overseas DC-4, trying to make the obligatory fuel stop in Gander's fiendish weather, rammed a Newfoundland ridge, with the loss of thirty-nine lives. Then TWA Captain Howard Tansey overshot the Shannon runway in fog, and twelve more died as his Constellation crunched down in a marsh. On some trips that winter, Pan Am Captain Chilie Vaughn said later, the crew outnumbered the passengers. And the ride hardly matched up with the glossy-finished advertisements for air travel. Flying high in northern latitudes, Pan Am Captain Hack Gulbransen took good care to wear his fleece-lined, buckle-up flying boots, and said later:

The floor behind the cockpit was a kind of skating rink. There was ice on it. And back in the passenger cabin the first thing the steward did was hand three or four blankets to each passenger. I'd go back for a visit and ask, "Is everybody all right?" And they'd be huddled up in blankets.

But neither then nor later did the airlines ever lose a plane over the Atlantic. The secret, of course, was four-engine airliners that could continue to fly when one and even two engines gave out. And by the summer of 1947, business began to pick up.

The passenger took his departure from a La Guardia terminal that was abustle with the casual confusion of the air world. The crowds hurried in and out the entrance in cotton dresses and business suits, hatless and carrying paper parcels as if setting out for a day's picnic in a beach wagon. There was little of the formality of leave-taking that marked the departure of an ocean liner or even a train. The traveler scuttled down a subway passage with crowds of other passengers. Then he made a brief dash in the sunshine to the boarding ramp, up the stairs, and into the dark cabin of the plane. Inside were soft carpets, upholstered armchairs, curtains, and only a little round peephole of a window to look out at the sky. Compared to the old trimotors with their big, sliding-pane windows, these airliners had shut out the sky.

So there was no blast of air, no deafening roar, no need to stuff cotton in the ears for these encapsulated passengers. They shed coats, unfurled newspapers. When the plane left the ground no one waved, no one held his breath. Then, when they were ready to unfasten their seat belts, the captain stepped through the door at the front of the cabin. Although intercoms on planes had come in during the war, none yet functioned well enough for use as a public address system. Poised and occasionally smiling, the captain gave the passengers a short résumé of the flight, talked reassuringly about the weather to be expected, and introduced the pursers and stewardess. Hal Blackburn of TWA said at the time:

Everybody on a plane is scared, because for us flying is unnatural. Probably not for our children, certainly not for our grandchildren. I know quite a lot of passengers who talk as if they were unafraid, but I doubt it. Let a radio antenna come loose and start slapping against the side of the plane, and you'll see them getting mighty concerned.

Around 6:30, when everybody was eating lamb chops on plastic trays, the captain made another visit to the cabin. This time he spent nearly an hour chatting with them, one after another. Blackburn said:

If you're worth your salt, you're thinking about those people back there. You've got to think of their stomachs and their ears — and their emotions too. If something unusual happens, I tell them about it. It's the unknown we're all scared of.

In those days pilots sometimes invited a passenger to visit the cockpit, and often offered the same privilege to children aboard. Airline captains had yet to hear of skyjackers, and the CAA had not yet ordered the doors to cockpits locked. But there were some surprises. Once as Blackburn was making an instrument approach to Zurich through fog and darkness, a drink-thickened voice behind him said, "Look here." Blackburn and the copilot instinctively turned around, and the frolicsome passenger snapped a flash photograph. Although momentarily blinded, Blackburn and his copilot regained their vision in time to continue the approach and land the plane without mishap.

After dinner's end somewhere over Nova Scotia, only tiny spotlights in the ceiling of the Constellation needled the darkness. Outside gray clouds swathed the plane until the engines suddenly slowed, signaling the descent to Gander. Then at 11:00, with all America behind, the passenger pushed back the seat to the precise angle of ease and tried to sleep. At 3:00 by his New York watch he woke from partial rest to brilliant sunshine. In 1927 Lindbergh had remarked, "There's only about two hours of complete darkness over the ocean." All the passengers stirred, coffee was passed.

At 6:30 passengers looked out and saw Ireland, offshore island of the Old World. On an ocean liner the passenger always had a sense of the element he was passing through. All those miles of real salt water helped bridge the intervening distance. But in an enclosed airliner he felt no sense of the element — it was as if he had passed through time, not air. The reassuring walls of time, the successive mileposts of sunrises and sunsets had been obliterated in the swiftness of flight. Looking down on Ireland's thatched roofs, sheep and cattle in the fields, haycocks, the early Constellation passenger could feel light-years away from America. "It's all a dream," travelers would say. The meal they ate at the long table in the Shannon lounge was ham and eggs, but not identified. Was it breakfast or lunch? By the New York-set watch it was 7:00 A.M. By Irish time it was noon.

The plane flew on, crossed the port of Cherbourg, passed the beaches of France still strewn with the wrecks of the Normandy landings. And the next thing the passengers knew, the voice was saying, "No smoking please. Fasten your seat belts. Passengers disembarking will please have their passports, their custom forms . . ." Crossing in the 1940s was a miracle that only the most hardened airman could have called routine. To the musician Maurice Dumesnel, it was "the great adventure." When he looked out into the midocean night at eighteen thousand feet, the stars were so bright they appeared as blotches of gold, and moonbeams cast a fairy light upon the endless carpet of clouds below. Only the Bach *Aria* or the slow movement of Beethoven's *Pathetique* Sonata could have expressed his feelings, he said later.

Pressurization allowed the Constellation's passengers to travel among the stars. But there were uncertainties about the severe interior pressures created on the skin of the plane. One night in 1947 TWA navigator George Hart was standing on a stool with his head up in the plane's Plexiglas astrodome taking star shots. Possibly his sextant knocked against the brittle Plexiglas. Whatever the reason, the Plexiglas blew out, and the pressure of the cabin air rushing out that small hole sucked Hart through, and almost caught up the stewardess who was steadying the stool. Only quick work by another crew member who caught hold of her leg saved her. Hart, ripped from the plane's capsuled warmth by the brief, violent force of what the engineers called explosive decompression, fell to his death in the cold Atlantic night.

Once TWA Captain Alexis Klotz was flying eastward when his flight engineer complained of trouble in getting the cabin pressurized. Then the pressure started building up fast. All of a sudden the hostess ran up to the cockpit to say that a woman passenger was stuck on the toilet and could not get off. Klotz understood in a flash that the woman's bottom had sealed what a mechanic had failed to seal properly before takeoff. He ordered the pressure turned off long enough for the poor woman to be released and got the crew to stuff the leaky toilet with newspapers so that pressurization could be maintained for the rest of the trip.

Pressurized planes were slower to be used on other international routes. In the Pacific it mattered less to the passenger because the weather was steadier. But in Latin America passengers had to take their lumps and like them. Crossing the Andes in 1947 from Chile to Argentina was still a hair-raising experience. Everybody sucked oxygen from tubes by their seats as the DC-3 rose to eighteen thousand feet, Willard Price said later, and women passengers frequently uttered little screams each time the plane jinked or twisted close to a granite precipice. Threading passes between Bogotá and Guayaquil, Dorothy Day clutched her six-month-old baby tight as the DC-3 rose and fell in the air currents. As usual, everybody got airsick but the baby. After one rocking, pitching, eighty-minute ride from Rio, Henry Hogg arrived at Sao Paulo so deaf he could not keep his appointments. Even after twenty-four hours, when he boarded the same delayed DC-3 for the trip onward to Uruguay, he could not hear his watch when he held it to his left ear. By the time the plane delivered him to Buenos Aires a day later, he noted in his diary, "I can now faintly hear my watch ticking, but the ear still rings." Hogg, a British import-export man, nonetheless noted, "A fine trip."

Even before the war, Pan Am had invited Americans to the first "air cruises" around South America. There were few takers. Now a cloud appeared, no bigger than a man's hand, that was soon to engulf international flying. Up to this time travel agents had pretty much devoted themselves to arranging rail and steamship trips. But about this time the tourist trade began to change. A few agents, persuaded by Pan Am, began to sell "package tours that whisk you to places that used to be a week away" — places like Bermuda, the Caribbean, Central America, Mexico, "even Europe." The first year after the war the trips had to be called off because Pan Am ran short of planes. In 1947 the lure was two-week air vacations that included "14 full days in Hawaii, Mexico, Cuba." No other form of transport could deliver such instant dreams.

Pan Am, still carrying the overwhelming proportion of U.S. overseas air travelers, announced the first around-the-world flights in 1947. The service couldn't quite circle the world, of course, because Pan Am was not allowed to fly across the United States. But you could easily go all the way simply by riding United or American or TWA cross-country. Los Angeles travel man Bert Hemphill thereupon organized the first Round the World Air Cruise. A dozen signed on. Starting out across the Pacific in DC-4 "sleeperettes" (reclining chairs laid flat during the sleeping hours), these hardy pioneers of airborne armies to come paid $3,750 to blaze the global sky track. They hit Peking the day the Communists took over, landed in Europe amid the Berlin Blockade, and got back to Los Angeles without a single forced landing and with only minimal delays fifty-two days after takeoff. Even without sticking to Pan Am's scheduled 138 hours, that was beating Jules Verne's eighty days with ease. Then in 1949 Pan Am introduced the double-decked Boeing Stratocruiser as an airborne challenge to ocean-liner luxury in transatlantic travel. Round-trip fares were $720, and members of New York's smart set took to flying to Europe just for the weekend.

Part V

Cruising
(The 1950s)

27

Air Coach Arrives

"*Here* they come," *Life* magazine proclaimed at the start of the summer of 1955. On the cover was a photograph of the new transatlantic liner S.S. *Independence* thronged with U.S. tourists departing for European vacations. Inside, along with pictures of gay crowds at dockside send-off parties, was a stunning photograph showing seven stately ocean-going palaces drawn up side by side at Manhattan piers to receive yet more thousands of Americans sailing for European holidays.

Precisely one year later, *Life* published a special issue — the "Air Age." Celebrating "man's new way of life in a world reshaped by the conquest of the skies," the most popular U.S. magazine threw its hat in the air for air travel — more Americans were now voyaging to Europe by air than by sea.

The reason for this sudden and dramatic change was that the airlines had changed. Starting around 1948, commercial flying began to turn into the kind of travel that the general public could afford. The big postwar Douglas and Lockheed airliners made mass air travel technically possible. Competitive pressures made it economically possible. Instead of vying just for Pullman passengers, the airlines went after the railroads' low-cost coach business.

The industry dubbed its new service "air coach." Many airline men did not like it. Back in 1940 United had toyed with low-fare service in second-class planes at off-peak hours in California, and dropped it. American's C. R. Smith said later, "Air coach was just a device for cutting fares," and others played it down as merely a "price-cutting innovation." But not Juan Trippe. Pan Am's leader said later that the introduction of air coach ranks after Lindbergh's flight and the onset of the jetliner as the third major milestone of airline history. "The importance of that change, which preceded the arrival of jets, was that for the first time the ordinary man began to fly with us," Trippe said.

On the international routes air coach, called "tourist class," not only lifted air transport over sea transport as the most popular means of crossing the Atlantic, it also enabled both Pan Am and TWA, for the first time, to stand on their own feet economically and get off government subsidy. And by dint of air coach fares, domestic airlines drew ahead of the railroads in number of passengers carried by 1955.

It was competition from the nonskeds that forced the airlines into their discovery of mass travel. The regular airlines had their schedules to maintain no matter how few riders showed up. They made a go of it by charging fares at about twice their costs, and turned a profit whenever their planes flew with more than half the seats filled. The nonskeds, under no such obligation, seldom sent their planes off until they were full. Full loads enabled them to charge lower fares and still make a profit. Gradually some of these nonskeds began flying passengers all the way across the country, first in DC-3s and then in DC-4 "air coaches." Offering a bargain ninety-nine-dollar fare — 30 percent less than the scheduled airlines — they usually found ways to delay their odd-hour takeoffs until every seat was full. At such prices, people who had never flown before bought tickets.

Airlines took note. Specifically, Capital Airlines, a lesser line long trodden under by Big Four competition, took note. In November 1948 Capital launched night coach service at much lower fares, and quickly built up a substantial trade between New York and Chicago, as well as other big cities. Seeing no recourse, the Big Four swallowed their sneers and followed. TWA, self-styled "airline of the movie stars," did a turnabout and offered transcontinental coach flights in early-model Constellations at bargain rates. Captain Russ Bowen said later, "We called 'em cattle class. We just lined the customers in — two double rows, stem to stern. Passengers could save quite a lot between California and Chicago. Utilization on those aircraft was tremendous — better than ever before or since."

Overseas, where flights were longer, the importance of filling those empty seats was even more compelling. When nonskeds began advertising flights between Puerto Rico and the mainland at fire-sale prices, Pan Am took the plunge and started tourist-class DC-4 service between New York and San Juan in September 1948. By seating passengers five abreast and cutting out galley equipment altogether, the airline made room for sixty-three passengers. Only one attendant was posted in the cabin. The only food provided for the twelve-hour trip was little packets of sandwiches and fruit that passengers could buy before going aboard. But the results were a revelation. At fares of $75 instead of the standard $133, the number of passengers quintupled in five months. This was in fact the start of the world's first migration by air. In the postwar years, 5 million Puerto Ricans moved to the United States. Two-thirds of them made the trip aboard Pan Am's cut-rate San Juan–New York service.

The DC-6 and Connie were first thought unsuitable for five-abreast seating — the DC-6 seemed too narrow and the Connie, while fat in the middle, tapered at both ends. So the manufacturers and airline engineers lavished their ingenuity on ways of crowding in more rows of seats. If "stretch" was a word constantly on their lips, another was "pitch." "Pitch" was their term for the distance from the back of one seat to the seat immediately behind it, and before they were finished protocols were drawn up that bound the airlines to conform to these measurements down to the last millimeter. On the first postwar planes the pitch was a wallowy forty-two inches. But TWA and American narrowed it to thirty-nine inches for their coach flights, and soon some drawing-board accordianist squeezed it down to thirty-four inches.

In 1950, on the strength of Juan Trippe's canny conviction that the San Juan success showed where the future lay, Pan Am ordered stretched DC-6s with the idea of commencing tourist flights on the Atlantic. The new planes turned out to be the sensationally profitable DC-6Bs — with one more row of seats and five-abreast seating, or a total of eighty-two seats for the crossing. Mobilizing Washington's support, Trippe proclaimed his proposed service was "simply an application of the tested American principle of mass production — achieving a wider and wider market by lowering the unit cost of the product." Perhaps more to the point, the growth rate of Atlantic travel, at first 15 percent a year, had fallen by 1951 to a mere 6 percent. This was the argument that brought his competitors within IATA into line. It looked as if the Atlantic airlines, now a dozen in number, had about reached the limit of people willing to pay the $711 first-class return fare to fly over the water.

So the Europeans fell in with Trippe's proposal. They agreed to start a new tourist class. The fare would be slashed, and to make this possible passengers would be seated much closer. A thirty-four-inch pitch between rows was prescribed, the airlines being left free to decide how many passengers should sit abreast. The free baggage allowance was cut from sixty-six to forty-four pounds. Meals were made simple and inexpensive. Free wines and spirits for tourist passengers were ruled out.

In spite of the detailed prescriptions, a row broke out the moment tourist flights started. The rules said that sandwiches could be served. The Scandinavian airline began offering the delectable open-face sandwiches, decked out with anchovies, herring, and fancy cheese, that are standard to Nordic cuisine. Thereupon TWA, stuck with the usual array of doughy indigestibles that everybody else called sandwiches, cried foul. TWA lodged a formal complaint with IATA. Fined $25,000, the stiffest penalty yet levied, the Scandinavians thought the publicity well worth it.

Sandwiches and all, tourist class brought a fantastic upsurge in Atlantic air traffic — up 17 percent by 1953, 9 percent more by 1954, then up 19 percent in 1955, and 20 percent more in 1956. By 1958 low-fare air travel constituted two-thirds of the North Atlantic business, and Pan Am found that between 67 and 77 percent of those flying tourist had never been in the air before. Having tapped a wholly new group and struck a mass market, Pan Am regained its proud position as the leading international carrier.

When smaller European airlines like the Swiss and the Belgians insisted that they lacked the equipment to operate separate first-class and tourist flights, IATA let them operate their planes part first-class and part tourist. This brought a further change. Pan Am's rival TWA not only followed the European example on the Atlantic, it also broke open the U.S. market by seating first-class and coach passengers on the same plane on its domestic flights. TWA sales chief Oz Cocke found this a way to respond positively to pressures from local communities and their congressmen, who kept demanding the popular coach flights from their cities even when traffic did not seem to justify both regular and bargain-price flight stops there. Thus it happened that, as soon as tourist flights got started, low-fare travelers everywhere rode on the same planes with first-class passengers. The airlines simply pulled a curtain across the aisle after takeoff to separate the sections. The bigger section had to be allotted to the low-fare riders at the back of the cabin. At home and overseas, lower-price air travel took the play. Democratizing air travel, coach service made the airlines *the* growth industry of the 1950s.

As air travel grew, tourism grew. Americans flying overseas had to have places to stay when they got there. In Europe this was not at once a problem, but in Pan Am's home ground in Latin America hotels were in short supply. Oddly enough, not Pan Am but the U.S. government first thought something should be done about this. Only a few steamship lines had ever put any money in hotels outside the United States. But in the war years Pan Am had got into Franklin Roosevelt's good graces building airports for him. One day the President remarked to Nelson Rockefeller, his youthful coordinator of inter-American affairs, that it would be a good idea to build hotels where none existed in Latin America and Pan Am should build them. When peacetime flying resumed, the Good Neighbors could pick up needed dollars if there were decent hotels in the places Americans might visit.

When Rockefeller passed along the boss's idea, Trippe replied that the U.S. hotel industry should do the job. Rockefeller checked and found no interest among hotel concerns. Then Trippe said, "If we must, how about Jesse Jones of the Reconstruction Finance Corporation coming through with some of those 4 percent loans of his?" The upshot was that Pan Am formed the Intercontinental Hotel Corporation (IHC) in 1946, and with the help of the new Export-Import Bank instead of Jesse Jones's RFC, set out to build some hotels.

Trippe borrowed the president of the Waldorf-Astoria to get the show started. Somehow the first hotels went up where there already were hotels — in Caracas and Bogotá. Trippe saw at once that it was good business and said with utter accuracy, "Some day it will make more money than Pan Am." By 1954 IHC was operating nine hotels around Latin America and turning its first profit. On the edge of the 1950s' tourist boom it opened still more, until by 1961 Pan Am's subsidiary ran some twenty-six hotels on six continents. By then the airlines were falling over themselves to get into the lucrative business, and there were Hiltons, Hyatts, and Sheratons to house deplaning Americans almost everywhere.

Even when air tourism surged, airline men held to their old adage that theirs was a commodity that had to be sold. In the 1930s they had used credit cards as a key weapon in capturing businessmen travelers. In the 1950s they went after the mass air travel market by inviting everybody to fly on credit. Pan Am sales chief Willis Lipscomb's catchy come-on was: "Fly Now, Pay Later." Under the scheme he put into action in 1954 with the aid of the Household Finance Corporation, all the customer had to do was walk up to a Pan Am counter and say where and when he wanted to go. The agent filled out the ticket. Then he trotted out the installment payment contract, which the prospective traveler signed on the dotted line just as if he were buying a car or a clothes dryer. It was another Pan Am coup. Of the first year's contracts, a delighted Trippe told shareholders, 70 percent "represent[ed] business which would not otherwise have been available to your company." Collection losses were negligible. All the other airlines rushed to adopt the scheme. Nothing signified better that air travel had become part of the American way of life than the fact that people now flew on the installment plan.

In 1950, twenty years after Walter Brown forced them to start carrying passengers so they would not be so dependent on mail pay, the airlines finally pulled free of Washington's supporting hand and began to fly without concealed subsidy from the U.S. Post Office. Not all managed to emancipate themselves at once. Most of the lesser trunk lines continued to receive what after 1950 was openly labeled as subsidy in addition to regular "service pay" for carrying mail. And the biggest domestic lines still needed a year or two to pay back the extra sums the Post Office Department advanced them to tide them over their hectic postwar years. But in 1952 TWA, making a sensational comeback from the brink, proudly announced that it was "off subsidy," and in 1957 even Pan Am, notwithstanding its unprofitable services to out of the way places like Africa and Alaska, informed stockholders, "For the first time in 30 years no subsidy was included in operating revenues."

For the airlines, the going was great. Besides the growth of business travel, citizens winning new leisure — shorter workdays, longer weekends, paid vacations — began spending some of the time and discretionary dollars they did not put into free-standing suburban houses and free-wheeling station wagons into flying holidays. Harvesting profits, the

Eddie Rickenbacker in 1939

airlines not only declared dividends but split stock, and sold out new issues every time they offered them. These were the years when pilots lived like country squires in places like Bucks County and Newport Beach, and raised cattle, bought oil leases, and sailed yachts while not off on their three-week trips to Bombay or Helsinki or Brazil. These were the years when stewardesses got their man.

The growth of air traffic brought new kinds of competition among the airlines. Northwest, leaping the Great Lakes as nothing on wheels could, became the nation's fourth transcontinental flying into New York and Washington. Western, almost devoured by United in the rocky postwar years, won the right to fly north from San Francisco to Seattle, served free champagne to riders, and ran United a good race for north-south travelers along the Pacific coast. The three big transcontinentals not only gained common access to both San Francisco and Los Angeles on the west and New York, Philadelphia, and Washington on the east but to practically every major city between. As air travel expanded the CAB gave the smaller trunks their chance to grow. Soon as many as five competing lines flew between many cities.

Up to a point this simply meant that the rival lines each got their chance to fatten up on a domestic market that was expanding at the rate of 15 percent a year. In these years all the trunk lines offered pretty much the same planes — four-engine DC-6s or Constellations on the longer runs, Convairs or Martins on the shorter hops. All could offer passengers about the same departure and arrival time. But there could still be differences, notably in the way they dealt with their riders.

In the earlier years of airline growth, each operator had its own special areas — Braniff in Oklahoma, Northwest in Minneapolis, American at Boston, United at Cleveland, and so on. More than others Eastern, dominating the eastern half of the country, had maintained a monopoly on many routes, and the ledgers showed it. This fueled sentiment for opening up these routes. So did Captain Eddie Rickenbacker's cavalier style of flying them economically without much concern for passenger well-being. At the start of the new era CAB granted first National and then Northeast the right to compete on Eastern's New York–Miami gold mine route. Big, tough, and efficient, Eastern more than held its own against these interlopers.

A more significant beneficiary of CAB efforts to open up Eastern's territory was Delta. So small it was put out of business when Walter Brown drew his 1930 airline map, Delta was one of the little fellows that resurfaced after the 1934 airmail contract cancellations. For a long time Delta lived strictly as a regional airline, flying between such cities as New Orleans, Birmingham, Atlanta. Then, in the reshuffle to give Eastern more competition, Delta gained a through route from Chicago to Florida in 1945, and began to attract a lot of passengers.

To a striking degree Delta's advance reflected the contrast between the airlines' top personalities. Captain Eddie, America's ace airman in World War I, knew all the tricks for making planes fly profitably. More than others, his pilots flew straight from airport to airport, through clouds and not around them, saving both time and money. Delta's C. E. Woolman, also an authentic airline pioneer, and a crop-duster before he ever carried a passenger, was as gentle as Rickenbacker was brusque. Though careful enough with his nickels and dimes, he never said an unkind word to anybody; the airline he ran offered courtesy with a Southern accent.

In 1946 a $5 million credit from Atlanta's Citizen and Southern Bank, largest yet put together by a southern lending institution, enabled Woolman to put high-flying DC-6s on the Chicago–Florida run and its other longer routes. Woolman continued to shuffle in and out of Washington hat in hand as representative of a poor-boy airline having trouble making it go. Except for the pilots' association, Woolman's airline had no unions edging the alacrity of its "Yes, suh" service. Customers noted approvingly that mechanics dropped their wrenches to hustle baggage out to the waiting room.

Such things counted because the nature of air transport was changing. In the new kind of operation the margin of opportunity lay in how attractive you could make it for people to fly your airline. Woolman was quicker and more effective at finding ways to serve passengers. In the next years the contrast sharpened between just flying airplanes on the one hand and currying the public on the other. Of course, not all the charm in the world could substitute for sound routes, and Delta's gains were invariably founded on yet further authorizations to fly to important places like Miami and Houston and finally New York. There are Georgians who say that the airlines "made" Atlanta. But it was Eastern that moved its headquarters from Atlanta to Miami, and Delta that left Monroe, Louisiana, to take Atlanta as its capital just when the rising flood of air travel demonstrated the centrality of that city. Atlanta, just the right distance between cities like New York and Houston and Chicago, became an aerial crossroads. Almost every big company opened offices in Atlanta, and Georgia filled with defense plants. C. E. Woolman's airline both fueled and fed on these big changes. But the rise of Delta to the point where, in September 1976, it unseated Eastern from its Big Four eminence was more than a symbol of Southern emergence. It was a phenomenon of the new kind of airline competition.

28

Nonstop at Last

In the 1950s the airlines of America reached for planes that would fly nonstop — nonstop across the country, nonstop across the Atlantic. The most aggressive of airline enterprisers, C. R. Smith, having uncharacteristically backed out of the Atlantic race, led the drive for the plane that would deliver his customers direct from coast to coast. Not content with the DC-6, which was ringing the cash register like no other airliner, he went to his old friend Donald Douglas for still more "stretch." Douglas gave it to him — in the DC-7, forty inches longer than the stretched-out DC-6B and with room to carry fifty-eight passengers. He was not in time to be first, however. Lockheed had stretched the Constellation into a Superconstellation nineteen feet longer than earlier models, with room for sixty passengers, and TWA put it into service between New York and California on October 19, 1953, a month before Smith launched his first nonstopper.

One thing made possible these feverish moves: the Wright turbocompound engine. Built for the Navy, the first really new engine to go on American airliners since World War II, this was the biggest, most complicated and most powerful piston engine ever hitched to a commercial plane. It was the piston engine's last hurrah.

Piston engines drive your automobile, but it has been a very long time since airplane engines bore any resemblance to what's under the hood of your car. From the time when the Wright Whirlwind propelled Lindbergh across the Atlantic, U.S. manufacturers preferred a radial design to the in-line type of engines they made in Detroit. Under the military's ceaseless demands for more power, the stubby little grinders with cylinders ringing the propeller shaft grew into snub-nosed monsters. By dint of fancy new alloys, higher-octane fuels, turbosuperchargers, and advances in propeller art, manufacturers brought these engines up to 2,000 horsepower on some of the fastest climbing wartime fighters. Then, having doubled the concentric rows of cylinders, Pratt & Whitney added so many more rows that people took to calling them corncob engines. And after the war, when U.S. airmen hesitated to shift to jets, both Wright and Pratt & Whitney wrung yet more horsepower out of their existing designs to give each successive model of DC-6 and Constellation a little more oomph. The Wright turbocompound, which powered the last and biggest piston airliners, marked the summit of this long and intensive evolution.

As far back as 1910 engineers had dreamed of building a compound engine — that is, an engine that captured exhaust heat for extra power to help turn the drive shaft. The turbosupercharger suggested one way. Harnessing energy from the hot exhaust gas to turn a shaft spinning the supercharger blower, it helped the engine perform better in the thin air at high altitudes. In the turbocompound engine Wright used three small turbines of the same type not to drive a supercharger but to put extra power straight into the main crankshaft. The return of energy from the exhaust gas was a form of jet propulsion: the engine's conventional piston power was supplemented by thrust — something like 130 pounds of it. On the face of it, such compounding was a formidable feat. This extra thrust was "free," obtained without any addition of weight — in effect, something for nothing.

In slightly different versions the turbocompound went into both the Superconstellation and the DC-7. Four of these 3,514-pound monsters, each generating 3,250 horsepower, had to be primed for takeoff. The three blowers of the exhaust-capture system were like nothing mechanics had seen before. "One hundred forty-four sparkplugs to change," wailed American's line chief Abe Hoyt. TWA's Operations Vice President Floyd Hall said later, "We had ten engine failures a day." TWA Captain Mo Bowen said he had never seen engines pulled out so often. With these enormous, Rube-Goldbergian power plants, TWA's Maintenance Vice President Ray Dunn said later, "we were taking piston engine technology as far as it could go."

But once the engineers sent these planes out to the passenger gates, they flew. Watching their turbocompounds like hawks, and switching planes at first inklings of trouble, American Airlines was able to make good on its nonstop claims. Eastbound flights kept their scheduled time of seven hours, fifty minutes, but westbound trips met head winds. Winter winds were particularly strong, sometimes as much as 100 to 150 knots. Under such buffeting American's DC-7s often took eight and a half to nine hours to reach the Pacific coast. Because a CAA regulation barred any crew from flying more than eight hours, the airlines had to ask Washington for permission to keep their crews aloft for nine hours in such cases. Permission was granted but not before the pilots struck American for more pay and pried away some of the DC-7's productivity gains for themselves.

Once the eight-hour limit had been relaxed, both the DC-7s and Superconnies began delivering the benefits of nonstop crossings to thousands of air travelers. But it wasn't easy. Like the DC-6 before it, the DC-7 labored hard to climb to the high altitudes it had been designed for. With thousands of pounds of fuel and sixty passengers aboard, it proved virtually impossible to reach 25,000 feet; and even after the plane had burned most of its fuel, the strain on the engines of operating such long distances at 350 miles an hour was simply unacceptable. For a time American instructed its pilots to fly no higher than 10,000 feet except when crossing the mountains. As it was, the huge turbocompounds worked so hard, American's Captain Ham Smith said later, that oil consumption got to be a bigger problem than fuel consumption. Many a time, Ham Smith said later, he feathered one of the four monsters before the end of the trip so he could be sure he had enough oil left to complete the journey.

Headaches and all, the big engines could haul more people longer distances, and that was what the airlines wanted in the booming 1950s. Once the DC7s and Superconnies had shown they could pound nonstop across the country, Pan Am and TWA were beating on the doors of Lockheed and Douglas to give them versions of these planes that could cross the Atlantic without stopping.

In the early 1950s TWA gained a slight edge in the Atlantic race when Lockheed produced a first, turbocompound version of the Connie that enabled it to advertise "fastest trips to Europe." But these planes still had to stop once on most crossings. Then Pan Am acquired the DC-7B, with more fuel capacity. On the first proving flight it flew from New York to Paris nonstop; but it had to stop at Gander flying back, and when it went into service in 1955, passengers remarked on the noise and vibration in the cabin. Thereupon Douglas agreed to a major design change. It revamped the plane's wing, adding four feet at each side of the fuselage. This did more than add space to store fuel. The wider span enhanced the efficiency of the wing as well. Another consequence was that the engines were moved five feet farther out from the fuselage. That considerably reduced the racket and vibration in the passenger cabin.

With the DC-7C Pan Am had the first true transatlantic airliner. It could fly nonstop across either way with a full load. As soon as it entered service in 1957, Pan Am lifted its share of all transatlantic air passengers to 30.6 percent and regained first place over TWA.

TWA fired right back. Lockheed revamped the Constellation wing and put on a goosed-up version of Wright's giant engine. "We aimed at making it a transoceanic high-speed transport that would overshadow all others," Kelly Johnson said later. The same drawing board that produced the U-2 and SR-71 for the Air Force in these same years added another twenty-seven feet to the wingspan of the durable 1940 design. "Hughes bought it," Johnson said later.

The DC-7C flew the Atlantic nonstop in 1957.

He liked the big wing, the big plane. But I got mental fatigue working nights on that plane. Hughes would clear out a ballroom at the Beverly Hills hotel. Then, when nobody could possibly get near to eavesdrop, we would sit at a card table in the middle of the floor. He'd ask all the technical questions he could think of — why was the wing contour the way it was? Why was the power-booster control located where it was?

He'd phone me and phone me. I had to ask Bob Gross to call him off.

Finally Lockheed's president told Hughes: "If you ask any more changes in a design already approved by your engineers, I'm going to cancel and you can sue me if you like."

In this version the Connie tipped the scales at 156,000 pounds — just about double the weight of the original model that Johnson designed seventeen years earlier. But it had no more seats than the preceding model, and Hughes's engineering aide Bob Rummel later said, "Economically, the last Constellation was a flop."

There was much more to the story than that. Back of the disappointment over the last and biggest Constellation were the frustrations over a failed attempt to put a new kind of power plant into the plane — and the perplexities over a decision commercial aviation had been putting off for years.

After World War II the question that overhung all others for commercial flying was when and how jet power would come to air transport. Compared with Germany and Britain, the United States had been very slow off the mark in turning to jets. America had concentrated hard on producing all possible piston-engine warplanes for victory. But after the war, when it alone possessed an undamaged industrial plant and plenty of unused resources, the United States still lagged behind in putting jet propulsion to commercial use.

As late as 1943 practically no one on this side of the water seems to have realized how tremendous an increase in speed the high-flying turbojet made possible. In the U.S. industry's experience, increases in airplane speed had been won slowly and at great cost: doubling the power of the piston engine had taken nine years, and was only accomplished by nearly doubling its weight, so that a 25 percent increase in speed was all that was achieved. One prominent engineer, hearing about the British turbojet, correctly calculated that the fuel consumption of such an engine would be hopelessly high at 400 miles per hour, the maximum speed then contemplated for fighter planes, and would become competitive only at 500 miles per hour or above. But 500 miles per hour — the speed at which jet engines topped pistons in effective power — he dismissed as quite unrealistic. That such speeds were unthinkable thus became the reason for not venturing into jets.

Airline chieftains, for their part, had little time to think about a breakaway to jets after 1945 because they were working night and day trying to get a little stability and profit into their business, and coax a reluctant public aboard their large and costly piston planes. They made some studies. They knew the Germans had cracked the biggest secrets of jet technology before going down to defeat. They were well aware the Nazis had evolved power plants for jet fighters that flew rings around Allied warplanes and had found the new sweptback shapes in wing design to sustain planes at the speeds jet propulsion made possible. But the British, who had pioneered jet technology, did not content themselves with studies. They had developed even more sophisticated jet engines than the Germans. With the knowledge that the United States had cornered the piston plane business, they prepared to harness their gas-turbine know-how to commercial use.

American manufacturers knew a bit about jets. Before the war ended, General Electric built gas-turbine engines to British designs, and Bell and Lockheed designed fighters that carried their thrust into the skies. But if the early efforts with jets told U.S. airline men anything, it was that gas-turbine engines were outrageous gulpers of fuel. This was, of course, perfectly acceptable in war. The U.S. military, especially after the cold war heated up, went ahead developing jet fighters and jet bombers. But U.S. airline men, who had to have the most economical plane possible, put jet transports out of their minds. They figured it would be years before anybody could come up with a jet plane that could fly as cheaply as their piston models.

The British had other ideas, several other ideas. First of all, they believed they could develop a gas-turbine-powered commercial airliner. This was their long-term project — they gave it five years. It would result in the De Havilland Comet, the world's first pure jet airliner. Secondly, they had a more immediate candidate for a turbine-powered commercial plane. It was a modification of the jet-propelled plane that was not quite so ambitious. A military prototype, a redesigned Gloster Meteor, had already flown in 1945. It did not fly quite so spectacularly fast, but it used a lot less fuel. This was the turboprop — so-called because part of the gas turbine's energy was employed to drive conventional-type propellers and the rest expelled out the back as a jet propellant.

In the years immediately after the war most aviation experts thought the turboprop would be the form in which jetpower would be put to civilian use. Jack Northrop, always a forward-looking plane maker, forecast that the turboprop would be carrying all the nation's airline passengers by 1960. Not only did it consume much less fuel than the pure jet, its propulsive efficiency — fundamental for the airlines' kind of flying — was greater. Moreover, the turboprop promised to match the range of speeds everybody was projecting for piston airliners for the next years. After a 1946 test run, the British knew already that they had an outstanding turboprop engine in the Rolls-Royce Dart.

But the turboprop was dogged by the kind of aggravating delays that afflicted British postwar recovery. The four-engine Vickers Viscount, ordered in 1947, did not enter airline service until 1953, when the Comet pure jet was already flying. One American airline showed interest in the British turboprop. Capital Airlines, formerly known as Pennsylvania Central, was the biggest of the nation's second-rank airlines. For years, because 85 percent of its routes were blanketed by the Big Four, Capital had been flying in the prop-wash of the transcontinentals. In their lean and hungry years, United, TWA, and American had all clawed their way to the top of the heap by introducing a superior new plane. At this point Capital decided to try it. British eagerness to crack the U.S. airplane market gave Capital its chance to zoom ahead of the pack. The British government export office guaranteed credit arrangements whereby Capital acquired fifty Viscounts.

On June 16, 1955, Capital's President Slim Carmichael sent off the first Viscount from Washington's National Airport. Just two hours and fifty-five minutes later the turboprop landed at Chicago, having flown the distance sixty-five miles per hour faster than either the DC-6 or Convair. Passengers liked the plane's big windows and cabin room. But the real novelty was the absence of piston-engine vibration. J. C. Hill, a Chicago lawyer, stood a quarter on edge on the armrest of his seat to verify Capital's publicity claims. "Best ride I ever had," he said. Within months Capital filled sixteen Viscount flights daily between New York and Chicago, outpacing the Big Four's DC-6s and Connies. That year Capital tripled its business.

But ever since Ludington Airlines went broke in 1932 trying to introduce mass air travel between New York and Washington, airline men knew how hard it was to make profits on short hauls. Speed, air's biggest asset, counts for less in the economics of short trips. Airlines with longer routes can be flexible and balance shorter trips with longer ones. Capital's average passenger trip was only 383 miles — half that of other airlines.

Given time, Capital might have won route extensions to help break out of this barrier as Delta and others did. But the Big Four brought on DC-7s and Superconstellations that could outleg Capital's turboprop and carry twice as many passengers besides. Capital's gamble failed. The airline fell behind on its five-year program of monthly repayments, and the British began to make noises about foreclosing. In 1958 a Viscount was involved in a midair collision in which twelve passengers died, and there were one or two other accidents. When jets entered the competition, Capital had nowhere else to turn. In 1961 it merged into United, which thus regained its position as biggest of all domestic airlines.

Rushing to catch up in jet technology, the Air Force and Navy sponsored a dozen tries at an American prop-jet engine. One was the early Pratt & Whitney design that was released for first commercial use by Lockheed and caused such disappointment at TWA. Touted to provide transocean range and 50 to 100 miles per hour greater cruising speed than any piston power plant, the engine fell far short of these claims when tried on a regular Connie. Kelly Johnson's Super-Superconstellation had to enter service with a conventional power plant after all.

For Lockheed the abortive attempt to put turboprops on the Constellation had consequences. When the jet age arrived, Lockheed did not turn to commercial jetliners as Boeing and Douglas did. Instead it went on with turboprop designs. Its engineers found a power plant to their liking in the Allison 501-D 15, a turboprop developed by General Motors' subsidiary in Indianapolis for medium-range Air Force bombers. They installed it in a transport for the military, the C-130 Hercules. Almost all the experts, including those in the services, had held that turboprops would have an important place in commercial aviation even if jet planes could be built to fly economically. In 1955 Lockheed offered its Model 188 Electra, and American and Eastern ordered seventy-five of them off the drawing board. Powered by four Allison turboprops, the Electra was smaller than the Constellation, with not quite as great speed but with just as great seating capacity. The airlines wanted the plane for their important medium-length routes — Chicago–New York, Washington–Atlanta, Dallas–Chicago.

Introduced at just about the same time as the first big jets, the Electra proved to be almost as transitory as the last Constellation. Fittingly enough, it was prop trouble that cut down this last of the major airliners to sport a propeller.

In the first two months of operation, two Electras shed their wings and crashed with all aboard. Instead of grounding the Electra, however, the CAA ordered the airlines to reduce its speed from 405 to 350 miles per hour. There were no more incidents. Lockheed undertook exhaustive wind tunnel and flight tests, and after nearly a year found the fault. When the Electra cruised at the design speed of 405 miles per hour in a narrow range of altitudes — between twenty-two and twenty-six thousand feet — vibration set in that caused a structural member supporting one of the propeller shafts to fracture. The propeller would start wobbling wildly, then the engine would shudder. This in turn shook the wing so violently that it began to wobble, then flap, and finally tear away.

Once the cause was found Allison added three more structural members supporting the propeller, Lockheed beefed up the wing, and the Electra flew on. But it was only a matter of time before the jetliners drove it from the skies. The turboprop was doomed, and for the same reason as the piston engine: the propeller on which it depended had reached the limit of its development.

In propellers it's not the number of revolutions per minute that count, it's the airflow over the tip of the propeller. Propellers are limited by their tip speeds — about a thousand feet per second as they carve out their helix of air. The din of small planes blatting deafeningly at low altitude above airfields betrays that their props are approaching the speed of sound — that is, their propeller tip speeds are getting close to the speed of sound.

At altitudes like twenty-eight thousand feet where modern airliners must soar to gain best speed, propellers run afoul of this speed of sound problem. In the thinner air at such heights, the speed of sound goes down. But in this thinner air, propellers have to work harder. Turning faster, they run into sonar threshold complications — turbulence and vortices created around the prop tips just where the greatest efficiency is needed.

The Electra lived with these limitations until 1977, when Eastern retired the last ones from its Boston–New York–Washington shuttle runs. But long before that time, jetliners without prop problems did the job much better. Lockheed, lost a bundle on its plane of transition. A decade went by before Lockheed, with its tri-jet L-1011, could get back into contention in the big airliner market.

29

Crowded Skies

T_o the best of my knowledge I never landed on a completed airport," said old-time TWA Captain Hal Blackburn. Cities simply couldn't update their airports fast enough to keep pace with ever bigger and faster planes. Try as they would to expand and expand, their airports were always five years behind the airliners.

Hayfield airfields? Those racetrack greenswards where Lindbergh liked to land? All were paved over by the end of World War II, by which time concrete had long displaced land as the costliest item in airport budgets. From the time of the Depression, Washington helped cities pay for their airports, and by 1945 the Feds were draping towers and runways with mysterious antennas, cables, and black boxes for electronic flying. To house the new four-engine giants, airline hangars ballooned to football stadium size. As for the airport terminal facilities, they were already being defined by the size of their parking lots.

When La Guardia Field opened in 1940, it was hailed as a modern communications miracle — just eight miles from downtown New York and yet one of the world's largest airports. Spreading over the waterside site of two earlier airfields, it seemed to have everything: a marine terminal just for flying boats; mile-long, paved runways for the landplanes; and a wide apron at the terminal where no fewer than fifteen "giant" DC-3s could line up abreast to discharge or receive passengers simultaneously. On a gracefully balconied outdoor restaurant overlooking the ramps, citizens could take lunch amid the excitement — the thought of noise "pollution" crossed nobody's mind then — of landings and takeoffs across Long Island Sound. Crowned by a tall control tower, the airport's main building had the look of a substantial municipal monument. Yet after no more than two years' operations, it was evident that the place could not handle the burgeoning traffic. Mayor La Guardia himself, intent on assuring New York's future as the country's air gateway, announced, "We are working on a new airport."

Idlewild, sixteen miles from Manhattan but still within the city limits, took shape on a vast landfill on the marshes of Jamaica Bay. They razed 1,100 buildings, pumped in 61 million cubic yards of sand, drained 5,000 acres into pipes as big as subway tunnels — and indeed, a new subway could have been built for less. By the time the runways were ready for planes in 1948, a new mayor threw up his hands and said it was simply too much for the city to manage. Idlewild and La Guardia, along with Newark Airport, were turned over to the interstate agency that had been formed to manage the metropolitan port of New York. Eight times the size of La Guardia, with two-mile runways and ten terminal buildings, a 520-room hotel, a church, three banks, an "Animalport," and 50,000 workers, Idlewild was a city within a city. Yet not even this huge complex could handle all the air traffic that buzzed in and out of New York in the 1950s. Neither Idlewild nor any other U.S. airport was ready for jets in 1956, when the first ones were due. In a scramble, Idlewild runways were lengthened, widened, and thickened (to thirty-two inches of concrete) to take the first 120-ton turbos that roared overseas two years later.

In the midst of these frenzies, La Guardia bounced back bigger and busier than ever. Like Washington's National Airport and Chicago's Midway, La Guardia met the one requirement that endured through these years of frantic change: accessibility. As business travelers abandoned Pullmans for airliners, these close-in airfields became hubs for the nation's fast-growing air commuter traffic. La Guardia's stately terminal building was leveled, long concrete fingers thrust out into the water to make room for big planes. The jetliners themselves were redesigned to land, where the airlines said jets never would, at La Guardia, Washington, Midway. Hal Blackburn said in 1978: "They were tearing up La Guardia the first time I flew there in 1940, and they are still at it."

The same scenes of "Dig We Must" were enacted everywhere else. Though no one ever thought to locate a jetport inside the thousand-year-old cities of Europe, the British and French shoveled away Roman roads and medieval keeps to lay long runways at London's Heathrow and Paris's Orly. That done, bulldozers and scaffolding found steady work at both these international gateways as taxiways, ramps, gates, customs and immigration sheds, police, fire, fuel, food, baggage, freight — all the appurtenances of a great port of entry — had to be built and rebuilt to keep up with each year's rising flood of air travelers.

Some cities couldn't keep pace. At Waco, Texas, loose stones on the runway damaged props and forced the closing of Rich Field. Portland, Oregon, and Cincinnati had to turn to adjacent states to make airports big enough for their zooming traffic. Pittsburgh, having cleared off a mountaintop for its airfield, spent the next twenty years filling in gulley after gulley to stretch its runways for each new stretched airliner. The shining steel and glass terminals that St. Louis and Milwaukee reared in the 1950s were too small for all the customers the day they opened. To make room for its International Airport, Los Angeles had to buy out 2,800 house-holders. To give successive generations ever more right of way, Minneap-olis–St. Paul all but shoved old Fort Snelling into the Mississippi River. Kansas City and TWA, which made its headquarters there, tried to get by with their bluff-ringed river bottom airport until 1951. Then the Missouri River flooded the field. Finding it could not live without air transport, Kansas City rushed to open a five-thousand-acre spread twenty miles out on the prairie. Newark Airport, closed in 1952 after three bad crashes, re-opened with terminals and runways three times as big only to be com-pelled by rising traffic to rip up and rebuild a few years later.

Chicago opened the world's biggest airport in 1955. Not only did O'Hare Field sprawl twenty-two miles from the downtown Loop, you might also have to hike a mile of corridors from one airline's gate to an-other's, and walk a quarter mile just from the ticket counter to where you boarded your plane. Far from deterring patronage, these distances simply dictated changes in where people met and moved around Chicago, and thereby demonstrated the pulling power of air transport on the nation's social and economic life.

So the new airfield was out in the cornfields? Expressways soon linked it with the city, interstate highways met on its margins, and motels with conference rooms rose all around it so the air travelers would not even have to go into town. Overnight 75,000 passengers were arriving and de-parting daily; O'Hare topped all the world's airports in traffic. America, the most restless of nations, had always changed trains at Chicago, but now it came and went on the labyrinthine taxiways of O'Hare. Such was the crush of travelers that the long-range plan for the new airfield had to be implemented at once. So Chicago, bulldozing furiously, led the way with some notable airport innovations. Paired, tangential runways ena-bled planes to take off and land simultaneously on all six strips. At the end of their corridor marathons, passengers found motorized "bridges." Thrusting from terminal wall to the doorways of airliners, these structures stretched like accordions to permit passengers to parade directly to and from their seats.

And still there was no keeping up with the rising tide of air travel. On some summer Sundays as many as 500,000 people jammed New York's Idlewild Airport. When airline limousines were stranded in rush-hour traffic and passengers missed their flights, airline men wailed that high-way congestion had become aviation's problem. But others pointed out that the big delays were *at* the airport. CAB Chairman Ross Rizley re-marked, "It takes longer to get your baggage at Washington airport than it does to fly from Chicago."

That statement even held true on sunny days. On-time airline arrivals were still far from regular in the 1950s. By then the impasse over the kind of instruments to use for bad-weather landings at the nation's airports had finally been broken, cutting the number of weather-grounded flights in half. And very high frequency omnidirectional radio ranges had largely replaced the four-course radio beams, ridding airliner communications of the old curse of static and affording pilots a wide choice of beams to fly on. But radars in towers were few at the start of the 1950s, and radars aboard airliners nonexistent. The *New York Times* pronounced "planning to take care of the new volume and speed . . . woefully inadequate."

So when the weather went sour there were stupendous traffic jams above big cities. September 15, 1954, became known as "Black Wednesday" at New York. In pea-soup weather 300 airliners droned round and round while one by one they felt their way down; no fewer than 45,000 passengers were delayed up to twenty-four hours, some as far away as Washington and Chicago. In 1955 United began installing the first radar in its airliners. But on June 21, 1956, there was an even bigger foul-up over the big eastern cities. Planes were stacked high around New York for fourteen hours, at Washington for twelve. Some thirty thousand passengers in the East and Midwest had to cancel their trips. It took Eastern Airlines three days to regroup its fleet.

At least no one was killed in either of these bad-weather backups, and certainly cancellations were better than crashes. In 1954 not a single fatality occurred on the airlines. But there was another aspect of the ever-rising volume of air traffic that caused more and more anxiety among those in charge of the nation's airways. This was the nightmarish chance of midair collision.

In a sense the worry was even greater in good weather because when the skies were clear the pilots flew "contact" — relying on their eyes, just like any cross-country motorist, to avoid bumping into another plane. By the 1950s planes flew so fast that not even a pilot's eyes could be quick enough. If his airliner approached a stationary object at its usual 300-miles-per-hour speed, a pilot had exactly twelve seconds from the moment he spotted it to take evasive action. If two airliners converged at such speeds, and the pilot saw the other at a distance of two miles, he had only one-tenth of a second to make up his mind. At one mile, he could not avoid collision.

Airliners flew along flyways ten miles wide, under orders to keep to the right and at least ten minutes behind other traffic, reporting their course and altitude at frequent intervals. By the mid-1950s, moreover, air traffic controllers were beginning to keep radar watch on planes on the approaches to large cities. But an average of four near misses, mostly involving small airplanes, were reported daily. Then in 1955 fifteen died when a TWA airliner was struck by a light plane at Cincinnati, and there was a stir. But the skies were still wide open, and largely unwatched, when luck ran out on June 30, 1956.

On that date occurred the worst crash thus far in U.S. airline history. In broad daylight a TWA Superconstellation with seventy aboard and a United DC-7 carrying fifty-eight collided at 21,000 feet above the Grand Canyon. Both planes fell into the Canyon within a mile of each other, and the doctor who went down to verify the 128 deaths found "a sight I don't want to have to see again."

How could so bizarre a conjuncture have happened? Both planes were flying east, setting their own course. But over the Canyon there were clouds. The TWA captain radioed he wanted to climb from 19,000 to 21,000 feet. Informing him that the United flight was crossing at that level, the Los Angeles controller said no. So the TWA plane asked to climb to "a thousand on top" — 1,000 feet above the cloud tops — and this was granted. Incredibly, the two transcontinentals collided. The investigators' verdict was that the two pilots had not seen each other in time. For the rest, they could only say that clouds got in the way, or that the pilots were busy inside their cockpits, possibly trying to swing their planes to give their passengers a better view of the Grand Canyon below.

The death of 128 people in the Grand Canyon disaster had decided results. First, the federal government ordered a lot of long-range radar equipment and instituted surveillance of all airway traffic at all times. Second, Congress in the Federal Aviation Act of 1958 took away the administration of airways and air safety from the Commerce Department and put it in the hands of a new Federal Aviation Agency (FAA), which was also charged with investigating air accidents. Third, President Eisenhower named General Elwood P. Quesada, a strong administrator, to manage the new setup. Quesada ordered new electronic gear for the airways. He tightened rules, requiring all airline pilots to quit flying at sixty. (In 1972, the FAA was merged into the Department of Transportation, and the task of investigating air accidents, important as ever, passed to an independent National Transportation Safety Board. Thus, only the five-man CAB, with its system of regulating air fares and routes, survived substantially unchanged through the half century that saw commercial aviation's fantastic blooming.)

But the arrival of jets imposed new strains, and in 1960 a more terrible midair collision occurred — right over the biggest city in the United States.

Again it involved TWA and United planes, but this time United's was a jet. At 10:33 on the dark morning of December 16, 1960, United's DC-8 from Chicago with eighty-four aboard rounded an airway intersection over New Jersey on its way to landing at Idlewild. Not yet under Idlewild's ground control, the big plane was still providing its own navigation as it flashed down through the overcast and crossed into the New York area at 6,000 feet. At this point it called Idlewild: "United 824 approaching Preston [the Jersey intersection] at 5,000 feet."

At that same moment an air traffic controller at La Guardia hunched over his radar scope as he nursed TWA's inbound Flight 266 from Columbus through rain, sleet, and snow toward his airfield. While he watched, an unexpected blip slid across his scope. He picked up his microphone to call the Superconnie with unaccustomed urgency. "Jet traffic off your right now at 3 o'clock at one mile." The TWA pilot, unable to see or be seen, acknowledged the message. A second later the controller shouted, "Transworld 266 — turn further left."

Hearing no answer and lacking means to query Idlewild at short notice, he watched the screen helplessly as the two blips continued for another second on collision course. Then there was one pinpoint of light where there had been two; the TWA signal vanished from the screen, and the second blip crept on its northeast course for eight miles. Then it too disappeared.

On a street of old brownstone houses in Brooklyn, a man looked up and saw "a large bolt of lightning." In a house on Staten Island across the Narrows from Brooklyn, a housewife heard a noise that sounded like "a thousand dishes crashing from the sky." As she watched, the TWA Connie carrying forty-four passengers and crewmen crashed to earth nearby. Seconds later, a broken and torn DC-8 jetliner, which had been United's proud Flight 824, fell out of the sky in Brooklyn.

Part of the jet cut through the roof of one house. Engines, fuselage, cargo, bodies, cascaded with thundering crunches into the street; rivulets of jet fuel skittered and splashed crazily and ignited into billows of flame, which in turn touched off the gasoline tanks of parked cars. Tenants fled from a row of burning brownstone rooming houses. The empty Pillar of Fire Covenant Church turned into an inferno. Two men selling Christmas trees on a corner, a snow shoveler nearby, and eight other Brooklynites, were killed instantly.•

From the torn DC-8 fuselage came piercing cries. A teen-age boy ran down the street screaming, "Oh, those people are burning to death." A passing priest ran into his church for holy oil, and ran out again to administer rites to the charred bodies. Seven alarms jammed the area with fire-fighting equipment. Ambulances lined up at makeshift morgues — a bowling alley and a vacant store — to transfer bodies to hospitals.

Miraculously, out of the carnage a boy was hurled to a soft landing in a snowbank. He was eleven-year-old Stephen Baltz, of Wilmette, Illinois, traveling alone to meet his mother and sister in New York. Two cops rushed to him, wrapped their coats around his flaming body, rolled him in the snow. Carried to a hospital, the boy said, "I remember looking out of the plane window at the snow covering the city. It looked like a picture out of a fairy book. It was a beautiful sight. Then all of a sudden there was an explosion. The plane started to fall and people started to scream. I held on to my seat, and then the plane crashed."

As hundreds of rescue workers fought through the smoking debris, others converged on Staten Island's tiny Miller Field, a small-plane airport. The TWA plane had fallen 150 feet away from houses and two schools. "It went down in a terrible way, one wing gone — and it turned over and over very slowly," a woman said. Seat-belted bodies were flung everywhere. Snow on the field was quickly crimson-soaked — and stretcher-bearing lines formed as in Brooklyn. But there were no survivors. Last to die was frail, fire-tortured Stephen Baltz. Fighting to smile for his father, Vice President William Baltz of Admiral Corporation, he "went to sleep" next day.

After exhaustive inquiry, the investigating board announced the probable cause of the disaster that had taken 139 lives. The jet, knifing down into the New York area, had "proceeded beyond . . . its airspace." In the seventy seconds before the collision, according to the flight recorder taken from its wreckage, the DC-8 reduced speed from 356 to 301 knots. Obviously it was still barreling along too fast to wait for clearance to cross the last thirty miles to its Idlewild landing.

Such tragedies leave lasting marks in the nation's air transport history. The Rockne crash of 1931 led to modern all-metal airliners. Out of the Cutting crash of 1935 came the charter on which air transport is founded, the Civil Aeronautics Act of 1938. And the Grand Canyon collision led to the legislative reforms of 1958. After the Staten Island disaster, worst so far, there was no such legislative sequel.

Of course Washington ruled at once that no plane should fly faster than 250 knots thereafter in terminal areas. And General Quesada lost no time in placing costly and sophisticated electronic gear at airports and along airways and tightening procedures and training to enable pilots and controllers to work at something more like jet speeds. By the end of the 1950s airline travel was already safer than going by train, and much safer than taking your car. But there were limits, and Congress gradually stopped pressing for schemes to tie airliners' automatic pilots to instrument landing systems that would automate bad-weather landings. Eastern Airlines later achieved automated landings with passengers. But nothing more was heard about automated flying after a nightclub comedian began parodying recorded in-flight announcements in 1960: "Nothing can go wrong, go wrong, go wrong. . . ."

After the Staten Island disaster the veteran government safety man Ed Slattery was asked if such collisions were not unlikely. "Unlikely?" he exclaimed. "It was unlikely that an Air Force bomber would ram the Empire State Building in 1945. It was even more unlikely that two airliners would smash out in the wild sky over Grand Canyon. Unlikely to happen, but it did."

30

The Travail of TWA

When in 1945 Juan Trippe proposed his "chosen instrument" world airline, to be owned proportionately by all the domestic companies, he conceded that by far the biggest stockholder would be Howard Hughes. Since Jack Frye had brought Hughes into TWA at the end of the 1930s, Hughes had steadily accumulated TWA stock until finally he held 78.23 percent. No other person in the airline world, including Juan Trippe, came anywhere near holding such a commanding position in any company. Welch Pogue, top government lawyer in airline matters then, said later that the question of Hughes's ownership did not even come up in the first years. It was not really until Hughes stepped in to arrest TWA's financial slide after World War II that his singular position came into question — and then it was quickly resolved in his favor.

Hughes's control of TWA was not direct; at no time was he an officer or director; he never attended a board meeting. His TWA shares were all held by the Hughes Tool Company. In 1948 the CAB formally approved the tool company's ownership of TWA shares so long as its sale of equipment to TWA did not exceed $300 in any one transaction. The prohibition was not without significance for Hughes, whose California-based Hughes Aircraft Company, founded in 1930 to build planes he could race, emerged about then as a major aviation manufacturer. Hughes's wartime try at plane-building had not come to much — two or three examples of the high-altitude reconnaissance plane in which he almost lost his life in 1946, the mammoth flying boat that took to the air just once in 1947. But when he lost out on plane contracts, Hughes Aircraft branched in another direction. Hughes, General Quesada said later, was the first to see that the military airplane would need lots more than the pilot. Hiring a hatful of talented scientist-engineers, his outfit made the sophisticated equipment that aimed and fired the guns on warplanes. When the cold war electronics boom broke out in Southern California in the early 1950s, Hughes Aircraft rode its crest. Before long the company was the nation's ninth largest defense supplier, turning out weapons-systems, missiles, lasers, and satellites for the Pentagon and CIA.

Hughes never got to profit much personally from this bonanza. As was his habit, he took the line that the company was his alone, and refused to give the scientist-engineers a share in the ownership. That brought such a blowup that Air Force Secretary Harold Talbott threatened to take away Hughes Aircraft's contracts. Rather than give in, Hughes gave all the company's stock to a newly formed Hughes Foundation for Medicine. Hughes became the foundation's sole trustee. But he could no longer dip into the aircraft company's resources for his own purposes. From then on he left Hughes Aircraft pretty much alone.

There never was a businessman like Howard Hughes. You have to go back to Henry Ford to find one so bent on going it alone. But at least you could find Ford, in the midst of his plants or in the midst of his family. Hughes at fifty was solitary, reclusive, and suspicious, only fleetingly seen in public. He was thinner than ever, with little left of that appealing vagrancy — the tieless and hatless rich boy, in soiled shirt, rumpled trousers, and sneakers — of the years when he briefly succeeded Lindbergh as America's most feted flier. He had no domicile, no office. Any face-to-face encounter with him took place in a rented house, or in a car parked on a deserted street, or in a hotel room bare of any evidence that the man lived there.

His hearing, damaged by three plane crashes and years of exposure to roaring airplane engines, had grown steadily worse. His hearing aid, specially made for him by Hughes Aircraft electronics specialists and carried in his shirt pocket, gave him terrible headaches. He stuffed towels in windows and doors against eavesdroppers, and then he required others to speak loudly so he could hear. He shied away from shaking hands lest he be contaminated by germs and catch some new ailment. Wherever he lived he bivouacked. On the bed lay a cardboard box with a couple of shirts in it, and alongside it dozens of manila envelopes — his filing system. Somewhere nearby was a slide rule, and sometimes a pad of yellow paper. He worked alone, by phone, mostly after midnight.

The fountain of Hughes's wealth remained Hughes Tool. From the early days when he bought out his uncles and became sole owner, the company, which sold oil-drilling bits, was a veritable gusher itself. Vic Leslie, who later worked for Hughes, got the first outsider's peek checking out a loan for the Mellon Bank in 1941, and found "a veritable gold mine, carrying down to net profit no less than 50 percent of sales."

"Flying is my life," Hughes said, and he could still fly. Ralph Damon, hired on the rebound from American after he quarreled with C. R. Smith over the sale of American Overseas to Pan Am, was running TWA quite capably in these years, and Hughes poured in money for bigger and faster

Constellations to enable his airline to keep up in the race for the fast-growing travel market. Hughes liked to check out the planes himself. To satisfy CAA requirements that a TWA pilot always be present in a TWA plane, Captain Lee Flanagin used to go along as safety pilot. Flanagin said later:

He was not the type of pilot you'd find on the airlines — a very good pilot, not too good on instruments. He had plenty of confidence. And he always worked through the manuals, maintenance as well as flight manuals, before he came out to fly.

The first night he asked me to be there at ten o'clock. We flew out to Palmdale, and he shot landings, 500 of them. It was six hours straight of touch-and-go, touch-and-go in figure eight patterns. I finally told him I couldn't go any longer, there was no more gas. Afterwards he said: "The airplane is unstable close to the ground. It's got too much dihedral in the wings."

Hughes flew them all — Lockheeds, Boeings, Douglases. Because of their old feud, however, he did not ask Donald Douglas for a crack at the DC-6. Instead he called C. R. Smith, who let him try his hand on one of American's first ones. Hughes also bought single examples of different types of planes and kept them hangared around southern California with crews standing by in case he should want to fly them.

TWA may have been one of the world's great airlines. But its profit-and-loss record read like the fever chart of a patient with an obscure infection. There was good reason to locate the source of this infection in Hughes's own self-interested desire to keep the airline weak and the profits low. In that way he could avoid having to pay penalty taxes on the huge accumulations of his profits from the tool company. And in that way he could be justified in the eyes of the CAB and the Internal Revenue Service in supplying unusual amounts of cash to TWA. A fair estimate of the total capital Hughes poured into TWA over the years might be $70 million, an investment incomparably greater than that of any other individual in the airline business. But the fact was that TWA had declared only one dividend — a 10 percent stock distribution in 1953 — during the two decades the line had been under Hughes's control. That was no way to run a public utility, or even a private corporation, and suggests manipulation for purposes other than those of the enterprise itself. Yet TWA as an operating airline, staffed by capable and devoted people and supplied with planes as good as anybody's, never ceased to hold its own. TWA's losses had sometimes been staggering — 18 million during the first three postwar years. Then had followed a slow recovery during Ralph Damon's presidency. Damon took office with a five-year contract that contained a no-interference clause. But in 1954 the contract expired, and Hughes continued it by a gentleman's agreement.

The very next day Hughes began interfering in the line's daily affairs, by long phone calls in the middle of the night, by countermanding orders delivered next day by Hughes lackeys. It was the time when all the airlines were pondering the big leap into jets. As TWA's financial angel, Hughes would order the planes, and leave it to the tool company to pay for them. As airline president, Damon also happened to be an old-time plane builder who understood aircraft design. Damon more than anybody was the man who kept TWA from taking the halfway step to turboprops. "No way," he cried in 1952 after studying the design plans that Hughes's old friend Bob Gross presented for Lockheed's Electra turboprop. "I'm an engineer, and I say that plane is not structurally sound." Later Lockheed brought in its scheme for putting turboprops in Connies. It fizzled, but by then Damon was no longer around. He stood up for two years more under the night-and-day pressure, then died on the job, victim of pneumonia and exhaustion in January 1955. Thereafter TWA went into another sinking spell, and lost $16 million in three years.

Hughes had led the airlines into long-range landplanes with the Connie. Of course he wanted to be first into jets, too, but in this he failed. Probably he failed because he tried to make an extremely complex judgment unaided. More accurately, he failed because he demanded one design to serve two purposes. He wanted a plane that would be able to cross the Atlantic nonstop as well as a plane that would fly TWA's domestic routes. He looked over the front runner, Boeing's 707, and decided it lacked the range for nonstop transocean flying. But he could not apply to Lockheed for something better because Lockheed, taking orders for the smaller turboprop Electra, had bowed out of the jet race. Nor could Hughes turn to the other big Los Angeles outfit — he had been persona non grata for years with Donald Douglas. So Hughes went to Convair in San Diego. Having captured the old DC-3 market with its highly successful short-haul airliners, Convair might, Hughes thought, be just the company to come up with a jetliner bigger and better than anybody's. Once again there were secret meetings in Los Angeles at which Hughes told Jack Naish and Jack Zevely of the Convair command about the dual-purpose design he wanted. Then Hughes and his aide Bob Rummel met Ralph Bayles, Convair's chief engineer. The scene was the grand ballroom of the Beverly Hills Hotel at midnight, with Hughes's Mormon guards at the doors and the three figures hunched over a card table out in the center of the bare floor.

But Hughes's 1940 headstart was not to be repeated. Convair's design, based as it was on the same power plant, was no better than Boeing's 707. And while Hughes hesitated, Juan Trippe beat him and everybody else to the history-making takeoff. In a bold and astute decision, Trippe ordered (for the staggering sum of $270 million) not one but two jetliners — the Boeing plane that would be ready first, and a somewhat larger Douglas design that would arrive later with the hoped-for nonstop transocean capability. American and United placed their orders soon after.

Trailing badly, Hughes tried to catch up — and got into the worst bind of his life. He first put in his order for Boeings a good four months late. Next, in an ill-advised ploy, he ordered 360 of the Pratt & Whitney engines needed to propel the Boeing jetliners. His thought, which proved unfounded, was that he could corner the supply of engines and trade them for higher positions on Boeing's jet delivery list. But that was only a start. Well after the first big wave of jet orders, Convair informed him that it had something new after all. Its engineers had drawn plans for a jetliner built around a hot new engine — the same General Electric J-79 engine that made Convair's B-58 the world's fastest bomber. Early in 1956 the Air Force declassified the new engine. Just two days later Convair's men invaded a party Hughes was holding at a beachside villa in Santa Monica. A beautiful girl fainted onto the lap of one of the engineers, but Hughes dropped everything to study the sketches for an airliner that, while smaller, could fly maybe thirty-five miles per hour faster than the speeds projected for the first Boeing and Douglas jets. This was the Convair 880, and Hughes, gambling that the last might yet be first, ordered twenty.

What was all this going to cost? Pan Am, American, and United had worked out the most elaborate long-term credit arrangements with banks and insurance companies to finance by far the largest outlays any airlines had ever undertaken. Hughes's style was breathtakingly different. He had simply placed his orders without checking numbers and costs with anybody, not even with his money man Noah Dietrich. When the contracts came in, Dietrich was dumfounded to find that the cost of all the new equipment Hughes had committed himself to pay for added up to nearly $500 million — more than the sum of Hughes's assets.

Though the old Hughes-Dietrich bonds had frayed badly by this time, Dietrich appears to have done the best he could. He devised a plan to use $100 million of tool company cash plus a $300 million convertible bond issue to be floated by New York bankers. Hughes stopped it — he wanted no diluting of his ownership. After that, Dietrich said later, "we just went round and round." At the end of a year in the red, TWA got a new president, Carter Burgess, a Virginian who had been assistant secretary of defense. In the eleven months that Burgess held office, he never even met Hughes. The airline continued to lose money. And on March 12, 1957, came the final parting of the ways with Dietrich. At sixty-seven Dietrich, at a phone in one room at the Beverly Hills Hotel told Hughes at a phone in another that he was through pulling financial rabbits out of hats for him unless Hughes gave him a piece of the tool company. All corporations made stock-option arrangements for their managers, but Hughes, with his obsession for sole ownership, paid men like Dietrich big salaries rather than part with any part of his property. "You're holding a gun to my head," Hughes said, and that was the last word between them. The following day Howard Hughes, for a quarter century Hollywood's most eligible catch, married his long-time actress friend Jean Peters in Tonopah, Nevada.

About that time Lockheed completed the first of the eleven Model 1649A Superconstellations that Hughes had ordered to tide TWA over the final transition to jets. Since he also wanted to look over the Vickers Viscount propjet in Montreal, he decided to give the big new airliner a test by flying it to Montreal. Lee Flanagin having been grounded with a heart ailment, Captain Ted Moffitt went along as safety pilot.

But in Montreal Hughes sent his safety pilot back and, instead of returning the new airliner, lived on it, surrounded by bodyguards. Much of the time he spent lying on a mattress on the plane floor poring over the manuals for the Viscount and the 1649A. In New York Carter Burgess began to fret over the airline's missing plane. He had not even managed a telephone conversation with Hughes, who sent him orders through such people as Rummel, Flanagin, and his Mormon aide, Bill Gay. Then Burgess learned that Hughes, with only a TWA flight engineer in the cockpit with him, had flown the airliner to the Bahamas to look over some real estate. Soon Gay and Jean Peters joined him. Weeks went by. Still there was no sign that TWA was going to get its fancy new plane back. Finally Burgess said, "I'm going to get it back." Unable to get through to Hughes, he presented his demand to Bill Gay, "Bring the airplane back or I quit." To Vice President Floyd Hall he said, "I stood at the gates of Armageddon. I told the man." Next day Bill Gay told him his letter of resignation was accepted.

For the next seven months TWA was without a president. And for a total of six months TWA was without its airliner, while Hughes remained in the Bahamas. Then the phone rang in Lee Flanagin's Los Angeles home. The message was, "Mr. Smith is calling you from the Bahamas." Flanagin's wife said, "Oh, it's Howard Hughes calling you from the Bahamas." What Hughes wanted to talk about was the weather at Los Angeles. He was coming back.

The flight engineer having long since returned to his TWA job, there was no one to help Hughes fly the big plane but Bill Gay. At Los Angeles Vice President Hall and other TWA officials waited anxiously at the airport. Flanagin was at home nearby when the Los Angeles tower telephoned and said Hughes was approaching the area in the 1649A and asking to speak to him. Flanagin said later:

There was heavy rain and he didn't have any approach charts with him. He asked me what he should do. I told him it looked good toward Long Beach, and to circle at Long Beach until the showers let up. I had the tower tell him to come along the coastline, and he could see to approach into the west.

By that time, it was ten o'clock at night. He made a satisfactory landing, and I met him. He asked me to go and apologize to the tower. They said, "Forget it."

Somehow the TWA officials missed Hughes. Hall said later:

Four cars rolled up to the plane that we hadn't seen. The people in the plane got out — Hughes, Jean Peters, Bill Gay and some others. And off he rode without a word with us.

Next morning he sent word across town, "Change all four engines." "Why?" I asked. "That plane hasn't more than 30 hours' flying time." Hughes's answer was, "I don't want any airplane that I've flown flying passengers without an overhaul."

The time was drawing near when all the new jets would be ready, and still Hughes had no arrangement to finance his payments for them. Just about the time *Fortune* put him on its list of the world's richest men, tool company earnings fell in a worldwide oil glut, and it came out in a Boeing submission to the CAB that Hughes had failed to make the first down payment on his contract. Equitable Life then let Hughes know there would be no renewal of a TWA loan unless he promptly set up long-term financial arrangements for the purchase of the jets.

In the absence of Dietrich, Hughes called on his old friend Bob Gross of Lockheed to help. The first thing Gross did was to find a new president for TWA. He was Charles S. Thomas, wealthy former secretary of the Navy and a former member of the Lockheed board. Although Hughes would not see Thomas, he was at least willing to make the phone call. Thomas told him he would give it a whirl. Then Gross flew east to canvass the banking situation for Hughes. Gross found First Boston Bank willing to devise a financial plan. Nothing came of it when Hughes declined to meet the bankers face to face. Meanwhile the first Boeing 707 jetliner was delivered, and Hughes managed to pay for it by borrowing from the tool company's banker Ben Sessel at Irving Trust.

By this time Hughes was postponing plane deliveries and putting off payments at both Boeing and Convair. For Convair the impact was so devastating that its parent, General Dynamics, ran up the biggest year's loss in U.S. corporate history. All Wall Street knew the fix Hughes was in. When the bankers closed in on Henry Ford back in the 1920s, Ford had wiggled out by shipping thousands of cars and dunning his dealers for cash payment in what amounted to a huge involuntary loan. But Hughes had nowhere to turn for cash except his own companies; of these Hughes Aircraft was now off limits, TWA was a money loser, and even Hughes Tool had suffered a drop in earnings. Things looked so bad that TWA President Thomas thought the airline should try to save itself. He got behind an earlier plan that resembled Dietrich's original lineup. Angry that Thomas should independently take a hand, Hughes turned against him. Thereupon Thomas quit.

That did it. Nobody in Wall Street was going to lend Hughes the huge sums he had to have when his airline didn't even have a president, and there was no assurance of sound and steady management for the future. Loans kept falling due, plans kept collapsing, the airline kept losing more money. The day finally came when Banker Sessel, flanked by the representatives of the insurance companies and investment houses, told Hughes's lawyers there would be no more extensions. The only way Hughes could get the long-term credit for all his planes was by agreeing to place his TWA stock in a voting trust and by letting Wall Street's nominees run the airline. It was brutal. They were not taking away his property, but they were taking away his power. With nowhere else to turn, Hughes bowed, and on the last day of 1960 control of TWA passed out of Hughes's hands.

31

Air Travel in the 1950s

W_{ell} into the 1950s you still found yourself flying, if you insisted on traveling to places like Dayton or Mobile or Missoula, in a buckety old twin-engine DC-3. Even on an average-length trip of six hundred miles in a four-engine plane pressurized to fly above the weather, you commonly made several stops along the way. And if the plane, the crew, and the airway were all now rigorously supervised, irregularity remained rife. The air after all was not man's natural element. Scheduled flight was but a quarter century old. In 1951, the airlines carried out a million scheduled flights — of which less than 50 percent were completed on time.

By the 1950s, however, the swiftness of air travel attracted huge numbers of customers. The CAB, instead of permitting new entries into the business, dished out additional certificates to existing companies. The result was that several competing airlines vied for passengers on practically every route in the country. In this way, all the airlines took part in the tremendous growth of air travel, which zoomed upward at the annual rate of 15 percent — changing the nation's style of living and doing business in the process.

For the passenger, part of this process was a change from the prewar everybody-is-a-VIP treatment toward a mass-handling experience. You could still pick up a phone and get a reservation confirmed for a plane journey without having to go and stand in line, as people did when traveling by train. The airline might take a little time clinching your seat for you: the electronic signaling system by which the agent now checked available space had to clear it with other cities served by the flight, and that sometimes took a few minutes. You could still catch the airline limousine downtown, but the limousine was now a bus, and it stopped at only the bigger hotels.

Most of the passengers were businessmen. The rise of fast air service impelled all sorts of giant corporations, from Prudential Life to General Foods, to decentralize their operations, opening regional centers in places like Boston, Atlanta, Houston, New Orleans. The 1950s were the years when the hotel boom started, and rent-a-car agencies opened at every airport to accommodate the managers, specialists, and corporate salesmen who trooped on and off airliners from daylight to dark. Often the airline credit card that paid for their tickets also served to pay for the rental car and hotel room. Others, convention-goers especially, combined business and pleasure: golf bags were weighed along with the briefcases and overnight bags, and the credit cards paid for the excess baggage too. And for the first time travelers began appearing at the boarding gates with tickets purchased through travel agencies — many of these obtained on credit, on the popular FLY NOW, PAY LATER installment plan. Amid the bustle and din of the terminal buildings, these well-heeled riders bought their copies of *Life* and *Look,* their paperbacks and their trip insurance — $25,-000 worth for a mere quarter, plus five cents more for the stamp to mail the policy to next of kin.

As the planes got bigger and faster, sleeper service vanished from the domestic lines and survived only minimally on overseas routes. Air travel had originally been modeled on the Pullman, and the Pullman's ideal was to match the service in a good hotel. In the spring of 1952 *Harper's* magazine essayist Bernard De Voto took a trip across the country by DC-6, and noted how for both airlines and their customers standards of comfort and pace inevitably clashed. Departure from Washington was delayed a half hour "while homing planes felt their way in." To De Voto it seemed, "Until you are airborne, you are just an annoyance. You are chivvied by announcers, ignored by everyone else, and given as little information as possible." Then, "once you are in the air you are treated with a deference elsewhere reserved for movie actors, surrounded with luxury above your station."

In a first-class DC-6 there were forty-four places to the Pullman's twenty-eight, but there were two pretty girls hovering over you as you sat up all the way to San Francisco. As late as 1950 airline captains still passed a little piece of paper back among the passengers to keep you informed of the plane's altitude, air speed, and location, and the temperature and weather ahead. However, the public address system, required in big planes for such vital crash messages as "Everyone leave by the rear door," had now been installed. They spouted "orations of welcome and rapture from which there appears to be no reprieve," De Voto complained.

The cramped spaces of airliner galleys ruled out any chance that meals could be cooked on board as they were on Pullman trains. Deep-frozen food still had not come to the airlines. Though United's chefs put aboard 944 different menus a year in the 1950s, De Voto was served the inevitable chicken. He bewailed the paucity, even during meals, of conversation aloft — "no parlor car exhibitionists, nearly everyone reading his paperback book until he falls asleep."

Surcease for the restless air traveler, however, came aboard in the 1950s. From the days when the Philco salesman bought a bottle and went wing-walking on Al De Gormo's Boeing biplane, captains and steward-esses had been under strict orders to bar passengers who tried to carry this type of insurance along. But in-flight drinks were served from the earliest days in Europe, and Pan Am was always ready to meet the foreign com-petition with wine and liquor on Atlantic crossings. Before Pearl Harbor, C. R. Smith asked American's lawyers to look into state licensing and the niceties of serving drinks on planes that passed above dry states. But the airlines, bent on gaining public confidence, held back. They wanted no outsiders yelping, "Drinking and flying don't mix." Neither the CAB nor the Air Transport Association, voice of the airlines, ever had anything to say on this delicate topic. Gingerly, as the last states moved to repeal their dry laws, the airlines experimented in the early 1950s by offering passen-gers on certain flights "complimentary drinks."

Then the first domestic, Colonial, broke the ice and began selling drinks on its flights to Bermuda, and, to the surprise of the others, made a little money. It was a time when the airlines were looking for extra income. Without exactly broadcasting it, American and TWA set their steward-esses to pouring drinks for their thirstier passengers — a quota of two per purchaser — before dinner. But United's President Pat Patterson held out. His daughter Patty had gone off and got herself a job as stewardess with American, where she served drinks and where, Patterson growled, "They're treating her like a barmaid." Patterson was damned if his airline would have its stewardesses performing such unladylike tasks.

United's service expert Don Magarrell fretted and stewed. "They were killing us in the competition for passengers," he said later. Finally, Ma-garrell devised a little tray that would hold a glass and a two-ounce bottle and fit right in where the casserole for the dinner entrée went. Then he put colored caps on the bottles so that the stewardess would not have to use her lily-white hands to pour any liquor at all. One day in 1955 with the whole top executive staff sitting expectant, Magarrell, pretending to be a stewardess, bent over and placed the meal before Vice President Bob Johnson, acting the part of a passenger. Magarrell said later:

The whole idea was that we were offering liquor in a dignified way that would not offend anybody's sense of what the girl-next-door stewardess should be doing.

I asked Bob if he wanted martini or manhattan. Then I went out in the next room and brought in the new little tray with the setups. "Very well," said I, when he was finished, "you are now ready for dinner."

And I took that tray away, and put his dinner into that depression.

There was silence in the room for some time. "Gentlemen," I said, "this is an easy and economical and pleasant way to meet the competition." Then Patterson said, "Oh, I guess we could do that."

For six months, before Pat came around, Magarrell had been offering "Our Old Fashioned Fruit Cocktail," with an ounce and a half of bourbon in it, to patrons of United's "Executive Flights" between New York and Chicago. Now, having officially hopped off the wagon, United began featuring "Extra Dry 10-to-1 Beefeater Martinis" on its afternoon flights. Controller Curtis Barkes soon said, "You're spending too much on liquor." Magarrell blindfolded his staff and was delighted to find that none could tell the difference, when he put them to the test, between eight different gins. He had found a low-price outfit in California that moved its gin around in railway tank cars. He got his next supply at half price, changed the liquor cards to read "Extra Dry 10-to-1 Mainliner Martinis," and served more of them than ever.

About that time Boeing's 377 Stratocruiser arrived, with its cozy little circular cocktail lounge with benches for fourteen below decks, and a space in the tail that converted into two berths, like a cabin cruiser — "great for honeymooners and gentlemen with their secretaries," Magarrell said. On United's long flights to Honolulu, the bar filled as soon after takeoff as the pilot turned off the "Fasten your seat belt" sign. On Northwest's flights across the north country, the Stratocruiser bar took the chill off the coldest evenings. And since the planes were pressurized to levels equivalent to no more than 5,500 feet, the flight surgeons gave special blessing to descents into Wellwood Beall's "conversation pit": a drink was relaxing, and it was good for travelers on long flights to get up and walk around.

If Americans learned about Einstein's $E=mc^2$ from the atomic bomb, it was probably the airplane that brought home to them his theory of relativity. His message that space and time were not what you thought did not square, of course, with prior experience. To be told that space was curved was fantastic; the thought that time, far from being absolute, was only a framework of man's devising to place himself in his surroundings, was enough to make newspaper readers of the 1920s clap their brows in disbelief. Was it possible, in the example the Sunday supplements loved to cite, that if a ten-year-old twin was sent off on a round trip to the nearest star at just under the speed of light, in the eight and a half years that would have elapsed during the journey, he would have aged only fourteen and a half months while his earthbound twin was getting on for nineteen?

True, the Newtonian concept of immutable time still worked for airplanes: at a 250-miles-per-hour speed, the old calculations differed from the new by only about one-billionth of a second. But planes traveled so much faster that they knocked over all previous experiences of duration. Flying put people in a mind to see that time after all was a pretty relative thing. On landing at Denver after only six hours on the way from Washington, De Voto remarked:

There's no answer to the speed of the plane — it's inconceivable. Logic dissolves when you summon up other schedules — 5 or 6 months by ox team, 7 or 8 weeks by steamboat and stage, 42 hours by streamliner and lesser trains.

The wonder of speed was all the greater in those years because you could see so much from the moderate heights at which you traveled. On a DC-6 crossing American Airlines Captain Paul Carpenter veered far north, as became possible with improved communications and radar controls in the 1950s, to avoid a weather front. Flying close to the Canadian line, he suddenly found himself looking down on the very ground he had traversed as a boy. He said later:

Every summer we would go from near Williston with horsedrawn carts full of hardware and other supplies to the farm my dad bought across the Dakota line in eastern Montana. One of those nights we pitched camp alongside Cow Johnson's soddy — sod hut. The rough road we followed split Brush Lake, milky and salt-caked on one side, crystal clear on the other.

In 15 minutes I flew over the 90 miles it took us three days to cross, and I could spot the same sloughs.

You got an even greater sense of annihilating distance flying the Atlantic. If you were a businessman, you found out that Amsterdam was not that much farther from New York by air than Atlanta, and when the Common Market opened up enticing commercial opportunities in Europe, you flew back and forth opening branches and buying into overseas companies. If you were a tourist, you discovered that you could spend all but a day or two of your three-week or four-week vacation on the other side, seeing London, Paris, Rome, swimming in the Mediterranean. If you were a student, you found that Icelandic, flying by such out-of-the-way places as Reykjavík and Luxembourg, could lift you across at a much lower fare in just a little more time.

You could fly so fast and so far because you flew high — this was the boon of the pressurized cabin. Creating a capsulized environment, in which passengers could ride in comfort through air too thin to breathe, was an American achievement. It was all the more impressive considering that Britain's first Comet jetliners, designed also to fly high to make the most of their revolutionary new means of propulsion, came to grief in successive 1954 crashes precisely because the pressurized cabins in these otherwise flawless planes blew apart in midair.

Neither in their pioneering years nor later did such a high-altitude disaster befall American airliners. In the prewar Stratoliners, the cabin sometimes became uncomfortably warm on hot summer days because the heat of compressing the air raised the cabin temperature and humidity still higher. In the later Constellations and DC-6s, turbosuperchargers driven by the plane's engines maintained an inside pressure equivalent to

8,000 feet no matter how high the airplane climbed, and kept the cabin well ventilated besides. In practice, these planes flew at 20,000 feet or less. Many times, when bucking headwinds on a westbound trip like De Voto's, they flew much lower. On the Boeing Stratocruiser, third of the postwar generation of U.S. airliners to arrive, improvements in the system made it easier to operate at high altitudes. This plane, once it had shed weight by using up most of its fuel, could and did climb to 25,000 feet for the last three hours of a New York–London crossing.

Between 1945 and 1955 passengers on these high-flying planes experienced some eighteen episodes of sudden decompression in flight. As luck would have it, the first case occurred on a British-operated Constellation whose astrodome blew off in October 1946 at 19,000 feet over the Atlantic. The air rushed out the opening with a roar, carrying along all the navigator's papers. BOAC Captain Ken Buxton slapped on his oxygen mask and brought the plane down to 7,000 feet within ten minutes. During the descent a steward had to give oxygen to several elderly women and a year-old baby. Everybody else was all right, although one passenger was trapped in a lavatory when negative pressure sealed the door shut. He banged and shouted. "The gentleman was, I think, a bit frightened," Captain Buxton said later.

In all the incidents only four lives were lost. One was George Hart, the TWA navigator sucked out when his astrodome blew out at 20,000 feet over the Atlantic a year later; one, a purser killed when a prop blade snapped off and tore through the cabin of a transatlantic Stratocruiser; and two, sucked out when Stratocruiser doors blew open — one a steward and one a passenger. In the case where the steward was lost, Pan Am's Stratocruiser was flying over Long Island at only 10,000 feet. In every other instance the pilots brought the planes down to ten or twelve thousand feet within ten or twelve minutes without injury to the others left aboard. On the whole, it was a great record and goes to explain why the scheduled airlines never lost a plane flying the North Atlantic. Comfortable as well as fast, high flying had everything to do with the great advance in public acceptance of air travel in these years.

Flying high meant less airsickness. Before airplanes were pressurized and flew above the weather, there wasn't much the airlines could do about this most important aspect of passenger comfort. They did what they could. After stewardesses reported that people in window seats seemed to throw up oftenest, American Airlines advised susceptible riders in its DC-3s and DC-4s to take aisle seats away from "visual stimuli outside the plane," adjust their seats to a reclining position, and fix their eyes as nearly as possible on the ceiling. In 1938 a young United physician heard that a new drug for pregnant women's morning sickness also kept them from getting carsick on the way to the doctor's office. This was Dramamine, and United lost no time spreading the word that it might help

queasy passengers more than the spirits of ammonia they had handed
around earlier.

Airsickness was a touchy subject, and not only for the airlines. Fear and
apprehension, of course, still had a lot to do with it, although, as one of
the industry's medical advisers said, "The average passenger, especially a
man in the presence of a stewardess, tends to conceal them." People became uneasy, the medics warned, at any alteration in the sound of the engines in flight. Even when the pilot throttled back after a climb to
cruising altitude, flight crews should provide a prompt explanation in the
cabin. To avoid downdrafts and "the sense that support is suddenly with-
drawn," airliners should detour around storms. In 1955 United installed
the first radars on airliners to enable pilots to spot thunderstorms at a distance and steer around them.

In 1953 Dr. Ross McFarland of Harvard, chief counselor on what came
to be known as "human factors" in air transport, ticked off the chief
causes of discomfort for airline passengers. Though high flying had
sharply reduced the number of cases of vomiting witnessed and reported
by stewardesses, airsickness still ranked number one on the misery list.
Well back in second place was aerotitis — the medics' name for the unpleasant effects of rapid pressure changes, including earache. Others were
extremes in cabin temperatures and drafts, oxygen lack at high altitudes,
and excessive noise and vibration that caused headaches and some temporary deafness. But Dr. McFarland had not a word to say about fatigue
induced by long air trips. That was a problem uncovered by the next big
leap in the cometlike trajectory of airline achievement.

Slow though they were at the start, airliners kept carrying people faster
and faster. During the 1950s they carried so many so far and so fast that
thousands and even millions of passengers' bodies were suddenly knocked
out of kilter, their minds were measurably skewed, and the transitions
they experienced between daylight and dark were so abrupt that they
couldn't help but realize that time itself was a sometime thing. At last
everybody got an inkling of what Einstein was talking about.

All this came about when, on the crest of the air travel boom, the airlines introduced the biggest of all their piston planes, the DC-7s and the
Super-Superconstellations. These planes were no faster than the planes
they replaced — the DC-7 was actually a bit slower than the DC-6. But
they had the capability, thought absolutely essential to competitive advance, of flying nonstop across the country and across the Atlantic. These
flights lasted ten to twelve hours and carried passengers through four or
five zones. This was a triumph of progress, a feat inviting comparisons
with De Voto's ox teams.

Then, just when the airlines seemed to have attained air travel's ultimate goals, passengers lifted so miraculously from Los Angeles to New

York, from New York to London, began complaining they felt terrible for days after they got there.

What nobody, neither the airlines nor their medical advisers, realized was that, in picking up people on one continent and depositing them ten or twelve hours later on another, they were upsetting some of the deepest rhythms of the human body. To collapse night and day in this abrupt fashion fouled up a whole array of biological clocks inside air travelers that as groundlings they had been only dimly aware of. The metabolic insult of swift flight across several time zones was so deep-thrusting that it disturbed not only bodily but also mental functions.

The most spectacular air traveler of the 1950s was John Foster Dulles. Aboard his government DC-7, the granite-faced old "cold warrior" set all sorts of records for one-man diplomacy, flying ceaselessly around the flanks of "world Communism" and shoring up the resolves of wavering allies. "There's a trough of low pressure moving rapidly eastward across the Atlantic, and guess who's in it!" read the caption of a British newspaper's cartoon alerting its readers to yet another visit from the high-flying U.S. Secretary of State. Little did Dulles or anyone else know what these long flights did to him. Returning from one of them to be told that Egypt had just bought arms from Russia, Dulles summoned the Egyptian ambassador to say there would be no loan for Egypt's Aswan Dam — a hasty act that provoked the nationalization of the Suez Canal and the Anglo–French–Israeli invasion of 1956. Before his death a year or two later, Dulles told columnist Marquis Childs he never would have made that mistake had he not been so tired after his long journey. The redoubtable secretary of state, and U.S. foreign policy, were victims of what later came to be known as "jet lag."

By the time the jetliners took to the air, FAA tests found out that air travelers had good reason to feel out of sorts after flights across several time zones. Professor F. Halberg of the University of Minnesota published findings in 1958 on what he called "circadian rhythms" — from the Latin *circa dies* for the time period of twenty-four hours. It seemed that in all human beings, no matter where they lived, heartbeat, temperature, perspiration, digestion, and urination rose and fell with daily cycles of activity and sleep corresponding roughly to the cycles of dark and light. And all these obscure rhythms were knocked out of whack when you flew across four or five time zones. You didn't only feel dreadful, your mental acuity also slacked off. After long trips, the tests showed, people did simple arithmetic more slowly; their responses to stimuli lagged. Yet the body was so flexible that after a few days the rhythms were reestablished and you felt all right again. After a return flight, there was a disturbance but not for so long. And it seemed that these subtle rhythms really kept time with the turning earth: individuals tested after flying all the way from New York to Chile — six thousand miles but all in the same time zone — experienced no foul-up in their internal clockworks at all.

When jet travel made the flying hangover commonplace, passengers learned how to live with it. All they had to do was slow down for a while after landing. Statesmen began breaking their long journeys with twenty-four-hour stops along the way. Businessmen took care to arrive a day ahead of time to rest up before negotiating a big deal. Tourists also were advised to take it easy after long flights and not rush right out to view the Taj Mahal. "Give the body a chance, let it come back to normal," said American Airlines medical director Dr. Ludwig Lederer.

After rushing back and forth on nonstop flights between Los Angeles and New York, one TV network man told Lederer he got so mixed up, so disoriented, he was ready to chuck his job. "Do you know any drugs for it?" he asked. Lederer said, "No, only sleeping medicines, but do you want some advice about how to lick your problem?" And Lederer told him that if he caught a plane in California at nine in the morning that landed in New York at 6:00 P.M., by his body clock that was only 2:00 or 3:00 in the afternoon. Lederer said, "You're a vice president, aren't you? If you want to call a staff meeting, why not call it that night at 8:00 P.M.? Let the New York people play up to you a little, and you stay on California time." The TV man told him afterward it worked like a charm.

Later Lederer said:

The older you are the worse jet lag affects you. The younger guy's rhythms bounce back faster. Don't rush out to catch planes. Catnap on the flight. Go easy on the food, and keep your clothing loose. Stretch, and move around. It helps circulation, especially in the legs — sitting for hours with legs bent at 90° isn't good. And when you get there, stay on your own time. Then, slowly each day, acclimate yourself to the time at your destination. We just have to learn to live in a relative way.

Part VI

Epilogue
(The 1960s and On)

32

Into the Jet Age

The man who led America into the age of jet travel was Bill Allen, Boeing's president. Wiry, a bit gnomish, with a high forehead and sharp brown eyes behind gold-rimmed glasses, Allen was a lawyer, no kind of airman at all. But he was a Westerner in an industry with a big westward tilt, and he had grown up with the business. A mining engineer's son, he was born at Lolo, Montana, right where Lewis and Clark crossed the Rockies on their way to the Pacific. Trained at Harvard Law School, he caught on with a Seattle firm that assigned him, as befitted the office's juniormost employee, to the Boeing account. In 1926 the Boeing business carried a retainer of fifty dollars a month.

Flying in open cockpit planes to negotiate airport leases and insurance contracts, Allen spent almost all the next years on the rising company's affairs. He drew up Boeing's first airmail bid, then set up Boeing Aircraft and Transport to do the job. When the New Deal forbade aircraft manufacturers to own airlines, Allen helped create United Airlines from Boeing's and other properties. But instead of joining United's move to Chicago, Bill Allen stayed with his Seattle firm and immersed himself further in Boeing's manufacturing growth. In the tremendous outpouring of U.S. air power in World War II, most of the bombers that hit Germany and Japan were Boeings. Near the end of the war President Phil Johnson, most dynamic of the engineers that built them, died suddenly, and a directors' committee of which Allen was a member took a long time searching for a successor. Finally the committee decided that Allen was the one man for the job. Protesting he was no engineer, he took over the job the day the war ended. He was then forty-five.

If America had a Krupp, Boeing was it. "What does a company like Boeing do between wars?" asked *Fortune* magazine. In the full knowledge that Boeing had lost money building commercial planes in the past, Allen decided that the company simply had to go ahead and build a commercial version of the B-29 that carried the A-bomb to Hiroshima. The result was the Stratocruiser, big, comfortable — and expensive. Allen was lucky to sell fifty-five. The company dropped $50 million on it. Boeing fell far behind Douglas and Lockheed in the race to supply commercial airliners.

But when the cold war came along and the only way the United States could send A-bombers all the way to Russia was by refueling them in flight, Boeing recouped. The Air Force bought 888 planes built like Stratocruisers but known as KC-97 tankers and rigged to tote aviation gasoline instead of passengers. This reduced the Stratocruiser loss to zero and gave Allen ideas for a commercial comeback.

As the cold war heated up the Air Force demanded nothing less than jet bombers for its "global" mission. As the nation's preeminent designer of bombers, Boeing built the first ones — speedsters with GE power plants and sweptback wings inspired by drawings that Boeing engineers had picked up at German plants after V-E Day. The idea of angling back the wing's leading edge so sharply was to fool the air and make it seem that the wing was not going so fast. Reducing drag, it enabled the plane to fly much faster before shock waves formed and thus reach speeds at which jet power paid off.

Having learned to harness the jet engine to the swept wing in the 500-mile-per-hour B-47, Boeing men like Maynard Pennell and George Schairer wanted to apply their designs to an airliner. Pennell said later, "On August 19, 1949, we made a presentation of three possible airliners, all jets. We singled out Pan Am. We tried to get Pan Am to sign a study agreement to pursue development of one of the international models." But no airline could fund anything so big and so advanced, and besides the Air Force had specified turboprops for the B-52 that Boeing was to build as America's next intercontinental bomber.

In September 1950 Allen traveled with Pennell to the Farnborough Air Show to see Britain's trail-blazing Comet jetliner shoot through the sky. "Do you think we could build one as good?" asked Allen. "Much better," replied Pennell. "We kept leaning on him," Pennell said later, "not that he wasn't interested." But Allen cannily held off, looking for a way to parlay a commercial design into something for the military too. Not until 1952 did Allen make his decision, and by then there had been a major technological breakthrough.

Boeing's old Pratt & Whitney partner at Hartford, making its own convulsive switch from pistons to jets under the stimulus of huge cold war contracts, had come up with a tremendous new jet power plant, better than anything built by the British or by GE or Westinghouse, the American turbine specialists. The new Pratt engine was at once so compact and so powerful that the military's interest in transitional turboprops suddenly evaporated. Placed in the F-100 interceptor, it carried fighter pilots for the first time faster than the speed of sound. And when the military ordered it inserted in Boeing's B-52 intercontinental bomber instead of a turboprop power plant, Allen's notions about the commercial jetliner finally fell into place. Surely, bombers built to fly to Moscow at jet speeds must be fueled in flight by tankers capable of the same speeds. In other words, if Boeing built an experimental jetliner, it was a good bet that the Air Force would want to order the same plane, or one much like it, to take the place of the outmoded KC-97.

But it wasn't that simple. For one thing the Air Force had already started talking to Boeing about putting jet engines into existing KC-97s. And though the Air Force was the country's, and the Congress's, darling in those days, there were limits on what it could ask. At that moment, particularly when it had just been voted a whole fleet of jet bombers, its chieftains could not bring themselves to ask for costly new tankers. On top of this, the airlines themselves had ordered another round of piston planes, and were both unwilling and unable to put up advance money for such a chancy venture as an experimental jetliner.

So Allen had to gamble. Military requirements having produced the right engine, he turned to his managers and asked a lot of hard questions. Their answers satisfied him that Boeing, though loaded up with immense bomber contracts, had the skills, the materials, the space, the plans, the budget to design and build a single prototype commercial plane. It could be done in two years, he was told, and would cost $16 million. On May 16, 1952, Bill Allen flashed the "let's-go," and Wellwood Beall, Ed Wells, Maynard Pennell and some four hundred engineers fell to work. Allen never let them lose sight of his dual-purpose aim. "We looked at the prototype requirements of both the commercial jetliner and the military tanker throughout," Wells said later.

Of course the experimental model would have the benefit of all the work on the bombers. It would have the magnificent swept wing, raked back to a thirty-five-degree angle, that delivered those 600-mile-per-hour speeds in the B-52. And so long as it kept the same 132-inch-wide body diameter of the old Stratocruiser, Ed Wells figured the plane would have the range to cross the Atlantic nonstop comfortably in either direction. But the commercial version must have three features that the airlines would insist on. Firstly, it must be a low-wing plane, so that the passengers would have the wing's protection under them in case of a wheels-up landing, a ditching, or a scrambled takeoff. Secondly, the jet engines must be hung separately under the wings because if they were paired in pods as in the bombers, safety inspectors would be sure to say if one engine caught fire the plane might suffer the loss of both. Thirdly, for stability in landing and taxiing, the plane must have tricycle instead of the bicycle landing gear specified by the military. Demanding performance before all else in its planes, the Air Force was willing to settle for less safety and comfort than the civilians.

A crucial element in Allen's calculations was that in creating the experimental plane, Boeing would have the right to use floor space and tools that had been bought by the government for the company's contract work on the bombers. In the spring of 1953 the Air Force, fearful that Boeing could not carry on a commercial program without jeopardizing the military program, withheld permission to rent the government facilities Allen had counted on. Thereupon Allen flew to Colorado Springs to buttonhole Air Force Secretary Harold Talbott. Talbott will be remembered as one of the Dayton promoters who used Orville Wright's name to swing profitable plane contracts in World War I, and who later, as a member of TWA's board, helped swing the contract for the first Douglas airliners. Talbott gave in, and thereby cleared the track for the first U.S. commercial jet.

At Seattle the life-sized wood mock-up was ready that fall. "We were more than fast — we had nobody's requirements to meet building that airplane," Ed Wells said later. At the last minute there was a structural problem with the huge tricycle landing gear. "First it had to be understood," Wells said, "loads were not going where our analysis said they should go. Then it was soon fixed." On May 15, 1954, the plane was rolled out of the plant, and old Bill Boeing himself, age seventy-two, paid his first visit to the premises in twenty years to glimpse the gleaming giant. The prototype of the 707 jetliner weighed 124 tons, 124 times the weight of Bill Boeing's first plane of 1916. Its engines generated 11,000 pounds of thrust, the equivalent of 110,000 horsepower, or 275 times greater than that of Bill Boeing's first mail plane of 1927. Its wingspan was 129 feet, its length 127 feet, and the production model carried up to 179 passengers — eighteen times as many as Bill Boeing's first modern airliner of 1932.

By this time the Air Force, just as Allen figured, had decided it needed a jet tanker. Lockheed won the design competition. But what was a paper design against a plane in being? On July 15, 1954, test pilot Tex Johnston took the 707 prototype aloft on its first flight, and reported they had a winner. It could climb and it could run — something like six hundred miles per hour. After that General Curt LeMay, the redoubtable, cigar-chomping chief of the Strategic Air Command himself, arrived from his Omaha headquarters to try it. On the first pass LeMay made a beautiful landing. The second time he bounced it forty feet into the air. "If I just learned to let these airplanes fly themselves I'd be a lot better off," he growled. He was enthusiastic. He went back to Washington and extracted an interim contract for twenty-one Boeing tankers modeled on the experimental jetliner.

In the sharing out of military contracts, Douglas had counted on getting at least a share of this business. But that interim order gave Boeing such an inside track that Douglas could not possibly match Boeing's terms on price and delivery. "We cried foul, but it did no good," Art Raymond said later. Bill Allen's gamble had already paid off. In 1955 the Air Force

The Boeing 707

signed with Boeing, and in seven years the company built 732 KC-135 jet tankers from the jetliner design.

With that large a military backlog Allen could now turn to marketing his 707 jetliner, and at his own pace. Events followed fast, but to explain what happened it is necessary to go back and see what the airlines had been up to.

The airlines had looked into jets after the war and decided they were out of the question, too costly to buy, and much too costly to operate. Jets guzzled fuel. They also lacked range. They were all very well for the military, who wanted performance and little else in their warplanes. But the airlines required planes that could fly straight and level, reliably and comfortably, for stated distances day after day. It was like comparing firetrucks and buses: the Air Force's fire wagons left their stations only occasionally and briefly. They experienced none of that monotonous pounding that airliners had to take.

So the commercial operators continued to load up with ever more piston planes, their only question being whether perhaps to install turboprop power plants instead of piston engines in their next generation of airplanes. In the early 1950s their talks with their main suppliers at Douglas and Lockheed centered around turboprops. In a much publicized speech at Syracuse, American's C. R. Smith declared that turboprops, not jets, were best suited to airline needs. At the time this belief created a procurement problem. The military had switched its support from turboprop power plants, and when the plane makers talked to the engine makers, the engine makers wanted to talk about jet engines because that was where their business was. In June 1954 when American signed up for Lockheed's Electra, the turboprop that powered it came from Allison in Indianapolis, which had never supplied an engine for an airliner before.

Pan Am, however, was the airline to watch. Its needs were different from those of such domestic outfits as American and Eastern. In fact, Pan Am's routes were a little like those of Fire Chief LeMay — long and overseas. Back in 1949, when Boeing had landed its first jet bomber contract, Allen's men found Pan Am interested in its early sketches for possible jetliners; it was the stony indifference at American and United that cooled off any disposition at Boeing to start into commercial jets at that time. When Britain brought out its Comet, the first of all jetliners, Pan Am's Andy Priester said the airline could not afford to pass it up. Juan Trippe placed orders for three of a longer-legged version, the Comet II, that might conceivably hurdle the Atlantic.

But the Comet came to grief, and it was another byplay with the British in 1955 that started Pan Am on the way to the age of jet travel. At the IATA meeting in London that spring, engineers from Rolls-Royce, the leading British engine manufacturer, cornered Trippe. On their own initiative, they said, they were developing a new turboprop engine for British European Airways. All Pan Am would have to do, they suggested, would be to take the DC-7s it had just ordered from Douglas, put this new turboprop engine in it — and the airline would have a superb new flying machine with plenty of range for its transocean needs.

Off went Trippe's procurement experts with the Rolls men to Santa Monica. After four days' palaver, the Douglas chieftains seemed to like the idea. Then San Kauffman of Pan Am asked Donald Douglas during the martini break if he had thought about when Rolls might have the engine ready. First thing after lunch Douglas said, "We haven't got around to delivery dates yet." Up to that point the Britons had said nothing about this subject. They asked for a recess. Kauffman said later, "Bear in mind that we knew the Boeing prototype was already flying." The British trooped back into the room. The new turboprop for the DC-7 could be delivered in the fall of 1959 — four years hence. "That's about when the Lockheed turboprop will be ready — you'll have a good competitor for the Electra," they said.

That night the Pan Am men called Trippe, told him about the exchange, and said they wanted to visit Boeing. Trippe said go ahead. At once Pan Am's Frank Gledhill phoned Wellwood Beall in Seattle. "Oh gee, you're just in time," Beall said. "I haven't ridden in the prototype myself yet. Come up and we'll give you a ride. We've got a flight scheduled Saturday."

Airline procurement men are not necessarily flyboys. Gledhill was an old mining engineer, Kauffman a groundling from Yale. The day they arrived something went wrong with the ailerons, and the flight was put off. Gledhill's reaction was to phone the home office to see if their insurance covered riding in an experimental plane. Kauffman said, "We don't need insurance, they're going to give us parachutes." Gledhill said, "At 30,000 feet? We'll freeze to death." Next morning they put on parachutes. A crew member pointed to a trapdoor in the plane's floor. He explained that just below it was an overhead bar, with another door below it to get out. "If you have to get out, you let yourself down through the trap door, grab hold of the bar, then let go for the jump." Looking at the bulky Beall, Kauffman said he doubted the hole was big enough. But Beall, an old airman, was not going to back out if his prospects from Pan Am went up. Pilot Tex Johnston roared, "What are you worrying about?"

They flew to 35,000 feet. "What speed, what power! I have never had such an experience in my life," Kauffman said later. But just then he had a job to do. It seemed to him pretty noisy in the bare, uninsulated plane. To satisfy himself where the noise came from, he went forward and got Johnston to cut the engines. It was engine noise, all right, but back in the cabin Gledhill and Beall got a bad scare thinking the engines had failed. The Pan Am men's judgment: "Terrific — if we can come up with economic operating figures that compare with other planes." Beall was ready with the answers. What the airline men had not reckoned on was the saving to be gained by the jet's much greater speed: planes that flew so fast could make many more trips, even if they used more fuel. Beall said, "We can match the DC-7, and we're shooting to equal the DC-6B."

Having seen the future work, Kauffman was all for making a deal at once. But Trippe and Gledhill had other ideas. "Let's be fair to the other guys," Gledhill said. So the Pan Am men flew right back to Santa Monica, and told all about their Boeing experience. "That," Kauffman said later, "was when Donald Douglas told us about the Rand report." It seemed that the RAND Corporation, the Air Force's Santa Monica–based research outfit, had told Washington that straight jets would never have the load-carrying capacity for long-range work. Douglas said: "We haven't looked at jets much because they burn too much fuel for over-ocean flying."

Gledhill took pains to stress that Pan Am would be very interested in a Douglas jet, and within weeks Donald Douglas decided for a gamble even riskier than Bill Allen's — to build a rival jetliner without supporting military orders, and on Douglas's own money.

Even while Douglas pondered his decision, the Pan Am men did not shut out other possibilities. They went across town to see Bob Gross at Lockheed, and asked if he was interested. Gross's answer, "No, I think three of us would kill each other. Our hands are full with the turboprop Electra." Kauffman said later, "Bob Gross was a pretty canny thinker. There *was* a third entry. Convair built a big jetliner — and nearly went bankrupt."

Convair plunged a bit later, egged on by Howard Hughes, and lost $270 million on its 880 jetliner. Douglas did not lose as much, but as Art Raymond said later, "It all but wrecked the company, and it all but wrecked me. Gone were the days of the DC-3 monopoly and overwhelming competitive advantage." From a standing start, without even a pause to build a prototype because Boeing was already so far ahead, Raymond's team produced, in the DC-8, a jetliner worthy of the proud Douglas name. Before it all ended, Douglas Aircraft had to accept merger into McDonnell-Douglas. But the Santa Monica company had a superb record for designing airliners and Gledhill, sticking by Douglas, said, "We'll get a better plane with two airplanes." Allen, who wanted to get the military tanker for the Pentagon started first, now gave his salesmen the green light to negotiate with Pan Am, and a head-to-head sales fight was on.

In October 1955, Trippe rocked the aviation world by announcing by far and away the biggest order ever placed by an airline: $265 million for the world's first fleet of jetliners. The division was a rude shock to Seattle — twenty Boeings, twenty-five Douglases.

After that, the other airlines could not sit on their hands long. Boeing's advantage was that it had already created a plane that flashed back and forth to Baltimore in less than eight hours — less than it took piston airliners to fly across the country one way. But Boeing would not be delivering the plane in commercial quantities for three more years, and Douglas, if successful with a jet design, might not end up so far behind. Douglas's advantage was that its plane, being later, could include the latest advances and accommodate the particular needs of airline customers.

So it was that Pan Am bought its DC-8s with the latest souped-up Pratt & Whitney engines capable of crossing the Atlantic nonstop in either direction. The original Boeing design with its Stratocruiser-sized fuselage, had had this capability. But then the Air Force, for its special purposes, had ordered the body widened by some eight inches. Later on, American and United came shopping. They demanded a further four-inch widening to permit six-abreast seating in the cabin. When United proceeded to order the DC-8, whose fuselage was 147 inches wide to the Boeing's 140, "we finally realized they were serious," Pennell said later. Allen caved in and agreed to add the extra four inches for American. That made the first

Boeing jetliner so thick-waisted for its engine power that it lost any chance of accomplishing the nonstop ocean crossings that were so important for Pan Am and its public.

From that point on, therefore, Boeing began offering two models of its 707 — one the original version and the other a larger plane with fifteen feet more wingspan and the bigger Pratt & Whitney engine. This larger, so-called intercontinental version became the best-seller, and outsold the DC-8 three-to-two. Pan Am was permitted to change its order so that all but its first six 707s were of this type. Outwardly they were practically indistinguishable, and when the DC-8 appeared, the Boeing and the Douglas turned out to be such look-alikes that most people could not tell them apart. But the DC-8's wing had five degrees less sweepback than the 707s, and this was enough to give the Boeing a twenty-mile-per-hour speed advantage.

As cities rushed to lengthen and thicken their airport runways, the airlines retrained their flight and ground staffs for the huge new machines and a large part of the nation's financial community was drawn into figuring ways of paying for the monster planes. Stuffed with electronic and hydraulic gear of staggering complexity and all kinds of backup systems to boot, the jetliners by all prior standards were prodigiously expensive — the DC-8 was termed a bargain because it cost $300,000 less than the $5.5 million 707.

This, said Eastern Chairman Floyd Hall later, was the time when the airlines finally came of age. Ever since they started carrying coach passengers, they had been growing faster than any other American industry. But now they became bigtime business. Without exception, the sums they had to lay out for jets exceeded their net worth. How to raise such sums was a challenge to lending institutions. Airlines, unlike manufacturing firms, possessed no inventories except unsold seats. Unlike railroads, they did not even own their right of way. Of course the day had passed when the president of Bankers Trust could turn away an airline loan request with the sneer that there wasn't room on his roof to land foreclosed airplanes. Yet to an uncommon degree financing of large-scale airline expansion required accepting new kinds of risks. Taking a page from rail history, banks and insurance companies began buying long-term equipment certificates on terms that enabled the airlines to amortize their expensive planes out of revenues over five-year, seven-year, and ten-year periods. But would the Boeings and Douglases go the way of the Comets? Would turboprops put jets in the ashcan? One of Juan Trippe's triumphs in leading the plunge into jets was in arranging for sale of ten-year debentures on his airline's equipment, without the necessity of starting repayment before the third year. It stood Trippe in good stead that of all airline chieftains, he was preeminently a child of Wall Street, at home all his life in the world of high finance. At the time he obtained financing for Pan Am's huge outlay for jets, he was a member of the board of Metropolitan Life and a member of its investment committee as well. The contrast with Howard Hughes, the reclusive outsider in Hollywood who lost control of TWA precisely because the financial community lined up against him, is instructive.

First to the Antilles, first across the Pacific, first across the Atlantic, first around the world and still, in 1958, first among airlines after being denied a place as America's chosen instrument in the world — it was another proud moment for Juan Trippe when at 7:20 P.M. on the evening of October 26, 1958, a Pan Am 707 poised for the first scheduled U.S. jet flight from New York to Paris. For that day the age of jet travel began. Only three days before, Britain's Comet jetliner, enlarged and revamped, had made it safely across from London. And two years before, the Russians, who had also modified cold war bombers to lug their businessmen around their huge country, had put into service a fifty-passenger jetliner between Moscow and Khabarovsk. But the Comet couldn't cut it, and the Russian TU-104 proved a fuel gulper. The new American jet transport, however, opened a chapter of uninterrupted success. As that chapter unfolded, it became evident that the first 707 flight took a place in history second only to Lindbergh's over the same track twenty-one years before.

Forty deluxe-class and seventy-one economy-class passengers — more passengers than had ever embarked on a scheduled flight before — were honored guests of Juan Trippe at a sendoff party in Idlewild Airport's international arrivals building. One was actress Greer Garson, crossing on holiday with her banker husband. Another was Mrs. Clive Runnells of Lake Forest, Illinois, who set the tone for the trip. Mrs. Runnells, dressed in matching brown, was crossing to Paris for lunch at the Ritz, then flying right back.

In order to be first with scheduled U.S. jet service, Pan Am had accepted the early model of the 707 that lacked range to fly the ocean nonstop. Actually these early 707s could cross both the continent and the Atlantic nonstop when flying eastward, with the wind at their backs. But in transatlantic trips the only alternative airport short of their destination was the ocean. So large a fuel reserve, therefore, had to be carried that they almost always put in for more fuel at Gander or Shannon or Iceland or the Azores. That was not all. One of the peculiarities of jet-powered engines as then designed was that it took them longer to accelerate to takeoff speeds than piston engines. This even more than their great weight was why the new planes needed two-mile runways. To make up for this deficiency Pratt & Whitney had added water injection to the engines. Shooting water into the stream of air entering the engine increased the air's density and lowered its temperature. That allowed more fuel to burn, thereby stepping up the force of the combustion during the dash down the runway. The water boost was so crucial that Pan Am Captain Jim Fleming later said that on takeoff he would rather have lost an engine than his water supply. Aboard the first 707s was no less than 686.2 gallons of water for squirting into the four engines to help on the liftoff. The plume of black smoke that the early jets left behind as they roared up from airports was mostly water.

The evening of the inaugural takeoff was dark and drizzly. Captain Sam Miller, chief pilot of Pan Am's Atlantic division, had prepared for the trip by checking out in Boeing's prototype over Seattle, piloting Pan Am's first plane on test runs to four European capitals, and appearing on "Meet the Press." Since local residents were raising cain about jet noise, Miller planned to take off over Jamaica Bay, where there were fewer houses. But noise was only one of the new matters pilot Miller had to attend to. The Boeing 707 was a flying machine of a wholly different order of complexity. It was fifteen times as powerful, twice as fast, and twice as big as the Stratocruiser, the largest commercial landplane up to then. Flying it, Miller said later, required "management of mass and speed." Before departure Miller and his copilot, Captain Waldo Lynch, had to work up a much more precise flight plan. On reaching their starting point, they had to check such strange new drills as setting exhaust pressure ratio for all four burners and making sure of water-flow to each engine. Draped before their cockpit positions were six-stage "takeoff computation" charts no airliner had ever needed before. Releasing the brakes, Miller let the 125-ton giant roll. He said later:

We were taking off light, with just enough fuel to get to Gander. The first part of the run was just like driving the plane, power-steering it. Then, at about 80 knots, the aerodynamics took hold, the directional and lateral controls began to work. Waldo called off the acceleration intervals.

"V_1!" That was my last chance to hit the brakes, stay on the ground. It was wet and slippery — not much chance. The speed built up.

"VR!" I lifted the nose wheel, "rotated" the plane up on its rear wheels to tilt for the climb. With water injection, our speed was rising 4 or 5 knots a second. We were only a little more than a mile down the runway.

"V_2!" When Waldo said that, the airspeed indicator read 170 knots [194 miles per hour] and we lifted into the air. At our angle of 13° we climbed right into the clouds.

Then Waldo reached his right hand over and pulled the handle to raise the landing gear.

The first U.S. commercial jet flight was off, climbing at 3,500 feet a minute while the passengers oohed and ahed at the absence of noise and vibration inside the big cabin. Two hours and forty-two minutes later they touched down at Gander, where high-pressure hoses pumped 17,000 gallons of JP4 fuel into the tanks. An hour or so later the passengers were dozing high over the Atlantic weather. Watched by radar in five countries, the big plane began its descent over Ireland, and whooshed down at Paris in early morning sunshine eight hours and fifteen minutes after leaving New York — having traveled 3,300 miles at an average 475 miles an hour.

From the first, the jet transport proved what the young jet bomber pilots had told Boeing's Maynard Pennell when he asked if they met any delays: "All you have to do is fill up the tank and go." National Airlines leased one of Pan Am's 707s and on December 10, 1958, started carrying people between New York and Miami. American put the first 707s on transcontinental flights January 25, 1959. No fuel stops were necessary in the six-and-a-half-hour trip. Poet Carl Sandburg held court among 110 passengers in the "moving room" at 35,000 feet above his native Illinois. Afterward he said he had been "whisked, streaked, zipped, flicked, sped, hurtled, flashed and shuttled" across the sky faster than Phaeton's chariot pulled the sun.

There were problems. American's Captain Ham Smith, who piloted Sandburg's plane, said later, "Every takeoff was marginal. We used every available foot of runway. At San Francisco the landing gear would some-times hit the [radio] localizer shack lifting off the end of the runway." Captain Fleming, one of twenty-one "administrative pilots" who kept Pan Am's six 707s flying while the company resisted union demands for a share of the new planes' productivity gains, said later, "What we needed was good plumbers. The water for injecting the engines had to be pumped from the fuselage. In cold weather the pipes would freeze, and then we had water running down the aisles." On a chill wintry departure from Paris, Sam Miller pulled down the deicing bar that shot hot air into all four engine nacelles, and two of the four burners abruptly flamed out. Somehow they got them relit, and flew on. At New York engineers swarmed over the plane and concluded that what could not have hap-pened had happened: the diversion of air had caused what they called a "compressor mismatch" inside the power plants. Everybody gave thanks that the flameouts had occurred in engines on opposite sides of the plane, or the plane would have veered out of control. The deicing bar was ripped out and controls reset so that engines could be deiced one at a time.

With its sweptback wing the jetliner had a tendency, if jarred by rough air aloft, to get out of control. In the DC-8, this could lead to "upsets" at high altitude. In the Boeing 707, it set the plane rocking from side to side in a hammocklike motion known as a Dutch roll. Unless quickly arrested, it could cause discomfort or worse. Only months after the 707s went into service Pan Am Captain Howard Cone on a practice flight over France got into a Dutch roll so violent that an engine flipped off. Cone landed the plane at London with severed fuel and hydraulic lines lying back over the wingtop like tangled spaghetti. Instructions went out to all airlines on how to avoid this tendency. A couple of years later United went to Seattle to buy a version of the 707 known as the 720. On the acceptance flight United's Captain Weldon Rhoades cut back both engines on one side in the belief that prompt use of rudder should keep the plane from rolling. Boeing's Tex Johnston jammed the throttles forward again and roared, "You can't do that." When Rhoades asked why, Johnston said, "Well, you're supposed to train pilots not to do that." Back on the ground, Rhoades refused to accept the plane, and United backed him up. Boeing then decided to add three feet to the vertical stablilizer to give the plane rudder enough to hold it against Dutch roll. The FAA came to life and ordered the same modifications made on all early 707s.

The early jetliners were so sturdily built that they never suffered grounding. The 707's closest call came while Pan Am's "21 Club" — the twenty-one administrative pilots — were still manning the Atlantic schedules. On that occasion Captain Waldo Lynch had gone back to visit his 110 passengers and left his copilot, Captain Sam Peters, in charge. Peters thought the automatic pilot was on, as is usual and almost necessary in flying the heavy jetliners. But it wasn't on. Another automatic unit had kicked it off and Peters, talking on the radio, didn't notice that the plane was edging over. From 30,000 feet the 707 nosed into a dive. The dive was so steep and the G forces set up were so great that Lynch had to get down in the aisle and crawl back to his seat. Together the two pilots, using all their strength, pulled the plane out at 6,000 feet. This bent the plane's wings, but it survived. Had it not, the 707 would have been grounded. Of Lynch, Bill Allen said later: "They made him a vice president and fined him $1,000." From that time forward the FAA forbade captains to go back and visit with passengers.

Somewhat later a Pan Am 707 was lost in a severe thunderstorm over Elkton, Maryland. Airmen have always sworn that planes, whether metal or fabric, were invulnerable to lightning damage. Sam Miller testified, "In all my flying I never had an explosion. A prong of lightning can pass right through an aircraft without damage because the plane is not grounded." But the CAB decided a bolt of lightning might have ignited fuel vapor in the wings. Boeing changed the venting system in the wings, but the JP4 fuel developed for the Air Force was identified as the real culprit. Though much like kerosene, JP4 also contained some of the more volatile light spirits common to gasoline. From that time on, though it costs a little more, commercial jetliners flew on plain kerosene.

On the whole, jet fuel economics turned out about as the airlines expected. C. R. Smith said later: "Those first jet engines gobbled a lot of gas. This reduced range and reduced payload. But we saved a lot on the price of fuel" — kerosene then cost ten cents a gallon delivered on board. But the jet engine had other merits, Smith said:

They eliminated complexities that went with piston engines. On those last piston engines you had a lot of cylinders — 28 to 32 of them — that produced a tremendous vibration. In the jet you simply had blades all turning the same direction, and no vibration. There were goddam few engine failures after we began flying jets.

That was probably the airlines' biggest surprise. The capacity of their engines to take hard, unremitting use also surprised Pratt & Whitney, who had piled up spare parts for the usual replacement demands and lost a bundle because they weren't needed. Where the turbocompound engines on the DC-7s and Superconnies had had to be pulled off every thousand hours if not sooner, the power plants on the 707s and DC-8s ran and ran without any attention from mechanics whatever. Before long the government authorities raised the number of hours between jet overhauls to 1,400 (in 1978:14,000 hours). Such outstanding power plant performance of course enabled the airlines to operate their planes more intensively than ever before. TWA, which received only one 707 at first put that sole jet on transcontinental service at once and it performed so well it saved the airline from what would have been a disastrous month's loss of revenue to its competitors. At Pan Am the men of the 21 Club were hard put to keep up with such stalwart machines. Jim Fleming said later, "In the first ten months I crossed the Atlantic a hundred times. It was a workout landing at places like Boston, Keflavik, even Prestwick in Scotland. I remember flying the New York–Paris–Rome trip and falling sound asleep over the dinner table in Rome after landing. We got so tired it took years to get over the effects."

Meanwhile Pan Am's 707s had only BOAC's outclassed Comets to contend with, and the company, charging a $10 premium to ride on the big birds, was racking up $50,000 with every round trip. Indeed, the jetliner passenger-mile costs came out just about as Wellwood Beall said they would — equal to those of the DC-7. A gap still separated 707 and DC-8 costs from those of the DC-6B, that most economical of all piston airliners, but that was about to be closed. Passengers flocked to fly jet — in 1959 51 million flew in the United States, 7 million overseas in the first year. So the airlines, with 197 707s and 150 DC-8s either flying or on order at the end of 1959, were on their way to the heights. The plane makers too, especially Boeing, were whistling a happy tune. So much had the cold war wrought for U.S. commercial aviation and its passengers.

33

Flying High

The age of jet travel may be said to date from 1959. But the day when the uncertainty, anxiety, and adventure vanished from air transport was not ushered in until a year or two later. Then, at last, the airliner became a magic carpet that carried you so fast and so securely that you forgot you were flying. That moment arrived with the fan-jet engine.

The first jetliners flew as fast and smoothly as most passengers could have asked. But they gulped fuel. They couldn't lift the loads that airlines' profit projections demanded. Even with water poured into the engines to boost their thrust, they needed every foot of the two-mile runways to hoist a hundred or so passengers off the ground. They also created worrisome noise problems. Wheeling and taxiing around airport terminals, they gave out a frightful banshee howl. Though the manufacturers fitted "sound suppressors" on their engines to hold the din to a supposedly supportable hundred decibels, the airlines clapped earmuffs on their ramp workers to cut hearing damage. Barraged by the cries of nearby householders, the airlines could think of nothing better than to order their pilots to climb more steeply after takeoffs. President Bob Peach of Mohawk actually suggested the country might have to breed a population less sensitive to noise.

More power and less noise — you might have supposed it impossible for the airlines to achieve both. Yet this is what they did with the fan engine.

Once again they could thank the British for the idea. Back in 1936 the same Frank Whittle who gave the world the jet concept had applied for a patent that admirably spelled out the principle. Only he called it a "bypass" engine.

The principle of the bypass engine, which Rolls-Royce developed after World War II, was that some of the air taken in at the front would not pass through the combustion process. Instead it would simply be compressed on its way through separate ducts to the rear. There, just in front of the jet nozzle, this bypass air would join and reinforce the jet of gas exhaust that, spat out to the rear, gave the machine forward thrust.

The British put only a moderate amount of air through their bypass ducts. Stressing mainly the cooling effect of this air on the power plant, Rolls-Royce's chief engineer Adrian Lombard estimated no more than a 5 percent efficiency gain for his Conway engine, which BOAC, Trans-Canada, and Air India selected for their Boeing 707s. But he also noted that the noise of the jet would be reduced by the proportion of bypass air in the blast that came out of the jet pipe.

When Lombard outlined these advantages at a New York meeting, two top Pratt & Whitney engineers violently disputed him. But General Electric took up the concept. Sensing the fuel-saving possibilities of enlarging the ratio of bypass air, GE constructed an engine with much more airflow and greatly lengthened compressor blades to handle it. They called it a "fan" engine. Their version was known as an "aft-fan" engine because the compressor blades were at the rear of the power plant. Pratt & Whitney, whose straight jet engines went into both the 707 and DC-8, remained a vehement holdout until Convair chose GE's aft-fan engine for its new jetliner. Then things happened fast.

Boeing was impressed by the new engine, Maynard Pennell said later.

American Airlines also made it plain that they wanted the fan for noise-reducing reasons as well as for fuel efficiency. We took this very seriously.

In January 1958 we told Pratt & Whitney in New York that if they didn't offer us the turbo-fan we would have no choice but to shift to the GE aft-fan. They acted. On a Saturday morning 4 or 5 days later they invited us to Hartford for a look. Wright Parkins was the guy, outspoken, competent, who turned around at our demand.

We saw an engine that they had put together on paper in those few days. The first two stages of the compressor — the fan stages — were literally taken from a nuclear-powered engine they had given up on after 8 to 10 years. These stages used titanium blades and had considerable running experience and good performance data. They put this in front of a J57 engine. Then they added the new turbine stages to provide the extra power for these large fans, and that was their answer.

What had happened? Pratt & Whitney had done it again. They had taken the principle that Whittle had enunciated and Rolls had cautiously executed and expanded it almost beyond recognition. If Rolls had thought to send a little bit of air through the engine uncombusted, Pratt & Whitney turned its model into a veritable fresh-air machine. Less than half the thrust would be produced by the combustion core of the engine. Instead the compressor blades at the front were to be lengthened to squeeze in huge mouthfuls of air, lengthened until they were almost as long as old-time piston-plane propeller blades. The air the British had thought to bypass through small ducts was now compressed and propelled backward by these big blades and did most of the work. The combustion core's role shrank to that of a supercharger raising the speed of the fan blades to the point where they compressed and expelled the air that provided 60 percent of the plane's forward thrust.

It was another triumph of America's cold war imperium — and it was immediately applied to the U.S. military airlift in the Asian, African, and Middle Eastern crises of the 1960s and 1970s. But the dazzling payoff was in commercial air transport. The fuel saved by using mostly air as propellant was of the order of 15 percent. Reequipped with fan engines, the same Boeing 707s and DC-8s could fly farther while carrying up to seventy-five more passengers. And the decibel count at the end of the runway for these reequipped jetliners dropped from one hundred to ninety-five.

After that it was jets — American jets — all the way on the world's air routes. Even in planes designed for short and medium hops, everything went the American way. In the late 1950s the French brought out a nifty twin-engine jetliner called the Caravelle with a marvelous innovation: the engines were banished to the rear. This uncluttered the wing, vastly improving aerodynamic efficiency. Inside the cabin, the engines were almost inaudible. United Airlines, one of the domestic Big Four, placed an order for twenty of these fetching foreign machines. Thereupon Boeing and Douglas simply designed one-hundred-passenger models, fastened Pratt & Whitney fans on their tails (three on the Boeing, two on the Douglas), and took away both the domestic and world markets with their medium-range 727 and DC-9.

Fan jets further changed flying styles. Even on trips as short as Houston to Dallas or Minneapolis to Chicago, the way to get the most out of these planes was to take them up to thirty thousand feet as soon as possible. Sometimes they appeared to climb almost straight out of sight from takeoff. The 727 was also designed for quick descents. But a pilot had to know how to handle what the experts were pleased to call its "sink rate." Four nasty crashes occurred at Chicago, Cincinnati, Salt Lake City, and Tokyo, all because pilots thought they could drop toward the field and flare into a flat glide as if they were floating down in a comfortable old DC-6. Instead they either fell short of the runway or smacked it so hard they broke the landing gear and the bottom of the plane and lost half their passengers — not by impact but by fire and smoke. Once this hazard was understood, however, the airlines piled on the short-range jets — Boeing sold 1,350 of these 727s, Douglas 570 of its DC-9s — to carry businessmen in and out of such close-in airports as Chicago's Midway, New York's La Guardia, and Washington's National.

As fan jets kept air travel growing by 15 percent annually, the airlines looked beyond the small, medium, and merely large, and reached for the biggest fan jet they could get: the wide-bodied jumbo. In 1965 Juan Trippe led the way by placing yet another record order — $525 million for a fleet of twenty-five Boeing 747 jetliners capable of carrying two and a half times as many passengers as the biggest 707s and DC-8s. Tall as a five-story building, the 747 was half again longer and wider than its predecessors. Passengers sat nine abreast. The plane was capable of flying across *any* ocean; it brought every place on earth within twenty hours' flying distance. Douglas and Lockheed built wide bodies too.

There, in the roller-coaster progress of the airlines, the tremendous surge of plane-building ended. The industry had overreached itself. The demand for air travel finally slacked off; there simply weren't enough people flying to fill all the 374-seat jumbos. All the airlines felt the pinch and some of them went deep into the red. In other ways the industry seemed to have come up against limits. The next faster model of transport plane, the supersonic, was so costly it could only be built for commercial use if the Air Force went on building bombers. An explicit and huge subsidy was required, and after the cold war receded Congress and the public had no stomach for that. At the other end of the spectrum, helicopters and vertical takeoff planes also turned out to be unprofitable for airline operation. But in the late 1970s, prodded by Congress, the airlines began to offer bargain fares. Once more planes filled to overflowing. Air travel resumed its fantastic growth.

34

Air Travel in the 1960s

Jets took getting used to: when a Pan Am 707, laden with junketing travel agents and journalists, taxied to the end of the runway at New York's international airport in late 1958, a woman from Seattle peered out, then asked nervously, "Aren't you even going to use your propellers on the takeoff?"

At first the fashionable people who had hitherto made their seasonal moves from Palm Beach and Newport to San Moritz and Antibes by ship greeted the jetliners enthusiastically. For the "jet set," as these people were instantly dubbed, the airlines cordoned off up to half the cabin as "first class," and tried gamely to simulate the spacious elegance of ocean liners and make customers comfortable with blankets and pillows as if they were in deckchairs. It became briefly chic to fly from New York to London for a weekend, and the *New Yorker* reported a young woman who flew to Paris to have her hair done before a ball at the Plaza. *Vogue* solemnly noted "the great change in the concept of travel dressing" without steamer and wardrobe trunks or even the clutter of train and piston-engine toiletry:

The pale grey suit, its fresh white collar and cuffs, the white straw hat it's worn with, all such were impossible, or at least impractical, before the jets with their clean, unhectic swiftness.

The jet set were soon buried in the flood of economy-class riders. Experienced travelers held, however, that an airline's service style was determined by what the airline's native first-class passenger has been brought up to expect. Thus Air France offered meals a thousand miles long. And long after its rivals had retreated to plastic, Scandinavian's china and cutlery won design prizes at Stockholm's Museum of Modern Art. Japan Air Lines flight attendants were so attentive that outlanders often mistook them for geisha girls. And TWA, serving steak and potatoes even on the Atlantic, was still a Kansas City airline: when it put on airs, the result, one traveler said, was "lobster tails à la midwest."

In the first two years of jet flying, travel by air almost doubled. Between 1955 and 1972, the number of airline passengers in the U.S. rose from 7 to 32 million — almost a fivefold increase. In those years, according to Gallup polls, the proportion of Americans who said they had flown at least once rose at last to 50 percent. The increase in population, in the moneyed class, in overall income played its part in this swift growth, which exceeded that of any other U.S. industry for the period. Tremendous promotion by the airlines and travel agents of economy fares and package vacations also had their effect.

Those who flew were people with incomes two or three times above the general run, and they fell into two groups. On the one hand were those with enough spare cash to take long trips, usually for holidays and often as families, perhaps once or even twice a year. These were identified by the airlines as "discretionary" travelers. The rest were those whose trips were paid for — people, in other words, traveling on business. Declaring that "one of the 20th century's greatest romances is between businessmen and the jet," *Time* magazine reported in 1964 that no fewer than 86 percent of domestic airline passengers were businessmen. Though that number was almost certainly high, it appears that close to two-thirds of passengers on planes flying in and out of New York at that time were business executives and others traveling on expense accounts. Not only that, these were by far the most frequent air travelers, which made for a paradox: at the very time air travel was turning into a mass phenomenon, most jetliner seats were occupied by a relatively small part of the flying public. The Port Authority of New York and New Jersey reckoned that 5 percent of the passengers took 40 percent of all air trips in 1969 (each member of this group made an average of ten or more trips a year). This meant that the basic pool of these air travelers worked out at no more than 1.5 million people.

Accordingly, the busiest routes were the trade routes, topped by New York–Boston, New York–Chicago, and New York–Los Angeles. In 1964 and 1965 the airlines speeded things up by introducing the Boeing 727 and Douglas DC-9, which could fly in and out of close-in airports such as Washington's National and New York's La Guardia. At once the convenience for briefcase commuters revived La Guardia and made that field busier than either Idlewild or Newark. Flying to places like Atlanta or Chicago, executives did not even go into town but convened their meetings at the airport or nearby hotels, and jetted home for supper. Visiting buyers and bankers arrived oftener in New York but left a lot sooner than in earlier times. With jets the length of hotel room occupancy in Manhattan dropped from 3 to 2.2 days per guest. In some firms travel budgets tripled. When companies built new plants they put them near airports; when executives transferred to a new city went house-hunting, they kept handy access to the airport in mind. Every morning and evening at every major airport in the land these merchant travelers ebbed and flowed.

Their cars, left for their return, clogged every airport parking space. So strong were these diurnal tides that night flying on the airlines shrank to a trickle.

You could see anybody at airports — the Beatles, the President, your mother, your hairdresser, your congressman hustling home to his constituency for the weekend. In an off year Richard Nixon rode Eastern's shuttle between his New York law practice and Washington. One day in 1962 W. Alton Jones, president of Cities Service, called his friend C. R. Smith in New York for a seat on an American flight to Chicago where he planned to pick up his company Convair for a trip to the West Coast. C.R. said, "You're a goddam fool to get your ass shook up in a rattletrap old Convair 440. Let me fly you out there on a jet all the way." So Jones, one of the country's leading oilmen and a great friend of President Eisenhower, went to Idlewild and caught American's nonstop 707 for Los Angeles. With ninety-four others he lost his life when the plane crashed in the bay on takeoff. In his luggage was found a bag stuffed with $55,000 in currency for use, the press then reported, in California's gubernatorial campaign.

On weekends and at hours between the commuting rush the airlines shifted to pleasure travel. Economies made possible by the size and superb serviceability of the new planes held domestic fares down to an average five cents a mile, a third of what they had been in baling-wire days. This brought out the kind of individual private travel that made these the airlines' most profitable years. As Pan Am sales chief Willis Lipscomb said, an unsold airline seat was lost forever. Every seat filled by a "discretionary" traveler represented found money and, since the company broke even on a half-filled plane, often pure profit.

The jets exerted a particularly powerful pull on the young, who felt none of the old hesitancies about taking to the skies. With jets the number of young people studying in Europe rose 150 percent; it soon became commonplace for American students to know several continents first hand before they were twenty-one. The junior year abroad was a jet-age phenomenon. The summer mass migration of youth to Europe began with the jets: 120,960 young Americans were recorded in 1960 as visiting foreign youth hostels, nearly all of them in Europe. To keep their jet momentum the domestic airlines offered bargain youth fares to young people willing to stand by for the last unsold seats. The results were impressive. Where only 22 percent of young people ages eighteen to twenty-four had flown before 1962, the proportion rose to 59 percent by the time the airlines dropped the discount a decade later.

For personal travel Las Vegas and Miami led all destinations. Jetliners disgorging fugitives from the snow country turned the Caribbean into an American lake every winter. Priorities for these tourists were sun, sand, sea, and sex. However short the vacation, jets could deliver them all. Pan

Am offered fourteen-day to twenty-one-day excursions between New
York and London, California and Hawaii, New York and Puerto Rico; one
year American Airlines rolled up more profits on its Caribbean flights
than on all the rest of its system put together. With almost any place on
earth less than twenty-four hours from any major U.S. airport, office secretaries thought nothing of vacationing in Vienna, Delhi, Lima. One young
American woman, flying out to India, wound up as queen of Sikkim; another, lifted by jet to a job in Amman, married the king of Jordan.

Within two decades 100,000 Americans had circumnavigated the globe
by jet. By 1970 as many Americans had jetted to Japan as flew to all Europe in the piston years. Then, air tourism being a two-way street, the
Japanese turned into the leading air tourists to the United States. Europe-
ans· flying on international vacations outnumbered Americans, lifting
tourist income of Spain alone to 43 percent of that country's foreign exchange earnings. And Lufthansa put prayer rugs on its Middle East flights
so that Muslims could perform their offices on the way to Mecca.

The jetliner, so big and steady you hardly knew you were in the air, invited Bea Lillie's ocean-liner sally, "When does this place arrive?"
Though seated ten abreast you could get up and walk around, or in the
jumbos, climb to an upper deck for a drink. You watched a movie, listened
to music, and scarcely glanced at the clouds seven miles below through
which the Ford Tin Goose once battled. Your food was frozen, thawed
below decks in microwave ovens, and served, in metronomic cadence, a
meal a minute. There were still bad moments when a plane might bump
into the jet stream and flight attendants and unbelted passengers were
flung without warning against the roof. There were occasional disasters,
although the Boeing 747 went four years before its first crash. There were
also skyjackings.

The first act of aerial piracy was pulled off back in the 1930s in Peru
when insurgents took over a Fokker biplane piloted by an American,
Byron D. Richards, and used it to shower Lima with pamphlets. But never
was there a U.S. skyjacking until the jet age. Then, on May 1, 1961, a man
armed with a knife and pistol forced the skipper of a National Airlines
Convair to abandon his Miami–Key West run and fly to Castro's Cuba.
The other fourteen passengers were returned to Miami, but the skyjacker
dropped out of sight. El Pirato Cofrisi was the name he had given to the
ticket agent — Cofrisi was an eighteenth-century Spanish pirate. El
Pirato really started something. More flights were forcibly diverted to
Cuba. After a Frenchman hijacked a Pan Am 707 to Cuba, was returned
to Mexico, and sentenced to eight years in jail, there was a lull. Then
came 187 skyjackings in rapid succession, capped by the seizure of a
Northwest 727 by a man who compelled its return to Seattle for $200,000
and a parachute by which he disappeared into the night as the plane flew
south over the Cascades. The airlines only stopped the epidemic by

searching every passenger before departure with electronic metal-detecting gear. Overseas, where airlines often were government-owned, skyjacking became a weapon of political terror.

Although tickets were sold for balloon rides out of Paris when the Germans besieged the city back in 1871, travel by air is a recent phenomenon. The first passenger on a powered flight, Leon Delagrange, went up only seventy-odd years ago. The first commercial passenger of record was Mayor A. C. Pheil of Tampa, who paid $400 for the privilege of being lifted in Tony Jannus's air ferry across the bay to St. Petersburg in the winter of 1913. One of the most traveled of all air passengers was Sam Irwin, an old Curtiss-Wright hand and still a flying salesman in 1979, when he had logged 3.4 million miles. Close on his heels was the globe-trotting aviation editor Wayne Parrish, who had racked up 3.3 million miles. Bert Hemphill of Los Angeles, dean of U.S. travel agents, organized the Century Air Travelers Club for those who had flown to a hundred or more countries and "island groups." His star member was undoubtedly Al Rabin, a Los Angeles leather dealer who claimed to have visited 240 countries and possessed a United Airlines 4-million-mile card to show for it. But the palm for most traveled air traveler went by 1978 to John Otto, diplomatic courier for the U.S State Department. Since 1946, flying with pouches for almost every embassy and legation, Otto had piled up a total of 5.25 million hours as a passenger — almost eighteen months off the earth.

It has always been possible to charter your own airliner if you have the means. In 1959 the Rockefeller family hired a Pan Am DC-8 to fly them to a son's wedding in Norway. When a Pan Am Stratocruiser was forced down at Belém airport in Brazil in 1951, Bolivian tin baron Mauricio Hochschild hired a fifty-seven-seat Panagra DC-6 to fly down and fetch him and his wife, and no other passenger, on to New York.

Chartering a plane for two was a baronial gesture. But chartering a plane for a group could be a bargain because in return for a sure full load an airline charged less per seat. In early times few groups dared try. In 1934 Larry MacPhail, chief of the Cincinnati Reds, took the chance and chartered two American Airways Ford trimotors to transport his baseball team to Chicago — though a coach, two pitchers, and three infielders insisted on taking the train. In 1940 United lifted the University of Michigan Big Ten champions to California to play at Berkeley. It took three DC-3s to haul Tom Harmon and his mates from Ann Arbor, but Coach Fritz Crisler said the airlift helped them win by five touchdowns.

In 1945, facing competition from the nonskeds and their war-surplus planes, United opened a charter department and went after the sports business. United's Bob Johnson said later: "We made it possible for base-

ball to go national. We made it possible for pro football to go national."
Up to then the big leagues extended only as far west as St. Louis, the limit
for overnight Pullman jumps from the big eastern cities. But when United
started flying baseball clubs to all their away games, the owners saw new
possibilities. In 1957 Walter O'Malley moved his Dodgers from Brooklyn
to Los Angeles — chartering United DC-7s for cross-country road trips.
Then came jets, which completely redrew the map of professional sports.
Baseball, football, and hockey leagues added "expansion" teams in cities
once thought too remote to be in the majors. Clubs like the Braves moved,
trying the loyalty of fans, from Boston to Milwaukee to Atlanta. Wherever they went, United charters flew them.

For airlines that, like United and Pan Am, maintained big fleets, charters made up as much as a tenth of their traffic. When 250 employees of
the De Kalb seed company in Illinois wanted a week's outing in Hilo,
United could lift them all there in a single plane. Such parties, and there
were thousands of them in the 1960s, were called "affinity groups." As
leisure-time travel grew and grew, the airlines also went after the pack-
age-tour business that had been pioneered, with some CAB encouragement, by the nonskeds. Through travel agents and travel "packagers,"
tourists could buy bargain trips, including hotel beds and ground transportation, with their air tickets to such spots as Las Vegas, Honolulu,
Miami, if they were willing to wait until there was a planeful ready to go.

When the airlines began offering big commissions to lure such business,
the package-tourist often found himself flying Pan Am or Sabena instead
of such charter-only carriers as World or Texas International. Before long,
when there weren't enough charter passengers to fill a plane, they were
simply switched into the empty seats on a scheduled airliner. Then the
passenger who paid the regular fare to London or Bermuda or Honolulu
found himself sitting beside package-tourists who had paid perhaps 40
percent less. Eventually the CAB was obliged to recognize what was
going on. The airlines were authorized to offer a number of seats on every
flight at lower fares. In this way the rise of charter flying compelled, in
1978, the first reduction in air fares since the jets arrived. The result was a
new surge in air travel.

35

The Imperial Airplane

The impact of the airplane, which arrived so suddenly, has been great, greater than we have understood. Up to 1945, when a Boeing bomber delivered World War II's final blow upon the people of Nagasaki, 95 percent of all planes ever flown were warplanes. Thirty-odd years later the scales have swung into better balance. Having found more terrible carriers of destruction, the great powers have all but stopped making warplanes. Commercial aviation has come into its own, and commercial aviation helps bring men together. Yet after surveying the damage the airplane did in its first seventy-five years, the least that can be said is that it came into our hands so suddenly that civilization was hard put to contain it.

By all the standards of human transport, the airplane came like a bolt through the blue. The *Queen Mary* and the Twentieth Century Limited were crowning achievements of years of steadily advancing efforts to refine and improve, respectively, sea and land transport. In less than sixty years the flimsy, floppy kite of the Wrights, seized upon to lift us in peace as well as war, usurped them both.

What "public convenience and necessity" were served by displacing the ocean greyhound that could deliver you in four days from New York to Southampton, or the iron horse that could carry you overnight in safety and some comfort between New York and Chicago? It's a measure of how the airliner has changed our thinking that this sounds like a silly question. For the airliner has swayed us to prize speed beyond everything else in travel. It has shrunk space, enormously enhanced our mobility, and caused us to conceive our existence in space-time terms. If the supersonic transport ever becomes economical, Americans will undoubtedly build and fly it — three hours, we'll say, not three thousand miles, to Paris.

Lugging people and mail more than cargo, leaping over mountains and water, the airliner has altered our trade routes. It has shaped our economy, binding the Far West, the Southwest, the Deep South, not to mention Hawaii and Alaska, to the rest of the country as never before.

Decentralizing the nation's corporations, it gives rise to so-called conglomerates; it has enabled specialized firms to expand nationwide.

The airliner changed politics, sports, even matrimony. Presidential candidates campaigned by plane, rallying their supporters at airports. On jet wings, baseball and pro football went national. And when the nation's young, taking advantage of air travel, enrolled in colleges at a distance from home, they subsequently found spouses and jobs in other parts of the country. Air travel abetted other population shifts — millions of Puerto Ricans migrated to the mainland by air. Within the country the airliner lifted individuals long distances to look for new jobs and couples to scout for places to retire.

For Americans the airplane's international impact was profound. After 1945, when air power ruled the world, the United States built an empire without annexing any land. Under an aerial nuclear umbrella, the nation established bases in forty-four countries around the globe. These bases Washington managed and largely manned by air. At the same time American businessmen, traveling the same airways, branched overseas and created multinational corporations that enlarged the American presence on every continent. With jets not only businessmen but also millions of tourists found it just as easy to go to Madrid as Miami and vacationed three to six thousand miles from home. And jets ushered in a new diplomacy whereby Presidents, turning up as far away as Peking in a few hours, conducted our foreign relations face-to-face with other chiefs of state.

With jets the airlines indeed became big business. Their immense capital requirements in equipping themselves with such big planes forced TWA out of control of a maverick like Howard Hughes; the same need for staggering sums drove all the airlines to the banks and insurance companies and into covenants that required the financier's approval for any deviation from prescribed ways. Some, like Pan Am, nearly went broke. The old-timers who had led the airlines to the heights stepped down — Patterson of United in 1966, C. R. Smith of American in 1968, Eastern's Captain Eddie Rickenbacker in 1963, and last of all, Juan Trippe, whose career in airline building went back to the 1925 beginnings, in 1969.

Their successors, flying a thousand airliners an average of ten hours every day of the year, lifting a million of us every twenty-four hours between fifty major cities and foreign ports, kept America on the go as never before. But compared with the brash and gingery gang that guided aviation on up from wing-walking days, they were an anonymous lot. Like all supercorporate managers, they were beset with the problems of dealing with employees and customers who insisted upon being treated as people, not units. Computers kept their accounts, classified their help, booked their passengers, scheduled their flights, and even steered their planes.

Walking off their jets to the shriek of spewing kerosene, you may have glimpsed a Connie relegated to a hangar; it was soon gone forever. And gone were the fright, the fun, the follies of piston-engine flying. The novelty of long-distance passenger flight had ceased to be novel. There was now no other way to go. The miracle of flight, which had always depended on machines, had been triumphantly turned into a repetitive and unexciting routine. The adventure was over.

Acknowledgments

The development of air transport has been so recent that many of the participants were still around to tell how it was. I wish to thank the following for granting me interviews: W. J. Addems, R. F. Ahrens, E. R. Alexander, W. M. Allen, J. W. Austin, J. C. Baird, A. C. Ball, Erwin Balluder, Curtis Barkes, W. E. Beall, P. E. Bewshea, H. F. Blackburn, R. W. Blake, J. G. Borger, G. S. Bowen, M. O. Bowen, R. A. Bowen, T. L. Boyd, John Boyle, Harl Brackin, J. W. Brennan, M. A. Brenner, G. H. Brink, O. F. Bryan, Walter Bullock, W. N. Bump, J. H. Carmichael, P. L. Carpenter, R. J. Celestre, Cyril Chappellet, C. M. Christenson, E. O. Cocke, S. R. Cohen, T. G. Cole, F. B. Collins, Charles Colvin, J. G. Constantino, E. R. Cutrell, Jack Dawson, A. De Gormo, Vincent Doyle, R. M. Dunn, E. J. Eidsen, I. G. Entrekin, C. S. Feeney, Miriam Filkins, T. J. Flanagan, Lee Flanagin, J. L. Fleming, A. E. Floan, J. O. Fortuna, J. C. Franklin, H. J. Friendly, Paul Garber, Stanley Gewirtz, Gordon Gilmore, W. C. Golien, Melvin Gough, H. G. Gulbransen, J. E. Guy, M. G. Hack, F. D. Hall, K. R. Harkins, H. R. Harris, H. L. Hibbard, W. J. Hogan, R. F. Holbrook, F. J. Hoyt, S. D. Irwin, C. L. Johnson, R. E. Johnson, S. B. Kauffman, R. D. Kelly, J. F. Lederer, L. G. Lederer, E. H. Lee, J. C. Leslie, H. C. Leuteritz, H. S. Levelle, H. H. Lippincott, W. G. Lipscomb, D. S. Little, J. T. Logan, Stanley Lowell, P. D. McTaggart-Cowan, D. F. Magarrell, H. N. Maletz, W. M. Masland, S. M. Miller, T. M. Moffitt, Allan Moore, C. S. Murphy, A. R. Ogston, G. R. Parkinson, W. W. Parrish, Nathaniel Paschall, W. A. Patterson, G. E. Pendray, M. L. Pennell, L. W. Pogue, D. L. Putt, A. E. Raymond, L. J. Raymond, J. T. Real, P. H. Redpath, D. H. Renninger, W. E. Rhoades, Mrs. Paul Richter, G. S. Rosen, D. D. Rosenfeld, Jacob Rosenthal, R. W. Rummel, T. J. Sanders, C. F. Schory, R. C. Scruggs, S. L. Seltzer, B. F. Sessel, Julian Singman, J. E. Slater, E. E. Slattery, Jr., C. R. Smith, D. C. Smith, H. C. Smith, R. L. Smith, Zelma Smith, R. D. Speas, C. R. Speers, Anthony Stadlman, I. K. Staggers, E. T. Stern, T. K. Taylor, J. H. Tilton, S. G. Tipton, D. W. Tomlinson, J. T. Trippe, Marie Trotta, V. S. Trygstad, J. McK. Tucker, William Utley, C. S.

Vaughn, T. K. Vickers, R. L. Wagner, T. F. Walkowicz, F. L. Wallace, J. B. Wassall, J. C. Waugh, E. C. Wells, H. C. Westwood, L. M. Wilson, M. M. Wilson, N. K. Wilson, R. D. Wonsey. I also received important letters from many of the above, and from C. V. Abbott, J. R. Alderson, Luis Alvarez, R. K. Baker, R. E. Bilstein, Walter Braznell, E. R. Breech, J. H. Clemson, H. M. Cone, J. R. Cunningham, R. G. Davidson, R. J. Dick, R. M. Emmert, Warren Erickson, M. S. Ersoz, W. S. Green, J. A. Guglielmetti, G. Guillot, J. B. Jaynes, Martin Jensen, R. S. Johnson, C. S. Kachline, C. E. Kaul, C. F. King, W. T. Larkins, W. M. Leary Jr., A. V. Leslie, W. D. Lewis, A. J. Lilly, C. E. Lovett, Jr., M. E. McAlpin, Frank McCarthy, Frank McCulloch, J. C. Mackey, J. L. Otto, F. C. Pogue, A. H. Rabin, J. H. Roe, J. H. Rosdail, J. S. Runnells, Paul Strohm, G. R. Turner, E. D. Weeks, Lillian Wheeler, H. C. Whitford, M. D. Wright, and B. S. Wygle.

Valiant in helping to locate air pioneers were Dave Frailey and Kay Hansen at American Airlines, Gil Perlroth at Eastern, Roy Erickson and Ken Ruble at Northwest, Will Player, Jim Arey, and Ann Whyte at Pan Am, Ken Fletcher and Parky Parkinson at TWA, Jim Kennedy and Chuck Novak at United, Forrest Mulvane and Linda Cole at Western, Peter Bush and Gordon Williams at Boeing, Ray Towne and Harry Gann at Douglas, Wayne Pryor at Lockheed, and Chapin Leinbach at the Air Transport Association. Of many in Washington agencies who helped, I wish to mention Ed Slattery, Dennis Feldman, Nick Komons, and Alfred E. Kahn, and special thanks to my old friend and co-worker, veteran of twenty-five years on the Washington aviation beat, Jerry Hannifin. Librarians without number lent a hand but I would particularly like to mention Philip Lagerquist at the Truman Presidential Library, Patricia Marshall at the American Institute of Astronautics and Aeronautics, and most of all, the able and devoted staff at the New York Public Library, where this book was chiefly researched.

William H. Forbis, Captain R. J. Selmer, USN Retired, and Jerome Lederer, president emeritus of the Flight Safety Foundation, read the entire manuscript; and Per F. Dahl, Thomas B. Davis, Henry F. Graff, John C. Franklin, Jerry Hannifin, Harvey H. Lippincott, David C. Little, Wayne W. Parrish, Hart Preston, Arthur E. Raymond, Richard W. Solberg, and Tirey K. Vickers read chapters and suggested many improvements.

Bibliography

MANUSCRIPT COLLECTIONS

Franklin D. Roosevelt Papers, Hyde Park, N.Y.
Harry S. Truman Papers, Independence, Mo.
Walter F. Brown Papers, Columbus, Ohio.

GOVERNMENT PUBLICATIONS

Air Mail Service: *Air Mail Service Schedule, 1926.*
        Monthly Report of Operations and Maintenance, 1918–1921.
        Statement of Performance from 1918 to 1923, 1924.
Archives, Department of State Files. Foreign Relations of the United States. Re Colombia, 1939. Re Brazil, 1939–1941.
Army Air Force, *Aviation Psychology Program Research Reports Nos. 1–19, 1947.*
Census Bureau. *Statistical Abstract of the United States. Historical Statistics of the United States, Colonial Times to 1970, 1975.*
Civil Aeronautics Authority: *Air Commerce Bulletin.*
        Annual Reports.
        Civil Aeronautics Bulletin.
        Flight Information Manual.
        National Airport Plan, 1950.
Civil Aeronautics Board: *Accident Investigation Reports.*
        Aircraft Accident and Maintenance Review.
        Aircraft Accident Reports.
        Annual Airline Statistics.
        Annual Reports.
        Economic Reports (Dockets).
        Handbook of Airline Statistics.
        Report of Air Carrier Financial Statistics.
        Résumé of U.S. Civil Air Carrier Accidents.
        Study of Aviation Insurance, 1944.
        Study of U.S. Air Carrier Accidents Involving Fires, 1955–64, 1965.
        Subsidy for U.S. Certificated Air Carriers, 1972, 1974.
Commission on Intergovernmental Relations. *Staff Report: Federal Aid to Airports, 1955.*
Congress. *Congressional Record.*
    House. Judiciary Committee. *Monopoly Problems in Regulated Industries, Part 1.* Hearings, 84th Congress, 2d session, 1957.
    Senate. Commerce Committee. *To Create the All American Flag Line.* Hearings, 79th Congress, 1st session (March 19–May 4, 1945).
        Subcommittee on Appropriations. *Hearings,* 83d Congress, 2d session, 1955.

Post Office Committee. *Revision of Air-Mail Laws.* Hearings, 73d Congress, 2d
session, 1934.
Special Committee Investigating the National Defense Program. *Investigation of
the National Defense Program* Hearings, 90th Congress, 1st session (July
28–Aug. 11, 1947).
Special Committee on Investigation of Air Mail and Ocean Mail Contracts.
Hearings, 1933–34, Parts 1–9, 1934.
Department of Commerce. *Annual Reports of the Secretary,* 1926–1933.
  Bureau of Air Commerce: *Consolidated Interline Safety Agreement,* 1936.
                     Air Commerce Bulletin.
Department of State. *Foreign Relations of the United States,* 1928–1945.
Department of Transportation: *Civil Aviation Research and Development.*
               *Policy Study: A Historical Study Derived from
Application of Technical Advances to Aviation,* 1971.
Federal Aviation Administration. *Flight, The Story of Electronic Navigation.* Oklahoma
City, 1974.
Federal Aviation Agency: *Airman's Information Manual, Part 1,* 1976.
                  Airport Paving, 1962.
                  Flight Standard Information Manual.
Federal Aviation Commission. *Report,* 1935.
Federal Trade Commission. *Aircraft Manufacturing Corporations,* 1940.
National Advisory Committee on Aeronautics. *Reports.*
National Air and Space Administration. *Aeronautical Engineering.*
President's Air Coordinating Committee. *Civil Air Policy,* 1954.
President's Aircraft Board. *Hearings,* 4 vols., 1925.
President's Special Board of Inquiry on Air Safety. *Report,* 1947 (mimeo).
Public Papers of Franklin D. Roosevelt.
Public Papers of Harry S. Truman.

BOOKS, ARTICLES, AND INDIVIDUAL MANUSCRIPTS

Aero Club of America. *Annual International Aeronautical Exhibition Prospectus No. 1,*
1912.
Aitken, James J., and D. L. Aitken. *White Wings, Green Jungle.* Mountain View, Cal.,
Pacific Press, 1966.
Allen, Francis R. *Technology and Social Change.* New York, Appleton-Century-Crofts,
1957.
Allen, Richard S. *Revolution in the Sky.* Brattleboro, Vt., Stephen Greene Press, 1964.
Allward, Maurice. *Safety in the Air.* New York, Abelard-Schuman Limited, 1967.
American Airlines. *Annual Reports.*
———. *Admirals' Club Handbook,* 1965 (mimeo).
———. *Corporate Structure and History,* 1935 (mimeo).
———. *History of American Airlines,* 1967 (mimeo).
"American Air Transport through the Magnifying Glass," *Interavia,* October 1946.
Arey, James A. *The Sky Pirates.* London, Ian Allan, 1972.
Armstrong, H. G. *Principles and Practice of Aviation Medicine.* Baltimore, Williams &
Wilkins Co., 1952.
Armstrong, William. *Pioneer Pilot.* London, Blandford, 1952.
Arnold, H. H. *Global Mission.* New York, Harper & Row, 1949.
Ashwood, T. M. *This is Your Captain Speaking.* New York, Stein & Day, 1974.
Aviation Corporation of the Americas, *Annual Reports,* 1928, 1929.

Baarck, H. *Medical Guide for Air Travelers.* Darmstadt, Garuda, 1974.
Bach, Richard. *Stranger to the Ground.* New York, Harper & Row, 1963.
Bache & Co. *Handbook of the Air Transport Industry.* New York, 1940.
Balchen, Bernt. *Come North with Me.* New York, E. P. Dutton & Co., Inc., 1958.
Ball, R. S. *A Chronology of Michigan Aviation, 1834–1953.* Lansing, 1953.
Banks, F. R. *Aircraft Prime Movers of the 20th Century.* New York, Wings Club, 1970.
Barclay, Stephen. *Search for Air Safety.* New York, William Morrow & Co., Inc., 1970.

Bassett, P. R. "Aeronautics in New York State," *New York History*, April 1962.
————. "Passenger Comfort in Air Transport," *Journal of Aeronautical Science*, March 1935.
Baxter, James P. *Scientists against Time*. Boston, Little, Brown and Co., 1946.
Beals, Carleton. *The Crime of Cuba*. Philadelphia, J. B. Lippincott Co., 1933.
Beard, H. C. *Wings*. London, Blackett, 1934.
Beaty, David. *Human Factor in Aircraft Accidents*. New York, Stein & Day, 1970.
————. *Water Jump*. London, Secker & Warburg, 1976.
Benford, Robert. *Doctors in the Sky*. Springfield, Ill., Charles C Thomas, Pub., 1955.
Bennett, D. C. T. *The Air Mariner*. London, Sir Isaac Pitman & Sons, Inc., 1943.
————. *The Complete Air Navigator*. London, Sir Isaac Pitman & Sons, Inc., 1937.
Bernstein, David. "Our Airsick Airlines," *Harper's*, May 1947.
Bernstein, Marvin D., ed. *Foreign Investments in Latin America*. New York, Alfred A. Knopf, Inc., 1966.
Bilstein, R. E. "Technology and Commerce: Aviation in the Conduct of American Business," *Technology and Culture*, July 1969.
Black, Archibald. *Civil Airports and Airways*. New York, Simmons-Boardman, 1929.
Black, Hugo, Jr. *My Father: A Remembrance*. New York, Random House, Inc., 1975.
Blomquist, A. E. *Outline of Air Transport Practice*. New York, G. P. Putnam's Sons, 1943.
Boeing Company. *Annual Reports*.
————. *Pedigree of Champions*. Seattle, 1963.
Bond, D. D. *Love and Fear of Flying*. New York, International Universities Press, Inc., 1952.
Borden, Norman W. *Air Mail Emergency 1934*. Freeport, Me., Wheelwright Bond Co., 1968.
Bowers, Peter. *Boeing Aircraft since 1916*. Fallbrook, Cal., Aero Publications, Inc., 1966.
Boyd, T. A. *Professional Amateur: The Biography of Charles Franklin Kettering*. New York, E. P. Dutton & Co., Inc., 1957.
Bramley, Eric. "Truman's About-Face Climaxes Bitter Pan Am–AOA Case," *Aviation Week*, Aug. 1, 1950.
Brewer, Stanley. *Impact of Mail Programs on U.S. Air Carriers*. Seattle, University of Washington Graduate School of Business, 1967.
Briddon, A. E., E. A. Champie, and P. A. Marraine. *FAA Historical Fact Book*. Washington, 1974.
British Ministry of Information. *Merchant Airmen, 1939–1944*. London, His Majesty's Stationery Office, 1946.
Brock, Horace. *Flying the Oceans*. Lunenburg, Vt., Z. Stinebour Publ., 1978.
Brooks, Peter W. "The DC-3 is Twenty-Five," *Flight*, Dec. 23, 1960.
————. *Historic Airships*. London, Evelyn, 1973.
————. *The Modern Airliner*. London, Putnam and Co., Ltd., 1961.
————. "Why the Airship Failed, " *Aeronautical Journal*, October 1975.
Bruno, H. A. *Wings over America*. New York, Robert McBride, 1942.
Buck, Robert N. *Flying Know How*. New York, Delacorte Press, 1975.
————. *Weather Flying*. New York, The Macmillan Co., 1970.
Bullock, Raymond H. *Airline Piloting*. Denver, World Press, 1947.
Burby, John. *The Great American Motion Sickness*. Boston, Little, Brown and Co., 1971.
Burckhardt, Robert. *The Civil Aeronautics Board*. Washington, Green Hills, 1975.
————. *The Federal Aviation Administration*. New York, Praeger Publications, 1967.
Burden, W. A. M. *The Struggle for Airways in Latin America*. New York, Council on Foreign Relations, 1943.
Burgess, C. P. "Dawn of New Era in Passenger Transportation," *Aerial Age*, March 1923.
Burton, C. L., L. W. Mayer, and E. H. Spuhler. "Aircraft and Aerospace Applications," in *Aluminum* II, American Society for Metals, 1960.
Butz, J. S., Jr. "Pratt & Whitney Evolves Turbofan from J-57 Program," *Aviation Week*, Jan. 26, 1959.

Caidin, Martin. *Barnstorming*. New York, Duell, Sloan and Pearce, 1965.
————. *The Silken Angels*. Philadelphia, J. B. Lippincott Co., 1964.
Caldwell, Cy. "Atlantic Fever," *Aero Digest*, May 1927.

Canby, Courtlandt. *History of Flight.* New York, Hawthorn Books, Inc., 1963.

Caves, Richard. *Air Transport and its Regulators.* Cambridge, Harvard University Press, 1962.

————. "Performance, Structure and Goals of C.A.B. Regulation," in Paul McAvoy, ed., *Crisis of the Regulatory Commissions.* New York, W. W. Norton & Co., Inc., 1970.

Chamberlain, C. D. *Record Flights.* New York, Beachwood Press, 1942.

Chamberlin, Edward. *The Theory of Monopolistic Competition.* Cambridge, Harvard University Press, 1933.

Champie, Ellmore. *The Federal Turnaround on Aid to Airports, 1926–1938.* Washington, DOT/FAA Historical Staff, 1973.

Chapman, C. E. *History of the Cuban Republic.* New York, The Macmillan Co., 1927.

Cherington, Paul W. *Airline Price Policy.* Boston, Harvard Graduate School of Business, 1958.

Chorley, R. A. "Seventy Years of Flight Instruments and Displays," *Journal of the Royal Aeronautical Society,* August 1976.

Churchill, Winston. *The Hinge of Fate.* Boston, Houghton Mifflin Co., 1950.

————. *The Grand Alliance.* Boston, Houghton Mifflin Co., 1951.

Clayton, Hoard. *Atlantic Bridgehead,* London, Ian Allan, 1968.

Cleveland, R. M. *Air Transport at War.* New York, Harper & Bros., 1946.

Cohen, S. Ralph. *IATA, the First 30 Years.* Montreal, IATA, 1949.

Collings, Kenneth. *Just for the Hell of It.* New York, Dodd, Mead & Co., 1938.

Colonial Air Transport. *The Modern Way to Travel.* 1929 (pamphlet).

Colson, C. N. "Modern Carpetbagger," *Flight,* June 27, 1935.

Comstock, Louisa. "Put Your Two Weeks on Wings," *Better Homes and Gardens,* September 1947.

Conn, Stetson, and Byron Fairchild. *Framework of Hemisphere Defense.* Washington, Office of Military History, Dept. of the Army, 1960.

Connecticut General Life Insurance Co. *Flight Forum* on "Aviation in the Transportation System," 1967.

————. *Flight Forum* on "Transportation Technology Problems in the Air," 1962.

Conoley, Ken. *Airlines, Airports and You.* Don Mills, Ont., Longmans Canada, 1968.

Cooper, James. "The First Pressurized Airliner," *New Horizons* (Garrett Corp.), Fall 1956.

Coordinating Research Council. *Aviation Fuel Safety.* Report No. 380. New York, 1965.

Corbett, David L. *Politics and the Airlines.* London, Allen and Unwin, 1965.

Corrigan, Douglas. *That's My Story.* New York, E. P. Dutton & Co., Inc. 1938.

Cott, Nate, and Stewart Kampel. *Fly without Fear.* Chicago, Henry Regnery Co., 1973.

Cottrell, Leonard. *Up in a Balloon.* New York, S. G. Phillips, Inc., 1970.

Cox, H. G. "British Aircraft Gas Turbines," *Journal of Aeronautical Science,* February 1946.

Craven, W. F., and J. L. Cate, eds. *Army Air Forces in World War II, Services Around the World.* Chicago, University of Chicago Press, 1958.

Crome, E. A. *Qantas Aeriana.* Sutton Colfield, England, Francis J. Field Ltd., 1955.

Crittenden, Ann. "Juan Trippe's Pan Am," *New York Times,* July 3, 1977.

Crump, Irving. *Our Airliners.* New York, Dodd, Mead & Co., 1940.

Curtis Publishing Co. *Saturday Evening Post Survey,* 1946.

————. *Travel Market of the United States,* 1953.

Curtiss, Glenn H. "The Future of Commercial Aviation," *Aviation,* Aug. 1, 1919.

Cutrell, E. R. "Instrument and Radio Flying," *Aviation,* June 1935.

Damon, Ralph S. *TWA, Nearly Three Decades in the Air.* New York, Newcomen Society, 1952.

Daniel Guggenheim Fund. *International Safe Aircraft Competition.* New York, 1930.

————. *Pioneering in Aeronautics: Recipients of the Medal, 1929–1952.*

Davidson, Gustav. *Dictionary of Angels.* Glencoe, Ill., The Free Press, 1967.

Davies, Joseph E. *Mission to Moscow.* New York, Simon and Schuster, Inc., 1941.

Davies, R. E. G. *Airlines of the United States since 1914.* London, Putnam & Co., Ltd., 1972.

————. *History of the World's Airlines.* London, Oxford University Press, 1964.

Davis, Edward W. *Pioneering with Taconite*. St. Paul, Minn., Minnesota Historical Society, 1964.

Davis, Kenneth S. *The Hero — Charles A. Lindbergh and the American Dream*. Garden City, Doubleday & Co., Inc., 1959.

Day, Karl S. *Instrument and Radio Flying*. New York, Air Associates, 1938.

De Havilland, Geoffrey. "Commercial Aircraft," *Flight*, April 18, 1935.

———. *Sky Fever*. London, Ian Allan, 1961.

De Leeuw, Hendrick. *Conquest of the Air*. New York, Vantage Press, Inc., 1960.

Dellinger, J. H., and Haradan Pratt. "Development of Radio Aids to Air Navigation," Institute of Radio Engineers *Proceedings*, July 1928.

Denmark, H. W. "Early History of Air Traffic Control in the Armed Forces," *Journal of Air Traffic Control*, October-December 1975.

De Soutter, D. M. *Aircraft and Missiles*. New York, John de Graff, Inc., 1959.

De Voto, Bernard. "Transcontinental Flight," *Harper's*, July 1952.

Dickinson, Thomas A. *Aircraft Construction Handbook*. New York, Crowell, 1943.

Dietrich, Noah, and Bob Thomas. *Howard: The Amazing Mr. Hughes*. Greenwich, Fawcett, 1976.

Dixon, Charles. *Conquest of the Atlantic by Air*. London, S. Low, Marston, 1931.

Donovan, Frank. *The Early Eagles*. New York, Dodd, Mead & Co., 1962.

Doolittle, J. H. "Early Blind Flying," *Aerospace Engineering*, October 1961.

Douglas, Donald. "The Modern Transport Machine," *Flight*, April 18, 1935.

Douglas, George W., and J. Miller. *Economic Regulation of Domestic Air Transport*. Washington, The Brookings Institution, 1974.

Douglas Aircraft Co. *The DC-8 Story*. Santa Monica, Cal., 1972.

Driscoll, Ian. *Winged Journey*. London, C. Johnson, 1947.

Duke, Donald. *Airports and Airways*. New York, The Ronald Press Co., 1927.

Duke, Neville, and Edward Lanchberry, eds. *Saga of Flight*. New York, The John Day Co., Inc., 1961.

Dunsmore, F. W. "Radio Beacons Non-Directive and Directive," *Radio News*, May 1924.

Durno, George E. *Flight to Africa, A Chronicle*. U.S. Air Transport Command, 1943 (mimeo).

Durr, Clifford J. "Hugo L. Black: A Personal Appraisal," *Georgia Law Review*, Fall 1971.

Eastern Airlines. *History of Eastern Airlines*. New York, 1976 (mimeo).

Eckener, Hugo. *My Zeppelins*. London, Putnam & Co., Ltd., 1958.

———. "The Zeppelin Crisis," *The Air Ship*, April-June 1938.

Eddy, Paul, Elaine Potter, and Bruce Page. *Destination Disaster*. New York, Quadrangle, 1976.

Edgerton, J. S. "Seeing America from the T.A.T.," *Aeronautical Review*, August 1929.

Ellis, Frank H. *Atlantic Air Conquest*. Toronto, Ryerson Press, 1963.

Ellis, Steve. "The Boeing 40: Transcontinental Pioneer," *Air Line Pilot*, November 1977.

Emme, E. M., ed. *History of Rocket Technology*. Detroit, Wayne State University Press, 1964.

Ethyl Corp. *Aviation Fuels and Their Effects on Engine Performance*. Detroit, 1951.

Evans, Charlie. "The Howard Hughes Story," *Houston Chronicle*, Oct. 10–14, 1971.

Evans, Edward, and Richard Hoyt. *Air Transport as an Aid to Business*. Columbus, Ohio, 1929.

Fahey, James C. *Ships and Aircraft of the U.S. Navy*. New York, Gemsco, 1941.

Falconer, B. L. *Flying Round the World*. Boston, Stratford, 1937.

Farley, James J. *Jim Farley's Story*. New York, Whittlesey House, 1948.

Faunce, Cy J. *The Airliner and its Inventor*. Columbus, Ohio, Rockcastel Publishing Co., 1921.

Fay, Stephen, Lewis Chester, and Magnus Linklater. *Hoax*. London, Andre Deutsch, 1972.

Finch, R. J. *The World's Airways*. London, University of London Press, 1938.

Finch, Volney C. *Jet Propulsion Turbojets*. Millbrae, Cal., National Press, 1948.

Flanagan, J. C., ed. *Aviation Psychology Program in the Army Air Forces*, Research Report No. 1, Washington, 1948.
Fokker, A. H. G. "Airplanes for Commercial Aviation," *Society of Automotive Engineers Journal*, September 1927.
———. *Flying Dutchman*. New York, Henry Holt, 1931.
Ford, H. S. "Walter Folger Brown," *Northwest Ohio Quarterly*, Summer 1954.
Fortune: "Colossus of the Caribbean," April 1931.
         "No. 1 U.S. Airplane Company," April 1932.
         "U.S. Aviation and the Air Mail," May 1934.
         "Success in Santa Monica," May 1935.
         "The Air Is How Safe?" April 1937.
         "Jet Propulsion: the U.S. is Behind," September 1946.
         "The Airline Squeeze," May 1947.
         "The Selling of the 707," October 1957.
         "The Problem of Howard Hughes," January 1959.
Foulois, Benjamin. *From the Wright Brothers to the Astronauts*. New York, McGraw-Hill, Inc., 1968.
Frank, J. P. *Mr. Justice Black*. New York, Alfred A. Knopf, Inc., 1949.
Frederick, John H. *Commercial Air Transportation*. Homewood, Ill., Richard D. Irwin, Inc., 1961.
Freud, Sigmund. *Leonardo da Vinci*. New York, Vintage/Random House, Inc., 1932.
Freudenthal, Elspeth. *The Aviation Business*. New York, The Vanguard Press, Inc., 1940.
———. *Flight into History*. Norman, University of Oklahoma Press, 1949.
Friedman, Paul D. "Fear of Flying: The Development of Los Angeles International Airport." Master's Thesis, University of California at Santa Barbara, 1978.
Friendly, Henry J. *Federal Administrative Agencies*. Cambridge, Harvard University Press, 1962.
Fromm, Gary. *Aviation Safety*. Washington, The Brookings Institution, 1969.
Fruhan, William E. *Fight for Competitive Advantage*. Boston, Harvard Graduate School of Business, 1972.
Fulton, J. F. *Aviation Medicine in Its Preventive Aspects*. London, Oxford University Press, 1948.
Fysh, Hudson. *Qantas Rising*. Sydney, Angus & Robertson, 1965.

Galland, Adolf. *The First and the Last*. New York, Holt, 1954.
Gamow, George. *Gravity*. Garden City, Anchor Paperback, 1962.
Gann, Ernest K. *Ernest K. Gann's Flying Circus*. New York, The Macmillan Co., 1964.
———. *Fate is the Hunter*. New York, Simon & Schuster, Inc., 1961.
———. *Sky Roads*. New York, Thomas Y. Crowell Co., 1940.
———. "Superfort with Seats," *Flying*, February 1974.
———. "The Tin Goose," *Flying*, August 1974.
Garland, G. D. *The Earth's Shape and Gravity*. Oxford, Pergamon Press, Inc., 1965.
Gartmann, Heinz. *Man Unlimited*. New York, Pantheon Books, 1956.
Gauer, Otto H., and George D. Zuidema, eds. *Gravitational Stress in Aerospace Medicine*. Boston, Little, Brown and Co., 1961.
Gay, Leslie, and Paul Carliner. "Prevention and Treatment of Motion Sickness," *Bulletin of the Johns Hopkins Hospital*, May 1949.
Genevoix, Sylvie, and Marianne Gosset. *H.R.H. ou une Amerique*. Paris, Plon, 1972.
Gibbs-Smith, C. H. *The Aeroplane*. London, Her Majesty's Stationery Office, 1960.
———. *Aviation*. London, Her Majesty's Stationery Office, 1970.
———. *Flight through the Ages*. New York, Thomas Y. Crowell Co., 1974.
———. *The Invention of the Aeroplane*. London, Faber and Faber, Ltd., 1966.
———. *New Book of Flight*. London, Oxford University Press, 1948.
———. *World's First Aeroplane Flights*. London, Her Majesty's Stationery Office, 1965.
———. *The Wright Brothers: A Brief Account*. London, Her Majesty's Stationery Office, 1963.
Gill, F. W., and G. L. Bates. *Airline Competition*. Boston, Harvard Graduate School of Business, 1949.
Gillies, J. A., ed. *Textbook of Aviation Physiology*. Oxford, Pergamon Press, Inc., 1965.

Glines, C. V. *Compact History of the U.S. Air Force*. New York, Hawthorn Books, Inc., 1963.
————. *The DC-3*. Philadelphia, J. B. Lippincott Co., 1966.
————. *Saga of the Air Mail*. Princeton, D. Van Nostrand Co., 1957.
Goldsmith, Margaret. *Zeppelin: A Biography*. New York, William Morrow and Co., 1931.
Gordon, Thomas. *The Airline Pilot: A Survey of the Critical Requirements of His Job*. Washington, CAA Research Division Report No. 73, November 1947 (mimeo).
Gorrell, Edgar S. *Measure of America's World War Aeronautical Effort*. Northfield, Vt., Norwich University, 1940.
Goulden, J. C. *The Superlawyers*. New York, Weybright & Talley, Inc., 1971.
Grace, Dick. *I'm Still Alive*. Chicago, Rand McNally & Co., 1931.
Gray, George. *Frontiers of Flight*. New York, Alfred A. Knopf, Inc., 1948.
Gray, H. Peter. *International Travel, International Trade*. Lexington, Mass., D. C. Heath and Co., 1970.
Grey, C. G. "Aerial Transport in the Light of History," *Aeroplane*, Mar. 16, 1921.
————. "How the Atlantic Air Ferry Began," *Aeroplane*, Aug. 18, 1944.
————. "On Commercial Aviation," *Aeroplane*, Sept. 27, 1922.
————. "On the Great Flights," *Aeroplane*, May 25, 1927.
————. "On Passengers and Paper," *Aeroplane*, Dec. 6, 1922.
————. "On Passenger-Worthiness," *Aeroplane*, May 11, 1921.
————., ed. *Jane's All the World's Aircraft*. London.
Grierson, John. "Lindbergh, A Pioneer Remembered," *Aerospace*, October 1975.
Grooch, William S. *From Crate to Clipper*. New York, Longmans, Green and Co., Ltd., 1939.
————. *Skyway to Asia*. New York, Longmans, Green and Co., Ltd., 1936.
————. *Winged Highway*. New York, Longmans, Green and Co., Ltd., 1938.
Gunston, Bill. "Big Fans for the Airlines," *Shell Aviation News* 444 (1977).
Gwinn, W. P. *Technology, Key to the Universe*. New York, Wings Club, 1973.

Hage, Robert E. *Jet Propulsion in Commercial Air Transportation*. Princeton, Princeton University Press, 1948.
Haggerty, James J. *Aviation's Mr. Sam*. Fallbrook, Cal., Aero Publications, Inc., 1976.
Hall, Donald A. "Technical Preparation of the Airplane 'Spirit of St. Louis,'" NACA Technical Note No. 257. Washington, July 1927.
Hall, Floyd D. *Sunrise at Eastern*. New York, Newcomen Society, 1965.
Hall, Melvin. *Intercontinental Air Transportation, A Short Survey*. Washington, 1940.
Hallion, Richard. *Legacy of Flight*. Seattle, University of Washington Press, 1977.
Hamilton, Edith. *Mythology*. Boston, Little, Brown and Co., 1942.
Hamilton-Standard Co. *Prop to Pilot*. East Hartford, Conn., United Aircraft Corp., 1948.
Hamstra, J. H. "Two Decades: Federal Aero-Regulation in Perspective," *Journal of Air Law and Commerce*, April 1941.
Harding, W. B. *The Aviation Industry*. New York, Barney, 1937.
Harris, H. R. *60 Years of Aviation History*. Windsor Locks, Conn., 1974 (mimeo).
Harvard City Planning Series, I. *Airport*. Cambridge, Harvard University Press, 1930.
Hatch, Alden. *Glenn Curtiss*. New York, Julian Messner, Inc., 1942.
Haven, Violet S. *Hong Kong for the Weekend*. Boston, Contemporary Features, 1939.
Heinkel, Ernst. *Stormy Life*. New York, E. P. Dutton & Co., Inc., 1956.
Henderson, Paul. "Problems of Commercial Flying," *Aeronautics Review*, Nov. 21, 1930.
Heron, S. D. *Aircraft Operation with Fuels of 100 Octane Number*. Detroit, Ethyl Corp., 1937 (mimeo).
————. *Importance of Octane Number in Aircraft Fuels*. Detroit, Ethyl Corp., 1936 (mimeo).
————. *Mutual Adaptation of Aircraft Fuels and Aircraft Engines*. New York, Society of Automotive Engineers Preprint, January 1947 (mimeo).
Hershey, Burnet. *The Air Future*. New York, Duell, Sloan and Pearce, 1943.
Herzberg, M. J., ed. *Happy Landings: An Anthology*. Boston, Houghton Mifflin Co., 1942.

Hertz Corp. *History of the Hertz Corp.* New York, 1977 (mimeo).
Heyman, Hans, Jr. *Civil Aviation and U.S. Foreign Aid.* Santa Monica, RAND Corp., 1964.
Hickerson, J. M. *Ernie Breech.* Des Moines, Meredith Press, 1968.
Higham, Robin. *Air Power: A Concise History.* New York, St. Martin's Press, Inc., 1972.
Hildred, William P. *Five Years of Transatlantic Airline Tourist Service.* Montreal, IATA, May 1957.
Hoare, Robert J. *Wings over the Atlantic.* Boston, Charles T. Branford Co., 1957.
Hoff, Nicholas. "Thin Shells in Aerospace Structures," *Astronautics and Aeronautics,* February 1967.
Holley, Irving B. *Ideas and Weapons.* New Haven, Yale University Press, 1953.
Holmes, Charles H. *Passport Round the World.* London, Hutchinson & Co., 1937.
Hoover, Herbert. *Memoirs, II.* Garden City, Doubleday and Co., Inc., 1952.
Hopkins, George E. *The Airline Pilots.* Cambridge, Harvard University Press, 1971.
Horner, H. M. *The United Aircraft Story.* New York, Newcomen Society, 1958.
Horsfall, J. E. "Lindbergh's Start for Paris," *Aero Digest,* June 1927.
Hubler, Richard G. *Big Eight.* New York, Duell, Sloan and Pearce, 1960.
Hucke, H. M. "Precipitation-Static Interference on Aircraft and Ground Stations," Institute of Radio Engineers *Proceedings,* May 1939.
Hudson, Kenneth. *Air Travel; A Social History.* Totowa, N.J., Rowman & Littlefield, Inc., 1972.
Hughes, A. J. *History of Air Navigation.* London, Allen and Unwin, 1946.
Hunsaker, J. A. *Aeronautics at Mid-Century.* New Haven, Yale University Press, 1952.
———. *Some Lessons of History.* New York, Wings Club, 1965.

Ingells, Douglas J. *The Plane That Changed the World.* Fallbrook, Cal., Aero Publications, Inc., 1966.
———. *The L-1011 and the Lockheed Story.* Fallbrook, Cal., Aero Publications, Inc., 1973.
International Council in the Aeronautical Sciences. *Proceedings of the Centenary Congress,* 1958.

Jablonski, Edward. *Sea Wings.* Garden City, Doubleday & Co., Inc., 1972.
Jaynes, Jack B. "Autobiography." 1975 (manuscript).
Johnson, C. M. *Development of the Lockheed Constellation.* Lockheed, 1944 (pamphlet).
Johnson, George. *The Abominable Airlines.* New York, The Macmillan Co., 1964.
Johnson, Robert E. *Airway One.* Chicago, United Airlines, 1974.
Johnston, Lawrence H. *GCA — Ground Controlled Approach,* Report No. 438. Cambridge, Radiation Laboratory, Massachusetts Institute of Technology, Oct. 1, 1943 (mimeo).
Jones, I. H. *Flying Vistas.* Philadelphia, J. B. Lippincott Co., 1937.
Jones, Thomas V. *Capabilities and Operating Costs of Possible Future Transport Airplanes.* Santa Monica, RAND Corp., July 16, 1963 (mimeo).
Jordan, William A. *Airline Regulation in America.* Baltimore, Johns Hopkins Press, 1970.
Josephson, Matthew. *Empire of the Air.* New York, Harcourt, Brace, 1944.
Judge, Arthur W. *Modern Gas Turbines.* London, Chapman & Hall, Ltd., 1947.

Kancher, Dorothy. *Wings over Wake.* New York, J. Howell, 1947.
Kane, Robert M. *Air Transportation.* Dubuque, William C. Brown Co., 1967.
Keats, John. *Howard Hughes.* New York, Random House, Inc., 1966.
Kelly, Charles J., Jr. *The Sky's the Limit.* New York, Coward-McCann, Inc., 1963.
Kelly, Fred C. *Miracle at Kitty Hawk: Letters of Orville and Wilbur Wright.* New York, Farrar, Straus & Young, 1951.
———. *The Wright Brothers.* New York, Harcourt, Brace, 1943.
Kelly, Raymond C. "Our Amazing Transportation System," *Journal of Aircraft,* December 1977.
Kelly-Rogers, J. C. "Commercial Flying on the North Atlantic from Infancy to Maturity," *Aerospace,* January 1976.

Kennedy, T. H. *Introduction to Economics of Air Transportation.* New York, The Macmillan Co., 1924.
Klaas, M. D. *Last of the Flying Clipper Ships,* memo in Pan Am Archives, February 1964 (mimeo).
Klotz, Alexis. *Three Years off This Earth.* Garden City, Doubleday & Co., 1960.
Knowlton, L. S. *Air Transportation in the United States.* Chicago, University of Chicago Press, 1941.
Kokenes, V. C. *On Wings of Faith.* New York, Random House, Inc., 1960.
Komons, Nick A. *Bonfires to Beacons.* Washington, DOT/FAA, 1978.
———. *The Cutting Crash.* Washington, DOT/FAA, 1973.
Kuter, Laurence S. *The Great Gamble.* University, Ala., University of Alabama Press, 1973.
———. "Truman's Secret Management of the Airlines," *Aerospace History,* September 1977.

La Croix, Robert de. *They Flew the Atlantic.* London, F. Muller, 1958.
La Farge, Oliver. *Eagle in the Egg.* Boston, Houghton Mifflin Co., 1949.
Lamplugh, A. G. "Accidents in Civil Aviation," *Journal of the Royal Aeronautical Society,* February 1932.
Lancaster, O. E., ed. *Jet Propulsion Engines.* Princeton, Princeton University Press, 1959.
Lancaster, Osbert. *Signs of the Times.* Boston, Houghton Mifflin Co., 1961.
Langewiesche, Wolfgang. *Stick and Rudder.* New York, Whittlesey House, 1944.
———. "What the Wright Brothers Really Invented," *Harper's,* June 1950.
Lanier, H. W. *The Far Horizon.* New York, Alfred A. Knopf, Inc., 1933.
Lansing, John B., and D. M. Blood. *The Travel Market, 1964–5.* Ann Arbor, Institute for Social Research, 1963.
———. *The Changing Travel Market.* Ann Arbor, Institute for Social Research, 1964.
La Pierre, C. W. *Power from Progress in the Air.* New York, Newcomen Society, 1954.
Large, Laura A. *Air Travelers.* Boston, Lothrop, 1932.
Larkins, William T. The Ford Story. Wichita, Longo, 1958.
———. The Ford Trimotor. London, Profile Publications, 1967.
Leary, W. M. *The Dragon's Wings.* Athens, Ga., University of Georgia Press, 1976.
Lederer, Jerome. *Safety in the Operation of Air Transportation.* Northfield, Vt., Norwich University, 1939.
Lederer, Ludwig G., and George Kidera. *Studies in Motion Sickness.* Chicago, American Medical Association, 1952.
Lederer, Ludwig G., and J. H. Tillisch. "Medical Criteria for Passenger Flying," *Archives of Environmental Health,* February 1961.
Le Shane, A. A. "Colonial Air Transport," *AAHS Journal,* Winter 1973, Spring 1974.
Leviero, Anthony J. "Aviation Profits Cut 89% in Year," *New York Times,* January 9, 1947.
Leyton, Burr W. *Wings Around the World.* New York, E. P. Dutton & Co., Inc., 1948.
Lindbergh, Anne Morrow. "Airliner to Europe," *Harper's,* September 1948.
———. *Listen, the Wind.* New York, Harcourt, Brace, 1938.
———. *North to the Orient.* New York, Harcourt, Brace, 1935.
Lindbergh, Charles A. *Spirit of St. Louis.* New York, Charles Scribner's Sons, 1953.
———. *Wartime Journals.* New York, Harcourt Brace Jovanovich, Inc., 1970.
Lingeman, R. R. *Don't You Know There's a War On?* New York, Duell, Sloan and Pearce, 1964.
Lippincott, Harvey. *Adaptation of Military Aircraft Engines to Commercial Transport.* United Technologies, 1978 (mimeo).
———. "The Navy Finds an Engine," *AAHS Journal,* Winter 1961.
Lipsner, B. B. *The Airmail: Jennies to Jets.* Chicago, Wilson & Follett, 1951.
Littlewood, William. *Perspectives in Air Transport.* New York, Wings Club, 1966.
Loening, Grover. *The Air Road Will Widen.* New York, Wings Club, 1969.
———. *Amphibian.* Greenwich, New York Graphic Society, 1973.
———. *Our Wings Grow Faster.* Garden City, Doubleday Doran, 1935.
Lockheed Aircraft Co. *Annual Reports.*
———. *Of Men and Stars.* Burbank, Cal., 1957.

Lomask, Milton. *Seed Money: The Guggenheim Story*. New York, Farrar, Straus and Co., 1964.
Lombard, A. A. *Low Consumption Turbine Engines*. AIS Preprint No. 560. New York, June 20, 1955.
————. "Thinking about Aircraft Engines," *Journal of the Royal Aeronautical Society*, May 1958.
Lowell, V. W. *Airline Safety Is a Myth*. New York, Bartholomew House, 1967.
Lundberg, D. E. *The Tourist Business*. Boston, Cahners Books, 1972.

Macdonald, A. F. "Airport Problems of American Cities," *Annals*, September 1930.
McFarland, M. W., ed. *Papers of Wilbur and Orville Wright*. New York, McGraw-Hill, Inc., 1953.
McFarland, Ross A. *Human Factors in Air Transport Design*. New York, McGraw-Hill, Inc., 1946.
————. *Human Factors in Air Transportation*. New York, McGraw-Hill, Inc., 1953.
————. "Human Factors in Relation to Development of Pressurized Cabins," *Aerospace Medicine*, December 1971.
Mackay, James. *Airmails 1870–1970*. London, Botsford, 1971.
Magoun, F. A. *History of Aircraft*. New York, Whittlesey House, 1931.
Magoun, F. A., and Eric Hodgins. *Sky High*. Boston, Little, Brown and Co., 1929.
Maguire, D. R. "Enemy Jet History," *Journal of the Royal Aeronautical Society*, January 1948.
Mance, Osborne. *International Air Transport*. London, Oxford University Press, 1943.
Mansfield, Harold. *Billion Dollar Battle*. New York, David McKay Co., Inc., 1965.
————. *Vision: The Story of Boeing*. New York, Duell, Sloan and Pearce, 1956.
Marcosson, I. F. *Colonel Deeds, Industrial Builder*. New York, Dodd, Mead & Co., 1947.
Martin, Albro. *Enterprise Denied*. Cambridge, Harvard University Press, 1971.
Marvin, Langdon P. "Air Mail Subsidy Separation," *Georgetown Law Journal*, Spring 1952.
Marx, Joseph L. *Crisis in the Skies*. New York, David McKay Co., Inc., 1970.
Masefield, Peter. *Aviation and the Environment in the 1970s*. New York, Littlewood Memorial Lecture, Nov. 18, 1971 (mimeo).
————. *Gateways to the World over the Years*. New York, Wings Club, 1972.
Masland, W. M. "Twenty Years of Achievement on the North Atlantic," *Interavia*, April 1966.
Maxtone-Graham, John. *The Only Way to Cross*. New York, The Macmillan Co., 1972.
Maynard, Crosby. *Flight Plan for Tomorrow*. Santa Monica, Douglas, 1966.
Miller, Francis T. *The World in the Air*. New York, G. P. Putnam's Sons, 1930.
Miller, Ronald, and David Sawers. *Technical Development of Modern Aviation*. London, Routledge and Kegan Paul, Ltd., 1968.
Mills, Stephen. *More than Meets the Sky*. Seattle, Superior Publishing Co., 1974.
Mingos, Howard. *Birth of an Industry*. New York, W. B. Conkey, 1930.
Molson, K. M. "Some Historical Notes on Development of Variable Pitch Propellers," *Canadian Aeronautics and Space Journal*, June 1966.
Mooney, M. M. *The Hindenburg*. New York, Dodd, Mead & Co., 1972.
Moore, Byron. *The First Five Million Miles*. New York, Harper & Row, 1955.
Moore, W. G. *Early Bird*. London, Putnam & Co., Ltd., 1963.
Morgan, David P. *Fasten Seat Belts*. New York, Arco Publishing Co., Inc., 1969.
Morgan, Neil. *Oh, How We Flew: Fifty Flying Years*. Los Angeles, Western Airlines, 1976.
Morison, S. E. *History of United States Naval Operations in World War II*. Boston, Atlantic–Little, Brown and Co., 1947–1962 (15 vols.).
Morris, Lloyd R. *Ceiling Unlimited*. New York, The Macmillan Co., 1953.
Morton, Jay. "Cleveland's Municipal Airport," *National Municipal Review*, September 1926.
Mosley, Leonard. *Lindbergh*. Garden City, Doubleday & Co., Inc., 1976.
Moscow, Alvin. *Tiger on a Leash*. London, Deutsch, 1961.
Mott, George, ed. *Transportation Century*. Baton Rouge, Louisiana State University Press, 1966.

Munson, Kenneth. *Airliners between the Wars*. New York, The Macmillan Co., 1972.
————. *Boeing*. Fallbrook, Cal., Aero Publications, Inc., 1972.
Murchie, Guy. *Song of the Sky*. Boston, Houghton Mifflin Co., 1954.

Nahl, Perham C. "Six Pilots: How They Carried Out 1926 CAM-9 Inaugural," *Airport Journal*, June 1976.
Naylor, J. L., and E. Owen. *Aviation: Its Technical Development*. London, Peter Owen/Vision Press, 1965.
Nevins, Allan, and F. E. Hill. *Ford, Expansion and Challenge, 1915–1933*. New York, Charles Scribner's Sons, 1957.
Newton, Byron R. "Watching the Wright Brothers Fly," *Aeronautics*, June 1909.
Newton, Wesley P. *The Perilous Sky: U.S. Aviation Diplomacy and Latin America, 1919–1931*. Coral Gables, University of Miami Press, 1978.
Nicklas, D. R. *History of Aircraft Cockpit Instrumentation, 1903–1946*. Wright Air Development Center Technical Report 57-301, April 1958.
Nicolson, Harold. *Dwight Morrow*. New York, Harcourt, Brace, 1935.
Nielsen, Thor. *The Zeppelin Story*. London, Wingate, 1955.
Nielson, Dale, ed. *Saga of the Air Mail Pioneers*. Washington, Air Mail Pioneers, 1962.
Norway, Nevil Shute. *Slide Rule*. New York, William Morrow and Co., Inc., 1954.
Northrup, John K. "Looking Ahead to 1960," *Aviation Week*, Nov. 29, 1948.

Oakland (Cal.) Port Commissioners Board. *Oakland Municipal Airport*, 1928.
Ocker, William C., and Carl L. Crane. *Blind Flying*. San Antonio, Naylor Co., 1932.
O'Connor, Harvey. *The Guggenheims*. New York, Covici Friede, 1937.
Ogburn, William F. *The Social Effects of Aviation*. Boston, Houghton Mifflin Co., 1946.
Ogston, A. R., and W. G. Dukek. *Milestones in Aviation Fuels*. New York, American Institute of Aeronautics and Astronautics (Paper No. 69-779), 1969.
O'Neill, Ralph A. *A Dream of Eagles*. Boston, Houghton Mifflin Co., 1973.
Oswald, W. B. "General Formulas and Charts for Calculation of Airplane Performance," NACA Technical Report No. 408, *NACA Annual Report 1933*. Washington, 1933.

Palmer, Henry R. *This Was Air Travel*. Seattle, Superior Publishing Co., 1960.
Pan American Airways. *Annual Reports*.
————. *Keeping Fit for Flying*, 1943.
Paradis, A. A. *200 Million Miles a Day*. Philadelphia, Chilton Book Co., 1969.
Parrish, Wayne W. *The Traveler-Journalist Views the World*. New York, Wings Club, 1975.
————., ed. *Official Aviation Guide, 1929–1938*. Washington, *Aviation Week*, n.d.
Patterson, W. A. *25 Years among the Stars*. New York, Newcomen Society, 1951.
Patton, O. E. *History and Development of Flight Recorders*. Washington, CAB, 1966.
Pearcy, Arthur. *The Dakota*. London, Ian Allan, 1972.
Peters, Michael. *International Tourism*. London, Hutchinson & Co., Ltd., 1969.
Peterson, Houston. *See Them Flying*. New York, Richard W. Baron Publishing Co., 1969.
Phelan, James. *Howard Hughes: The Hidden Years*. New York, Random House, Inc., 1976.
Phillips, Almarin. "Air Transportation in the U.S.," in W. E. Capron, ed., *Technological Change in Regulated Industry*. Washington, The Brookings Institution, 1971.
Pillai, K. G. J. *The Air Net*. New York, Grossman Pubs., Inc., 1969.
Player, Willis. "A Lindbergh Remembrance," *Pan Am Clipper*, May 1977.
Poole, Frederick A. *Records of an Airplane Passenger in 1928 and 1929*. Chicago, privately printed, 1929.
Port Authority of New York and New Jersey. *Transatlantic Air Travel, 1956–1971*. New York, 1975.
————. *World Air Route Planning*. New York, 1945.
Praeger, Otto. *Address to Rotarian Club*. New York, 1920 (mimeo).
Pratt & Whitney Story. East Hartford, Pratt & Whitney Division, United Aircraft, 1952.
Price, Willard. "The Andes Flight," *Reader's Digest*, October 1948.
Pudney, John. *The Seven Skies*. London, Putnam and Co., Ltd., 1959.

Puffer, Claude E. *Air Transportation.* Philadelphia, Blakiston, 1941.
Pusey, Merlo. *Charles Evans Hughes.* New York, The Macmillan Co., 1931.

Rae, John B. *Climb to Greatness: The American Aircraft Industry, 1920–60.* Cambridge, MIT Press, 1968.
Randall, S. J. "Colombia, the United States and Interamerican Rivalry, 1927–40," *Journal of Interamerican Studies and World Affairs,* August 1972.
Ray, James R. *Story of American Aviation.* New York, Winston, 1946.
Raymond, Arthur E. *Who? Me? Autobiography.* November 1974 (mimeo).
Redden, Charles F. "Commercial Aviation," *Aviation,* Jan. 2, 1922.
Redpath, Peter, and James M. Coburn. *Air Transport Navigation.* New York, G. P. Putnam's Sons, 1943.
Reichers, Lou. *The Flying Years.* New York, Holt, 1956.
Rentschler, F. B. *An Account of Pratt & Whitney Aircraft Co., 1925–50.* Hartford, 1950.
Reynolds, Ruth. "Howard Hughes, the Man Who Has Everything," *New York Daily News,* Jan. 19, 1936.
Reynolds, Z. Smith. *Log of Aeroplane NR-898W.* Privately printed, 1932.
Richmond, Samuel B. *Regulation and Competition in Air Transportation.* New York, Columbia University Press, 1961.
Rickenbacker, E. V. *Rickenbacker.* Englewood Cliffs, Prentice-Hall, Inc., 1962.
Robbins, Christopher. *Air America.* New York, G. P. Putnam's Sons, 1979.
Robinson, Arthur L. "Human Powered Flight: Californians Claim Kremer Prize," *Science,* Sept. 16, 1977.
Robinson, Douglas H. *Giants in the Sky.* Seattle, University of Washington Press, 1973.
Rochester, Stuart I. *Takeoff at Mid-Century.* Washington, DOT/FAA, 1976.
Rodgers, William. *Think: A Biography of the Watsons and IBM.* New York, Stein & Day, 1971.
Rodman, Selden, ed. *The Poetry of Flight.* New York, New Directions Publishing Corp., 1941.
Rohrbach, Adolph. "Economical Production of All-Metal Airplanes and Seaplanes," *SAE Journal,* January 1927.
Rolt, L. T. C. *The Aeronauts: A History of Ballooning.* New York, Walker & Co., 1966.
Rosen, George S. *History of Aircraft Propellers.* Windsor Locks, Conn., 1977 (mimeo).
Rowe, Basil L. *Under My Wings.* Indianapolis, Bobbs-Merrill Co., Inc., 1956.

Saint Exupery, Antoine de. *Night Flight.* New York, Century, 1932.
——. *Wind, Sand and Stars.* New York, Reynal & Hitchcock, 1939.
Sandburg, Carl. *Complete Poems.* New York, Harcourt, Brace, 1969.
——. *The People, Yes.* New York, Harcourt, Brace, 1936.
Scamehorn, H. L. *Balloons to Jets.* Chicago, Henry Regnery Co., 1957.
Scandinavian Airways System. *The Making of SAS.* Oslo, 1973.
Schiff, Barry J. *The Boeing 707.* New York, Arco Publishing Co., Inc., 1967.
Schlaifer, Robert, and S. H. Heron. *Development of Aircraft Engines and Fuels.* Boston, Harvard Graduate School of Business, 1950.
Schlesinger, Arthur. *Coming of the New Deal.* Boston, Houghton Mifflin Co., 1959.
Schreuder, O. B. "Medical Aspects of Aircraft Pilot Fatigue with Special Reference to the Commercial Jet Pilot," *Aerospace Medicine,* April 1966.
Schwartz, Bernard. *The Professor and the Commissions.* New York, Alfred A. Knopf, Inc., 1959.
Scruggs, W. O. "American Investment in Latin America," *Foreign Affairs,* January 1932.
Scullen, George. *International Airport.* Boston, Little, Brown and Co., 1968.
Sealy, Kenneth R. *Geography of Air Transport.* London, Hutchinson & Co., Ltd., 1966.
Serling, Robert. *The Electra Story.* Garden City, Doubleday & Co., 1963.
——. *The Only Way to Fly.* Garden City, Doubleday and Co., 1976.
——. *The Probable Cause.* Garden City, Doubleday and Co., 1960.
Sevareid, Eric. *Not So Wild a Dream.* New York, Alfred A. Knopf, Inc., 1946.
Shamburger, Page. *Tracks across the Sky.* Philadelphia, J. B. Lippincott Co., 1963.
Shay, Arthur. "Idlewild, Airport of Tomorrow," *Interavia,* October 1946.
Shell Oil Co. *Civilian Wings for Everyone.* New York, 1940.

Shepherd, Christopher. *German Aircraft of World War II*. London, Sidgwick & Jackson, 1975.
Sherwood, Robert. *Roosevelt and Hopkins*. New York, Bantam Paperback, 1946.
Shrader, Welman A. *Fifty Years of Flight*. Cleveland, Eaton Manufacturing Co., 1953.
Sheridan, Hy. "A Pilot Looks at the Passenger," *Saturday Evening Post*, Nov. 29, 1947.
Siegel, Peter V., "Time Zone Effects," *Science*, June 13, 1964.
Sikorsky, Igor. *Recollections and Thoughts of a Pioneer*. 1964 (pamphlet).
———. *Story of the Winged S*. New York, Dodd, Mead & Co., 1938.
Simonson, G. R., ed. *History of the American Aircraft Industry*. Cambridge, MIT Press, 1968.
Smith, C. R. *Safety in Air Transport over the Years*. New York, Wings Club, 1971.
———. "What We Need is a Good Three-Cent Airline," *Saturday Evening Post*, Oct. 20, 1945.
Smith, Dean C. *By the Seat of My Pants*. Boston, Little, Brown and Co., 1961.
Smith, Henry Ladd. *Airways*. New York, Alfred A. Knopf, Inc., 1942.
———. *Airways Abroad*. Madison, University of Wisconsin Press, 1950.
Smith, Wesley L. *Air Transport Operation*. New York, McGraw-Hill, Inc., 1931.
Soule, Hartley A. *Flight Tests of the DC-3 Airplane*. NACA memo, October 1937 (mimeo).
Speer, Albert. *Inside the Third Reich*. New York, The Macmillan Co., 1970.
Sperry Rand Corp. *The Gyroscope Through the Ages*. New York, 1971.
Stark, Howard. *Instrument Flying*. Pawling, New York, Stark, 1934.
Stewart, Oliver. *Aviation, the Creative Ideas*. London, Faber and Faber, Ltd., 1966.
———. *First Flights*. London, Routledge and Kegan Paul, Ltd., 1957.
Straszheim, M. H. *International Airline Industry*. Washington, The Brookings Institution, 1969.
Stroud, John. *The World's Airliners*. London, The Bodley Head, Ltd., 1971.
Strughold, Hubertus, ed. *International Symposium on Physics and Medicine of the Atmosphere and Space*, 1960.
———. *Your Body Clock*. New York, Charles Scribner's Sons, 1972.
Stuart, Frank S. *Modern Air Transport*. London, John Long, Ltd., 1946.
Sutton, Graham. *Mastery of the Air*. New York, Basic Books, Inc., 1965.

Taylor, Frank J. *High Horizons*. New York, McGraw-Hill, Inc., rev. ed. 1964.
Taylor, J. W. R., ed. *The Lore of Flight*. Gothenburg, Tre Tryckare, 1970.
———. *Passengers, Parcels and Panthers*. New York, Roy Publications, Inc., 1956.
Thomas, Lowell. *European Skyways*. Boston, Houghton Mifflin Co., 1927.
Thomas, Lowell, and Edward Jablonski. *Doolittle*. Garden City, Doubleday & Co., 1976.
Thomas, Lowell, and Lowell Thomas, Jr. *Famous First Flights that Changed History*. Garden City, Doubleday and Co., 1968.
Thruelsen, Richard. *Transocean*. New York, Holt, 1953.
Time: "Merchant Aerial," July 31, 1933.
    "Clipper Skipper," Mar. 28, 1949.
    "The Mechanical Man," July 19, 1948.
    "Mr. Horsepower," May 28, 1951.
    "Gamble in the Sky," July 19, 1954.
    "Shootout at the Hughes Corral," Dec. 21, 1970.
Tinnin, David B. *Just About Everybody against Howard Hughes*. Garden City, Doubleday & Co., 1973.
Tipton, Stuart G. *The Airport Situation*. Washington, ATA, 1967.
Thayer, Frederick C. *Air Transport Policy and National Security*. Chapel Hill, University of North Carolina Press, 1965.
Tomlinson, D. W. *The Sky's the Limit*. Philadelphia, McCrae Smith, 1930.
———. *Legacy of Leadership*. New York, Walsworth, 1971.
Trieber, L. A. *Literature of Flying and Fliers*. Sydney, Les editions du courier australien, 1945.
Trippe, Juan T. "Air Transportation and International Trade," Address in Philadelphia, Sept. 25, 1951.
———. *Charles A. Lindbergh and World Travel*. New York, Wings Club, 1977.

―――. "Ocean Air Transport," *Journal of the Royal Aeronautical Society*, September 1941.

Trissandier, Gaston. *Histoire des ballons et des astronautes célebres.* Paris, Launette, 1887–1890 (2 vols.).

Trotta, Marie. "U. S. Air Travel Market, 1955–1972" in George P. Howard, ed., *Airport Economic Planning.* Cambridge, MIT Press, 1974.

Tucker, Ray. *Mirrors of 1932.* New York, Brewster, Warner and Putnam, 1933.

Tunner, William H. *Over the Hump.* New York, Duell, Sloan and Pearce, 1964.

Turner, Louis, and John Ash. *The Golden Hordes: International Tourism.* New York, St. Martin's Press, Inc., 1976.

Turner, P. St. John. *Pictorial History of Pan Am.* London, Ian Allan, 1973.

TWA. *Annual Reports.*

―――. *Legacy of Leadership.* New York, Wadsworth, 1971.

Tyler, Poynts, ed. *Airways of America.* New York, H. W. Wilson, 1958.

United Aircraft and Transport Corp. *Annual Reports.*

United Aircraft Corp. *Annual Reports.*

United Airlines. *Annual Reports.*

Vaeth, J. Gordon. *Graf Zeppelin.* New York, Harper & Row, 1958.

Van Zandt, J. Parker. "Looking down on Europe," *National Geographic*, March 1925.

Varney, Alex. *Psychology of Flight.* New York, D. Van Nostrand Co., Inc., 1950.

Veale, Sydney E. *Airlines and Airways of Today.* London, Pilot Press, 1947.

Villard, H. S. *Contact: The Story of the Early Birds.* New York, Thomas Y. Crowell Co., 1968.

Von Karmann, Theodore. *The Wind and Beyond.* Boston, Little, Brown and Co., 1967.

Wacht, Walter F. *Domestic Air Transportation Network of the U.S.* Chicago, University of Chicago, Department of Geography, Research Paper 154, 1974.

Wagner, Ray, and Heinz Nawarra. *German Combat Planes.* Garden City, Doubleday & Co., 1971.

Walsh, J. E. *One Day at Kitty Hawk.* New York, Thomas Y. Crowell Co., 1975.

Warner, E. P. *Early History of Air Transportation.* Northfield, Vt., Norwich University, 1938.

―――. "General Design of Commercial Aircraft," *Aviation*, Nov. 13, 1922.

Watson-Watt, Robert. *Three Steps to Victory.* London, Odham, 1957.

Watt, Sholto. *I'll Take the High Road.* Fredericton, Brunswick Press, 1960.

Wead, Frank. *Ceiling Zero, A play.* New York, Samuel French, Inc., 1935.

Wedemeyer, Albert C. *Wedemeyer Reports.* New York, Henry Holt, 1958.

Weiss, David A. *Saga of the Tin Goose.* New York, Crown Publications, Inc., 1971.

Wells, Edward C. *Aerospace Engineering.* New York, Wings Club, 1974.

Wenneman, J. H., *Municipal Airports.* Cleveland, Flying Review Publ., 1931.

West, Bruce. *The Man Who Flew Churchill.* Toronto, Totem Books Paperback, 1976.

Westwood, Howard C., and A. E. Bennett. "Footnote to the Legislative History of the Civil Aeronautics Act of 1938," *Notre Dame Lawyer*, February, 1967.

Wheat, George S., ed. *Municipal Landing Fields and Airports.* New York, G. P. Putnam's Sons, 1920.

Wheatcroft, S. *Air Transport Policy.* New York, Humanities Press, 1950.

White, Clayton S., et al. *Aviation Medicine.* New York, Pergamon Press, Inc., 1958.

Whiteside, T. C. D. *Problems of Vision at High Altitudes.* London, Butterworth, 1957.

Whitnah, D. R. *Safer Skyways.* Ames, Iowa State University Press, 1966.

Whittle, Frank. "Advent of the Aircraft Gas Turbine." Stuart Lecture, Nottingham, England, 1946.

―――. *Jet.* New York, The Viking Press, 1952.

Who's Who in World Aviation. Washington, American Aviation, 1955.

Wiggin, Charles L. *The First Transcontinental Flight.* New York, Bookmailer, 1961.

Wigton, Don C., ed. *From Jenny to Jet.* Los Angeles, F. Clymer Publications, 1963.

Wik, Reynold M. *Henry Ford and Grass Roots America.* Ann Arbor, University of Michigan Press, 1972.

Wiley, Frank W. *Montana in the Sky.* Helena, Montana Aeronautics Commission, 1966.
Wilkinson, P. H. *Aircraft Engines of the World.* Washington, Paul H. Wilkinson, 1965-70.
Williams, Geoffrey. *Fasten Your Lapstraps.* London, Max Parris, 1955.
Williams, J. E. D. *Operation of Airliners.* London, Hutchinson & Co., Ltd., 1964.
Wilson, George L. *Air Transportation.* New York, Prentice-Hall, Inc., 1949.
Wilson, J. R. M. "Turbulence Aloft 1938-53." Washington, DOS/FAA, unpublished MS, 1977.
Wood, John W. *Airports.* New York, Coward-McCann, Inc., 1940.
Wright Aeronautical Corp. *Annual Reports.*
Wright, Monte D. *Most Probable Position.* Lawrence, University Press of Kansas, 1972.
Wright, Orville. *How We Invented the Airplane.* New York, David McKay Co., Inc., 1953.
———. "How We Made the First Flight," *Flying,* December 1913.
Wright, Orville, and Wilbur. "The Wright Brothers Aeroplane," *Century,* September 1908.
Wright, Wilbur. "Western Society of Engineers Paper," *Flight,* Oct. 2-Nov. 27, 1909.
Wykes, Alan. *Air Atlantic.* New York, David White Co., 1966.

Young, Clarence M. *Airport Management in Europe.* New York, 1929 (pamphlet).
Younger, John E., and N. F. Ward. *Airplane Construction and Repair.* New York, McGraw-Hill, Inc., 1931.
Younger, John E., N. F. Ward, and A. F. Bonnalie. *Airplane Maintenance.* New York, McGraw-Hill, Inc., 1937.

Zim, Herbert. *Man in the Air.* New York, Harcourt, Brace, 1943.

Index